SYMPOSIUM ON
AERODYNAMICS & AEROACOUSTICS

ADVANCED SERIES ON FLUID MECHANICS

SYMPOSIUM ON AERODYNAMICS & AEROACOUSTICS

Tucson, Arizona March 1 – 2, 1993

Editor

K.-Y. FUNG

University of Miami, USA

World Scientific

Singapore • New Jersey • London • Hong Kong

Published by

World Scientific Publishing Co. Pte. Ltd.

P O Box 128, Farrer Road, Singapore 9128

USA office: Suite 1B, 1060 Main Street, River Edge, NJ 07661

UK office: 73 Lynton Mead, Totteridge, London N20 8DH

SYMPOSIUM ON AERODYNAMICS AND AEROACOUSTICS

ISBN 981-02-1732-3

Printed in Singapore by JBW Printers & Binders Pte. Ltd.

Preface

On March 1, 1993, Dr. William R. Sears celebrates his eightieth birthday. This event has been marked by award presentations, invited lectures, symposia, and reunions organized by his friends, colleagues and former students as tributes to his profoundly influencing works and personal qualities.

His colleagues at the University of Arizona (Thomas F. Balsa, C. F. Chen, K.-Y. Fung and Edward Kerschen) organized a symposium to bring together fellow researchers, colleagues, and friends of Bill to focus on four areas of Aeronautical Sciences to which he has made major contributions. These are Wing Design, Unsteady Aerodynamics and Separation, Aeroacoustics, and Self-Correcting Wind Tunnels. The symposium speakers were chosen and invited by the Program Committee composed of Sanford Davis, Fereidoun Farassat, Marvin Goldstein, A. R. Seebass and Ernest Smerdon. Eleven lectures were given in three sessions on March 1-2, 1993 in Tucson, Arizona. Each represents the latest development, a historic account, or survey of the status in that area of research. This volume contains the written versions of these lectures.

The supports from NASA Ames, Langley and Lewis Research Centers, as well as that of the College of Engineering of the University of Arizona, for defraying the costs of the symposium and publication of its proceedings are gratefully acknowledged.

K.Y.F.
University of Miami

A Biographic Note

Dr. William R. Sears, a member of the National Academy of Sciences and the National Academy of Engineering, joined the University of Arizona in 1974 after twenty eight years of service at Cornell University. While at Cornell, he established the famous Graduate School of Aeronautical Engineering and guided the education of a generation of outstanding engineers. During the early sixties, as John L. Given Professor of Engineering, Bill became the first Director of the Center of Applied Mathematics at Cornell. After joining the University of Arizona and a career of theoretically-inclined research since his appointment at Caltech as a wind-tunnel assistant to Clark B. Millikan, Bill returned to experimental studies in order to develop and refine the basic science and practical aspects of adaptive-wall wind tunnels. It is unanimously agreed that Bill's many scientific contributions are firmly etched into the archives of aeronautics.

During some of the darkest days of World War II, Bill left his position as assistant professor at Caltech to become Chief of Aerodynamics and Flight Testing in a new aircraft company, Northrop. Here he worked on several development projects, the most famous of which is the turbo jet powered flying wing - a concept that has reemerged in one of the premier military airplanes of today.

Indeed, Bill is one of the most prominent of the scholars who participated in the golden age of aeronautics and were personally touched by the legends of that era. After graduating from the University of Minnesota in 1934, he went to Caltech to study under Theodore von Kármán. This interaction, together with the intellectual climate at the Guggenheim Aeronautical Laboratory at Caltech and the blossoming of a personal relationship with Mabel Rhodes (Mrs. Sears) shaped our colleagues as we know him today: an exceptional scholar and a wonderful human being.

T.F.B.

Contents

Chairman: R. Shevell

Section III: ADAPTIVE WALL WIND TUNNELS
Chairman: S. Davis

Section I

AEROACOUSTICS

Some Aspects of the Aeroacoustics of Extreme-Speed Jets

Sir James Lighthill

ABSTRACT

The Lecture begins by sketching some of the background to contemporary jet aeroacoustics. Then it reviews scaling laws for noise generation by low-Mach-number airflows and by turbulence convected at "not so low" Mach number. These laws take into account the influence of Doppler effects associated with the convection of aeroacoustic sources.

Next, a uniformly valid Doppler-effect approximation exhibits the transition, with increasing Mach number of convection, from compact-source radiation at low Mach numbers to a statistical assemblage of conical shock waves radiated by eddies convected at supersonic speed. In jets, for example, supersonic eddy convection is typically found for jet exit speeds exceeding twice the atmospheric speed of sound.

The Lecture continues by describing a new dynamic theory of the nonlinear propagation of such statistically random assemblages of conical shock waves. It is shown, both by a general theoretical analysis and by an illustrative computational study, how their propagation is dominated by a characteristic "bunching" process. That process — associated with a tendency for shock waves that have already formed unions with other shock waves to acquire an increased proneness to form further unions — acts so as to enhance the high-frequency part of the spectrum of noise emission from jets at these high exit speeds.

1 INTRODUCTION

I warmly appreciate the invitation to give a Keynote Lecture at this auspicious Symposium on Aerodynamics and Aeroacoustics being held here today and tomorrow to mark the 80th birthday of our very eminent colleague Bill Sears. I pride myself enthusiastically on having been Bill's scientific colleague for well over half of his life, having indeed had already the signal pleasure 42 years ago to be a house-guest of Bill and Mabel in the charming surroundings of their then home at Cornell.

Bill and I share, too, a particular piece of scientific background in having both been early devotees of aeroacoustics, and this is why, after careful consideration, I chose to devote this Keynote Lecture to the Aeroacoustics portion of the Symposium's "double-barrelled" subject field. Yet before embarking on aeroacoustical matters I would like first to express, on behalf of everyone active in aerodynamic studies, our profound appreciation of the immense and wide-ranging contributions which Bill has made to aerodynamics. It was of course these that I gratefully acknowledged four years ago at Cornell in my W.R. Sears Lecture "Aerodynamic Aspects of Animal Flight.". Today, on the other hand, I'm giving a quite different Lecture (although still with the word "aspects" in the title!) — on "Some aspects of the aeroacoustics of extreme-speed jets."

In a way this Lecture represents a further development from my Wright Brothers Lecture [1] entitled "Jet Noise". That was a Lecture in which I tried to put across to a general audience a series of new advances in Jet Noise science which had emerged from a blend of my own earlier studies [2,3] with some excitingly novel extensions made by Shon Ffowcs Williams [4].

In this Lecture I shall briefly remind you of some of those investigations from the early 1960s. Then I shall go on to describe some recent researches — on the aeroacoustics of what I may perhaps designate as "extreme-speed" jets — to which I have been stimulated by some major new challenges of the early 1990s.

One of the most exciting of these is posed by the US High Speed Civil Transport project (HSCT), an extremely promising plan for a supersonic transport aircraft ingeniously designed to minimise the level of supersonic-boom annoyance. However the corresponding problems of reducing engine noise for such an aircraft to within acceptable limits raise some thorny questions and may, in particular, demand that a fundamental aeroacoustic analysis of jets at relatively high speed be undertaken.

In a purely aeroacoustical context the appropriate definition of the Mach number M of a jet is the ratio of its exit speed U to the speed of sound c_0 in the atmosphere into which it is radiating. Now, the general trend of aeroacoustic Mach numbers for civil aero-engines (in other words, for the engines of those aircraft which face the greatest aeroacoustic challenges) has been a downward trend for very many years, and this has allowed engines to become simultaneously quieter and more powerful because in a wide range of Mach numbers (see Figure 1 below) the ratio η of acoustic power radiated to jet power delivered varies as M^5. At lower jet Mach numbers, then, noise radiated can be less even though jet power is greater.

Evidently, it has only been the introduction of aeroengines of ever larger and larger diameter that has permitted such increased engine powers to be delivered at the necessary low noise levels by means of this downward trend in jet Mach number. For a supersonic aircraft, on the other hand, the use of extra-wide engines is out of the question because their supersonic-boom emissions, and also the associated shock-wave drag, would be unacceptably great. Such considerations rule out any similar reduction in jet Mach number in this case.

It follows that work on engine-noise reduction for the HSCT project has been calling for many new fundamental studies of the generation of noise by extreme-speed jets; that is, by jets at relatively high Mach numbers. This is the range of values of M in excess of 2 for which (see Figure 1) the proportion η of jet power that is converted into noise approaches an asymptotic value of 0.01 or a little less. Moreover the noise field has become highly directional, because the turbulent eddies that generate sound are themselves being convected at supersonic speeds so that they emit their own supersonic booms in the Mach direction defined by an eddy convection velocity and the atmospheric speed of sound [5].

In an actual aero-engine installation, of course, there may be a great difference between the pure noise field of the extreme-speed jet itself and the overall noise radiated from the installation taken as a whole. However, an essential pre-requisite for designing that installation so as to bring down ground noise levels is to understand as well as possible the primary noise field generated by the extreme-speed jet. This is why so many stimulating lectures contributed to the important Workshop on Computational Aeroacoustics which I had the pleasure of attending last year at NASA Langley Research Center [6] were devoted to different aspects of that primary noise generation.

In today's Keynote Lecture, the main material which I shall present describes certain new researches into which I was drawn as a direct result of listening to all those stimulating contributions which, as I have indicated, were concerned with how the extreme-speed jet generates a noise field emitted largely in the Mach direction. In fact the qualifying phrase which appears in my title "Some aspects of the aeroacoustics of extreme-speed jets" is mainly intended to recognize that noise-generation mechanisms in such jets have long been studied and, above all, are being investigated actively today. Accordingly I shall refer only briefly to those generation aspects of the problem and concentrate rather on some other aspects which appear, relatively speaking, to have been neglected.

These aspects of the aeroacoustics of extreme-speed jets on which I shall concentrate are related to those effects of nonlinear sound propagation which immediately start to modify the noise field once it has been generated. Such nonlinear propagation effects may readily be expected to be important from that analogy with supersonic-boom generation which I already mentioned; and which tends to suggest that each supersonically convected eddy will emit in the Mach direction a boom-like signal that should include one or more conical shock waves. Accordingly a random sequence of eddies should generate a random, thickly packed assemblage of conical shock waves, and this idea is supported by various experimental, theoretical and computational studies.

My lecture, then, is primarily concerned with the nonlinear acoustic propagation of such random assemblages of conical shock waves once they have been formed. Thus it includes a study of the inherent tendency of the shock strengths to decay (through conical spreading and internal dissipation), as opposed by increases in strength that occur whenever adjacent shock waves unite. A certain "bunching" tends to arise, because a union of two adjacent shocks is found to increase the likelihood of further union with other neighbouring shocks. The high-frequency part of the noise spectrum is made more intense than would otherwise be the case by these bunching tendencies.

A completely general theoretical analysis is used to demonstrate the universal tendency towards bunching. Then a computational study is carried out in order to exemplify details of the process, which is expected to be important for the appreciation of how Mach wave fields are modified in the region of conical propagation that surrounds an extreme-speed jet.

All of this new material concerned with nonlinear propagation is preceded, however, by a simplified summary account of how jet noise is generated. This account (broadly along the lines of my Wright Brothers Lecture) shows how some sort of continuous transition can be discerned between more familiar processes of jet-noise generation at relatively low Mach number and that generation of a random assemblage of conical shock waves which sets in at the higher jet Mach numbers.

Next, a well known transformation of coordinates [7,8] is used to reduce the problem of conical propagation, concerned with how a temporal waveform varies with distance r along the Mach direction, into a plane- wave problem. In this latter problem the time t replaces $r^{1/2}$, and attention is focussed on how a given spatial waveform varies with time.

The spatial waveform that needs to be studied in this latter context turns out to be a spatially unlimited assemblage of random "sawtooth" waves, of the type (see Figure 2) into which a general plane sound wave of large amplitude would evolve during a certain time t. It consists of randomly located shocks (with random strengths) separated by expansion waves in which the slope of excess signal velocity as a function of position takes the value $1/t$..

I shall present a new and quite general theory of the nonlinear dynamics of random sawtooth waves. It shows how the inherent tendency of the shock strengths themselves is to decay like $1/t$ but that this tendency is opposed whenever shocks unite with adjacent shocks. Moreover those "bunchings" of unions to which I already referred combine in a sort of "snowball" effect to make major modifications in the waveform.

After a transformation back into the original variables describing conical propagation, t is replaced by $r^{1/2}$ but predicted shock strengths have to be divided by a further $r^{1/2}$ factor. Thus the inherent tendency of the shock strengths is now to decay like $1/r$ but we shall yet again find that this tendency is powerfully opposed by those "bunching" effects whose special relevance, in a noise context, is to intensify the high-frequency part of the jet noise spectrum.

2 SCALING OF AERODYNAMIC NOISE AT LOW MACH NUMBER

An airflow of characteristic velocity U and length-scale L, with high enough Reynolds number $\rho UL/\mu$ (where ρ and μ are the air's density and viscosity), is a turbulent airflow. The chaotic sound field which it radiates through the surrounding atmosphere (with undisturbed sound speed c_o) is known as aerodynamic noise.

This sound radiation — apart from any effects of solid boundaries (see below) — is precisely that which would be generated by quadrupole sources of strength T_{ij} per unit volume [2,8], where T_{ij} stands for the difference between the momentum flux in the real airflow and that in a simple acoustic medium with sound speed c_o. The most important term in T_{ij} is the convective flux $\rho u_i u_j$ of a momentum component ρu_i carried by a velocity component u_j.

Turbulent airflows at low Mach number U/c_o are compact sources of aerodynamic noise because typical frequencies ω in the turbulence scale as U/L (Strouhal scaling) and therefore the compactness ratio $\omega L/c_o$ is small [9]. It means that differences in phases of emission for sounds reaching a distant observer are small enough for the whole flow field to radiate effectively as a single source.

This source may be of dipole type, with dipole strength F_i, in cases when the vector F_i represents a force acting between the turbulent airflow and a solid body immersed in it [10,11]. Typically, F_i scales as $\rho U^2 L^2$, so that its rate of change scales as $\rho U^3 L$; and then the radiated power

$$<\dot{F}_i^2>/12\pi\rho c_o^3 \tag{1}$$

scales as $\rho U^6 L^2/c_o^3$: a sixth-power dependence on flow speed. Also, the acoustic efficiency η, defined as the ratio of radiated power to a rate of delivery (proportional to $\rho U^3 L^2$) of energy to the flow, scales as $(U/c_o)^3 = M^3$.

On the other hand turbulent airflows at low Mach number in the absence of any such solid body radiate effectively as a single quadrupole source, with total strength Q_{ij} scaling as $\rho U^2 L^3$ (because the strength T_{ij} per unit volume scales as ρU^2). The acoustic intensity at a distant point, whose vector separation and scalar distance from the source are x_i and r respectively, is then

$$<\left(\ddot{Q}_{ij}x_i x_j r^{-2}\right)^2>/16\pi^2 r^2 \rho c_o^5 , \tag{2}$$

where the second time-derivative $\ddot{Q}_{ij}$ scales as $\rho U^4 L$. Accordingly, the radiated power scales as $\rho U^8 L^2/c_o^5$ (an eighth-power dependence on flow speed [2,9]) and the acoustic efficiency η scales as $(U/c_o)^5 = M^5$.

At low Mach numbers, therefore, such aerodynamic noise radiation of quadrupole type is unimportant whenever dipole radiation due to fluctuating body force (with efficiency proportional to M^3) is also present [13]. In other problems, however, such as the noise of a jet (with practically no fluctuating body force), the quadrupole source becomes dominant. For example, a turbulent jet of exit speed U radiates with an acoustic efficiency of order $10^{-4}(U/c_o)^5$.

3 AERODYNAMIC NOISE AT NOT SO LOW MACH NUMBER

The chaotic nature of turbulent flow implies that velocity fluctuations at points P and Q, although they are well correlated when P and Q are very close, become almost uncorrelated when P and Q are not close to one another. Here we recall that the correlation coefficient C between velocities $\mathbf{u}_p$ and $\mathbf{u}_Q$ is defined as

$$C = < \mathbf{v}_P \cdot \mathbf{v}_Q > / < v_P^2 >^{1/2} < v_Q^2 >^{1/2} \tag{3}$$

in terms of the deviations $\mathbf{v}_p = \mathbf{u}_p - <\mathbf{u}_p>$ and $\mathbf{v}_Q = \mathbf{u}_Q - <\mathbf{u}_Q>$ from their means. When two uncorrelated quantities are combined, their mean square deviations are added up:

$$< \mathbf{v}_P + \mathbf{v}_Q >^2 = < v_P^2 > + < v_Q^2 > \text{ if } C = 0 \tag{4}$$

(because the term $2<\mathbf{v}_P \cdot \mathbf{v}_Q>$ vanishes).

Theories of turbulence define a correlation length ℓ, with C not far from 1 ($\mathbf{u}_P$ and $\mathbf{u}_Q$ well correlated) when PQ is substantially less than ℓ, and C not far from zero ($\mathbf{u}_P$ and $\mathbf{u}_Q$ almost uncorrelated) when PQ substantially exceeds l. Roughly speaking, different regions of size ℓ ("eddies") generate uncorrelated sound fields, and the mean square radiated noise is the sum of the mean square outputs from all of the regions [3,9].

Typical frequencies ω in the turbulence are of order $\omega = v/\ell$, where v is a typical root mean square velocity deviation $<v^2>^{1/2}$. For each region of size ℓ, therefore, the compactness condition that $\omega l/c_o$ be small is satisfied when v/c_o is small.

Compactness, then, requires only that a root mean square velocity deviation v, rather than a characteristic mean velocity U, be small compared with c_o. The associated restriction on U/c_o is less and can be satisfied at "not so low" Mach number.

On the other hand, the sound radiated is no longer that of a single quadrupole source. Rather it is a combination of uncorrelated radiation patterns from different regions of size ℓ, each with an intensity field (2) where $Q_{ij} = \ell^3 T_{ij}$; while by equation (4) the intensity fields of different regions may simply be added. Therefore, on division by the volume ℓ^3 of a region, we obtain the intensity pattern radiated per unit volume of turbulence as

$$\ell^3 < \left(\ddot{T}_{ij} x_i x_j r^{-2}\right)^2 > / 16\pi^2 r^2 \rho c_o^5 \, . \tag{5}$$

4 DOPPLER EFFECTS ON FREQUENCY, VOLUME, AND COMPACTNESS

Moreover, the radiation from any single eddy is subject to a modification as a result of the eddy being convected at "not so low" Mach number (the Doppler effect). A rather familiar element of Doppler effect on the radiation pattern of a moving source of sound is its shift in frequency, but there are changes also in its effective volume, and in its compactness.

All of these Doppler effects on the sound received by an observer at a far- field location depend not on the speed V with which the source moves but on its velocity component w in the direction of the observer. In the case of an observer located on a line making an angle θ with a source's direction of motion at speed V, this velocity component w has the form

$$w = V \cos\theta \, . \tag{6}$$

Then, while sound radiation of frequency ω travels a distance $c_o T$ during a single period $T = 2\pi/\omega$, its source moves a distance wT nearer to the observer. Thus the wavelength λ (distance between crests) is reduced to

$$\lambda = c_o T - wT = 2\pi(c_o - w) / \omega, \tag{7}$$

and the frequency heard by the observer (2π divided by the time λ/c_o between arrival of crests) is increased to its Doppler-shifted value: the "relative" frequency

$$\omega_r = \omega(1 - w / c_o)^{-1} = \omega\left[1 - (V / c_o)\cos\theta\right]^{-1} ; \tag{8}$$

though this, of course, may represent a decrease where the angle θ is obtuse.

The corresponding change in effective volume results from an effective change in the source's dimension ℓ in the direction of the observer [9]. Because the near side N of the source region is closer by a distance ℓ than its far side F, there is a certain "lag" τ between the times of emission from F and N of sounds that reach the observer simultaneously. Then in the time t for sound from F to arrive it travels a distance $c_o t$, but the corresponding emission from N starts after a time lag τ during which its distance from F in the direction of the observer has increased from ℓ to $\ell + w\tau$, and the sound emitted then travels a distance $c_o(t - \tau)$. The condition for both sounds to reach the same point at time t may be written

$$c_o t = \ell + w\tau + c_o(t - \tau), \text{ giving } \tau = \ell / (c_o - w) . \tag{9}$$

The source's effective dimension in the direction of emission is therefore altered to

$$\ell + w\tau = \ell(1 - w / c_o)^{-1} = \ell\omega_r / \omega : \tag{10}$$

a change by the same Doppler factor ω_r/ω that modifies the frequency. Indeed, the eddy's effective volume during emission is also increased by the Doppler factor ω_r/ω, because dimension in the direction of the observer is so increased whilst other dimensions are unaltered.

Because the effective eddy volume ℓ^3 appears to the first power in expression (5), whilst effective frequency occurs to the fourth power in the mean square of the second time derivative T_{ij}, these Doppler effects produce an overall modification by five Doppler factors [1,4],

$$(\omega_r / \omega)^5 = (1 - w / c_o)^{-5} = [1 - (V / c_o)\cos\theta]^{-5}, \tag{11}$$

in the intensity pattern (5) radiated per unit volume of turbulence. This gives a first indication of an important preference for forward emission from turbulence convected at "not so low" Mach number.

On the other hand, it is essential to recognize how, as V/c_o increases, Doppler effect tends also to degrade the compactness of aeroacoustic sources in relation to forward emission. Not only does $\omega\ell/c_o$ increase in proportion to Mach number, but an even greater value is taken by $\omega_r\ell/c_o$, the ratio which must be small if convected sources are to be compact. A very marked restriction on the extent (11) of intensity enhancement for forward emission as V/c_o increases is placed by these tendencies [1,4,13]. They can develop, indeed, to a point where the compact-source approximation (of low-frequency acoustics) may appropriately be replaced by its opposite (high-frequency) extreme: the ray-acoustics approximation. Thus, for supersonic source convection $(V/c_o>1)$, the relative frequency (8) becomes infinite in the Mach direction

$$\theta = \cos^{-1}(c_o / V) \; ; \tag{12}$$

moreover, it may be shown that radiation from the source proceeds along rays emitted in the Mach direction [5].

As this is a paper which devotes special attention to such radiation in the Mach direction, some immediate comments on its nature may perhaps be made. Equation (6) shows that the source's velocity of approach w towards an observer positioned at an angle (12) to its direction of motion is the sound speed c_o. On linear theory this means that different parts of a signal are observed simultaneously — which, of course, is the well known condition of stationary phase satisfied on rays [8]. Sounds emitted by a source approaching at a speed w exceeding c_o would be heard by the observer in reverse order (so that "pap pep pip pop pup" became "pup pop pip pep pap"!) but when $w = c_o$ all the sounds (vowels and consonants!) would be heard together as one single "boom".

From these preliminary comments the importance of two influences treated in later sections, that place limits on the signal propagated along rays, will already be clear. These are (i) the duration δ of well-correlated emission from turbulent eddies and also (ii) nonlinear propagation involving a departure of the signal velocity from c_o.

5 UNIFORMLY VALID DOPPLER-EFFECT APPROXIMATIONS

The correlation duration δ for convected turbulent eddies is defined so that velocities at times separated by substantially more than δ are almost uncorrelated while there is good correlation between velocities at times separated by substantially less than δ. This definition in terms of time separations (for the moving eddy) is directly parallel to the definition of ℓ in terms of spatial separations.

Combined use of correlation length ℓ and duration δ affords an approximation to the radiation pattern from convected eddies that has some value at all Mach numbers. Thus it is a uniformly valid approximation, spanning the areas of applicability of the compact-source and ray-acoustics approximations.

Figure 1 uses space-time diagrams where the space-coordinate (abscissa) represents distance in the direction of the observer. Diagram (a) for unconvected eddies approximates the region of good correlation, which must have spatial and temporal dimensions ℓ and δ, as a simple smooth curve; actually, an ellipse with ℓ and δ as its axes. Diagram (b) shows how the region of good correlation is sheared (sheared, in fact, by a distance w per unit time) for convected eddies approaching the observer at velocity w.

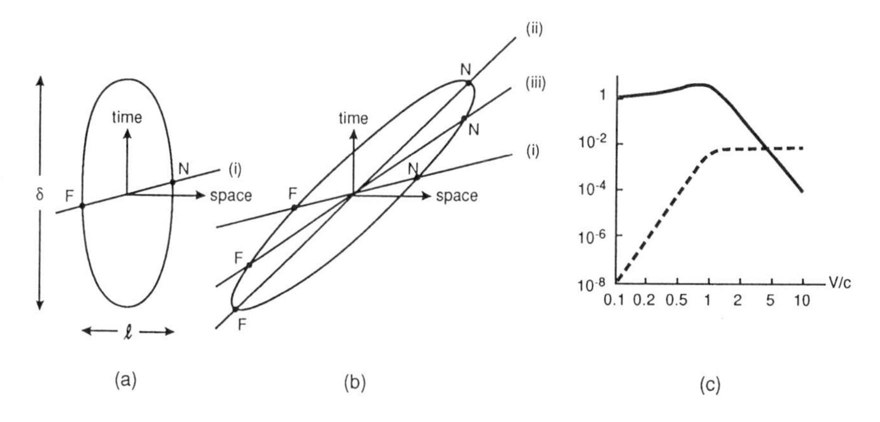

Figure 1. A uniformly valid Doppler-effect approximation.

Diagram (a). Space-time diagram for unconvected "eddies" of correlation length ℓ and duration δ.

Diagram (b). Case of "eddies" convected towards observer at velocity $w=V\cos\theta$, being Diagram (a) sheared by a distance w per unit time. Here, straight lines sloping by a distance c_o per unit time represent emissions received simultaneously by the observer. Case (i): w/c_o small. Case (ii): $w/c_o=1$. Case (iii): intermediate value of w/c_o.

Diagram (c).. Solid line: average (spherical mean) of the modification factor (13). Broken line: acoustic efficiency, η, obtained by applying this factor to a low-Mach-number "quadrupole" efficiency of (say) $10^{-3}(V/c_o)^5$.

Application to jets: For a jet of exit speed U, a typical eddy convection velocity V takes values between $0.5U$ and $0.6U$. In order of magnitude terms, then, the efficiency η makes a transition, at about $U/c_o=2$, between values around $10^{-4}(U/c_o)^5$ and an asymoptotically constant value of a little less than 10^{-2}.

Signals from a far point F and a near point N, in either case, reach the observer simultaneously — as do signals from other points on the line FN — provided that this line slopes by a distance c_o per unit time. Then diagram (b) distinguishes three cases as follows:

(i) in the compact-source case w/c_o is small, and the space component of FN in diagram (b) is $\ell(1 - w/c_o)^{-1}$, just as in equation (10) for the "usual" Doppler effect (which neglects the upper bound δ on correlation duration);

(ii) in the ray-acoustics case $w/c_o = 1$, and the space component of FN is $c_o\delta$;

(iii) in the intermediate case w/c_o is but moderately less than 1, and the space component of FN is ℓ multiplied by an enhancement factor

$$\left[(1 - w/c_o)^2 + (\ell/c_o\delta)^2\right]^{-1/2}. \tag{13}$$

Evidently, this form (13) of the enhancement factor comprehends all three cases and represents the effective augmentation of source volume due to convection.

The enhancement factor (13) needs to be applied, not only to the volume term ℓ^3 in the quadrupole field (5), but also twice to each of the pair of twice-differentiated terms inside the mean square; essentially, because time- differentiations in quadrupole fields arise from differences in the time of emission by different parts of the quadrupole source region — and the time component of FN in diagram (b) is simply the space component divided by c_o. As before, then, five separate factors (13) enhance the intensity field; and, with w replaced by $V\cos\theta$, expression (11) for the overall intensity modification factor is replaced by

$$\left\{\left[1 - (V/c_o)\cos\theta\right]^2 + (\ell/c_o\delta)^2\right\}^{-5/2}. \tag{14}$$

This is a significant change wherever $1 - (V/c_o)\cos\theta$ is relatively small, and it tends to limit the predicted preference for forward emission [1,4].

It also gives an improved description of the influence of Doppler effect on the overall acoustic power output from convected turbulence. For example, the solid line in diagram (c) gives a log-

log plot of the average value (spherical mean) of the modification factor (14) as a function of V/c_o on the reasonable assumption that $\ell = 0.6\ V\delta$. It will be noted that, as V/c_o increases, this average modification factor rises a little at first, but falls drastically like $5(V/c_o)^{-5}$ for V/c_o significantly greater than 1.

On the other hand aerodynamic noise at low Mach number (see above) has an acoustic efficiency η scaling as $(U/c_o)^5$ where U is a characteristic velocity in the flow. Actually, it would be permissible to take that characteristic velocity as the eddy convection velocity V; although, if this were done in the case of a jet, it would be important to recognize that V is not the jet exit speed itself but takes values between 0.5 and 0.6 times the jet exit speed. Thus, an order of magnitude $10^{-4}(U/c_o)^5$ for η in terms of jet exit velocity U corresponds to an order of magnitude $10^{-3}(V/c_o)^5$ in terms of the eddy convection velocity V.

The broken line in diagram (c) shows how this acoustic efficiency η, supposed to take the value $10^{-3}(V/c_o)^5$ for small V/c_o, is modified after multiplication by the average modification factor (solid line). This modification causes η to approach a constant value of about 0.005 (aeroacoustic saturation) at high Mach numbers, when slightly less than 1% of jet power is radiated as aerodynamic noise.

Supersonic jets are called "properly expanded" when they emerge — from appropriately shaped nozzles — as parallel flows. This is a contrasting case to that of a supersonic jet emerging as a non-parallel flow, which is necessarily followed by a sequence of shock waves (leading to augmentation of aerodynamic noise) in the characteristic "diamond" shock-cell pattern.

Aeroacoustic saturation similar to that indicated in diagram (c) is observed for properly expanded supersonic jets, with acoustic efficiency becoming close to an asymptotically constant value of a little less than 10^{-2} when the jet Mach number U/c_o exceeds about 2. These are the "extreme-speed" jets, with V itself (the eddy convection velocity) exceeding c_o, for which aerodynamic noise is directed [5] along rays inclined at the Mach angle (12). Before dedicating the remainder of the lecture to the nature of aerodynamic noise in this extreme-speed limit, I have attempted through the above discussion to exhibit that continuous trend of changing noise patterns which links it to the problem of aerodynamic noise at low Mach number.

6 NONLINEAR PROPAGATION OF NOISE FROM EXTREME-SPEED JETS

Extreme-speed jets have just been defined as those properly expanded jets, with speeds more than twice the atmospheric sound speed c_o, for which the eddy convection velocity V is itself supersonic and noise is primarily emitted in the Mach direction (12). Significant influences on this type of aerodynamic noise are exerted, not only by correlation duration (see above) for a supersonically moving eddy, but also by important consequences of nonlinear propagation.

Indeed, in a rather obvious analogy with a supersonically moving body, we may expect each eddy's sound field not only (i) to be emitted in the Mach direction but also (ii) to take the form of a "supersonic boom" which, from nonlinear propagation effects, incorporates one or more shock waves. On the other hand, because it is a chaotic sequence of supersonically moving eddies that generates such waveforms, the jet's near noise field must consist (as is, indeed, observed) of a thickly packed random assemblage of conical shock waves.

Thus although, as just remarked, there may be some sort of continuous trend in the mechanisms of generation of aerodynamic noise between classical processes at lower Mach numbers and such extreme-speed jets, nevertheless studies of its propagation become quite different in this latter case. As a marked contrast to approaches that use the full nonlinear flow equations in the jet and simple linear equations of propagation outside it (generation being associated with differences between the equations), it becomes necessary to recognize important nonlinear effects on propagation itself.

In the literature of aerodynamic noise for extreme-speed jets there already exist studies of the generation process that go far beyond the perfunctory sketches which I have included in this lecture. On the other hand, a properly detailed investigation of nonlinear effects on its propagation appears to have been neglected. For this reason, I have chosen to devote the rest of this lecture to such a detailed study of the nonlinear propagation of a thickly packed assemblage of conical shock waves once it has been formed.

Briefly, this detailed study demonstrates how the inherent tendency of the shock strengths to decay (through conical spreading and internal dissipation) is counteracted in part by increases in strength that occur whenever adjacent shocks unite. A certain "bunching" tends to arise, because union of two adjacent shocks is found to increase the likelihood of further union with other neighbouring shocks. The high-frequency part of the noise spectrum is made more intense than it would otherwise be by these bunching tendencies.

A well known transformation of coordinates [8] — suggested by the above-noted analogy with supersonic-boom theory [7] — is used (see Appendix) to reduce the problem of conical propagation, aimed at analysing how a temporal waveform varies with distance r along the Mach direction, to a plane-wave problem. In this latter problem the time t replaces $r^{1/2}$, and attention is focussed on how a given spatial waveform varies with time.

The spatial waveform assumed in this latter context is however immensely more complicated than the simple "N-wave" form appearing in supersonic-boom theory. Instead, it is a spatially unlimited assemblage of random sawtooth waves of the type into which a general plane sound wave of large amplitude would evolve during a certain time t.

This assemblage comprises randomly located shocks (with random strengths) separated by expansion waves in which the slope of the quantity "excess signal velocity" (excess, that is, over the undisturbed sound speed c_o) as a function of position takes the value $1/t$. The inherent tendency of the shocks themselves is also to decay like $1/t$; but, as already described, this tendency is opposed whenever shocks unite with adjacent shocks.

After an inverse transformation into the original variables describing conical propagation, t is replaced by $r^{1/2}$ but predicted shock strengths have to be divided by a further $r^{1/2}$ factor. Thus the basic tendency of the shock strengths is now to decay like $1/r$; but, yet again, the opposing tendency described above as "bunching" may prove to be important for the analysis of how aspects of aerodynamic noise are modified in the region of conical propagation that surrounds an extreme-speed jet.

7 RANDOM SAWTOOTH WAVES IN TRANSFORMED COORDINATES

The classical transformation described in the Appendix addresses the problem of how a thickly packed random assemblage of conical shock waves will evolve with increasing r (distance in the direction of propagation) by reducing it to an interesting, yet hitherto neglected, problem of plane-wave propagation. This plane-wave problem, which will now be defined, treats the evolution in time of a spatially unlimited assemblage of random sawtooth waves.

Because a random sawtooth wave is primarily important as a form of acoustic noise, any useful specification of such a wave must be one which facilitates identification of its noise spectrum.

Now, noise spectra are determined in practice — whether in a physical or in a numerical experiment — from signal records of very great yet finite length which are Fourier analysed by F.F.T. techniques.

Moreover any such Fourier analysis of a signal in an interval of great yet finite length expresses it essentially as a Fourier series, where successive terms in the series describe oscillations with very closely neighbouring frequencies. The noise spectrum is proportional to the squares of their amplitudes, regarded as a function of frequency, and the very close spacing between neighbouring frequencies allows it in practice to be depicted as a continuous curve.

Actually, the Fourier series in question represents an exactly periodic function, with period equal to the length of the interval. Specifically, it represents the unique periodic function (with that period) which coincides with the signal record within the interval.

These considerations lead us, in any study of spatially unlimited waves of random sawtooth type, to focus attention in practice on a random sawtooth wave which is specified on an interval of great but finite length L; while, for completeness, its form outside that interval is defined by requiring it to be a periodic function with period L. There is no implication here that our interest is really confined to strictly periodic functions; on the contrary, we are interested in waves that, besides being spatially unlimited, have an everywhere random character. But we recognize that any physical or numerical experiment will confine attention to an interval of great but finite length L within which Fourier analysis of the signal gives a representation of it as an exactly periodic function of period L, and it is this representation which we find fruitful to analyse.

Plane waves travelling in the x-direction in a homogeneous medium with undisturbed sound speed c_o may in nonlinear acoustics be conveniently described in terms of an independent variable

$$X = x - c_o t \qquad (15)$$

and a dependent variable equal to the excess signal velocity

$$v = u + c - c_o , \qquad (16)$$

where u is the fluid velocity and c the local sound speed (equal to $c_o + 0.2u$ in air). In a frame of reference moving at speed c_o, as defined by the space coordinate X, any value of v is propagated at a velocity equal to v itself.

This is the result which, whenever v is continuous, may be represented by the familiar partial differential equation

$$\partial v / \partial t + v \partial v / \partial X = 0, \tag{17}$$

with its characteristics of slope v in the (t,X) plane. However, the system necessarily tends to develop shocks, which will here be treated as discontinuities. The speed of a shock is equal to the average of the smaller and greater values of v which appear ahead of and behind it; accordingly, the shock absorbs characteristics ahead of it by running into them, while absorbing characteristics from behind as they run into it.

It may be shown (for example, by differentiating (17) with respect to X) that the reciprocal slope $(\partial v / \partial X)^{-1}$ of continuous parts of the wave must increase at unit rate along any characteristic. It means that any negative value of the reciprocal slope must after a finite time increase to zero, corresponding to infinite slope. This, of course, is when a shock appears, with its subsequent propagation governed by laws quoted above.

By contrast, any positive initial value of the reciprocal slope must grow indefinitely at unit rate, which after a time t adds a term t to that initial value. When t has become large, this added term t has become dominant over the initial value (at least, if this is not too big), so that to a close approximation the slope $\partial v / \partial X$ itself takes the value $1/t$. Indeed, a classically familiar argument why this must prove to be rather a good approximation is that, when t has become large, all characteristics with relatively bigger initial values of the reciprocal slope have disappeared through running into shocks; essentially, because of their tendency to be close to characteristics with negative initial values of reciprocal slope (from which shocks necessarily develop). The continuous parts of the waveform are dominated, therefore, by characteristics on which the slope has become close to $1/t$.

These are the reasons why an initial random acoustic wave of large amplitude develops after time t into what we define as a random sawtooth wave. This comprises shocks at positions

$$X = X_n \qquad (-\infty < n < \infty) \tag{18}$$

with X_n a decreasing function of the integer n; where the values of v behind and ahead of the nth shock are

$$v_n \quad \text{and} \quad v_n - z_n , \tag{19}$$

and where we shall describe as the "strength" of the nth shock the discontinuity z_n between the values of v behind and ahead of it. Moreover, the distribution of v has the slope $1/t$ between shocks, so that the value $v_n - z_n$ of v just ahead of the nth shock is related to its value v_{n-1} just behind the $(n-1)$th shock by the equation

$$v_n - z_n = v_{n-1} - (X_{n-1} - X_n)/t . \tag{20}$$

The decision to assume the wave periodic with period L, which has already been carefully explained, has certain consequences for the quantities (18) and (19). At any particular time t there will be a certain specific number N of shocks within a single period; here, N is a function of the time, taking integer values, which is reduced discontinuously by 1 at every instant when a union of two shocks occurs. The periodicity then implies that

$$v_{n+N} = v_n \ , \ z_{n+N} = z_n \ , \ X_{n+N} = x_n - L; \tag{21}$$

so that equation (20), summed between $n = 1$ and $n = N$, gives

$$\sum_{n=1}^{N} z_n = L/t . \tag{22}$$

This equation, indicating a balance between the net compressive and expansive effects in the sawtooth wave, ensures that the period L remains unchanged because the sum of the shock strengths on the left-hand side is found (see below) to vary as $1/t$..

Figure 2 illustrates the form of a random sawtooth wave at a particular time $t = T$, with the number N of shocks inside the period $0 < X < L$ equal to 25. The randomness of this wave derives from the fact that a random number generator was used to determine all the strengths z_n of the 25 shocks and all the spacings $h_n = X_{n-1} - X_n$ between shocks, subject to

(i) the need, which equation (21) demonstrates, for the sum of the spacings h_n to be L;
(ii) the corresponding condition (22) on the sum of the strengths; and
(iii) restrictions on each value of h_n or of tz_n to lie between $0.01L$ and $0.09L$.

Thus the spacings and strengths, subject to (i) and (ii), take random values between these upper and lower bounds. (Here, while the upper bound is intended to reflect some sort of limitation on sizes and strengths of noise sources, the lower bound is imposed mainly for convenience of graphical representation.) The above values fix uniquely a plot of v against X with zero integral over a period, and the evolution of this particular sawtooth wave is exhibited in Section 11 below.

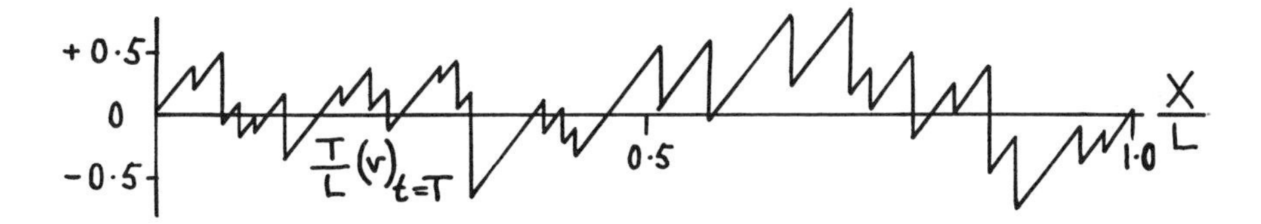

Figure 2. A random sawtooth wave, with the number N of shocks inside the period $0 < X < L$ equal to 25; and with their spacings and strengths determined, subject to conditions (i), (ii) and (iii), by a random number generator.

For a random function v of X defined in an interval $0 < X < L$ (here, a particular single period for the function), its noise spectrum as defined so that $P(k)dk$ represents the contribution to the function's mean square from wavenumbers between k and $k + dk$ is classically given as

$$P(k) = \frac{1}{\pi L} < \left| \int_o^L v e^{-ikX} dX \right|^2 > \tag{23}$$

where the angle brackets denote a statistical mean value. The high-wavenumber behaviour of this noise spectrum, for a random sawtooth function v dominated by discontinuous changes z_n at points X_n as in (18) and (19), is

$$P(k)=\frac{1}{k^2\pi L}<\left(\sum_{n=1}^{N} z_n e^{-ikX_n}\right)\left(\sum_{n=1}^{N} z_n e^{ikX_n}\right)>=\frac{1}{k^2\pi L}\sum_{n=1}^{N} z_n^2 \ , \tag{24}$$

so that it depends on the sum of the squares of the shock strengths z_n.

It is this dependence which makes unions of shocks important. At any such union, the sum of the shock strengths themselves remains unchanged; however, the sum of their squares is necessarily increased, so that the high-frequency part of the noise spectrum is intensified.

8 SAWTOOTH-WAVE EVOLUTION UP TO WHEN A FIRST UNION OCCURS

The dynamics of the shocks is analysed next up to when a union first occurs. At each instant the basic law

$$\dot{X} = \mathrm{v}_n - ½z_n \ , \tag{25}$$

which expresses the velocity X_n of a shock wave as the average of the values (19) of v behind and ahead of it, implies also that

$$\dot{\mathrm{v}}_n = (-½z_n)/t \tag{26}$$

because, during a small time δt, the shock absorbs into itself from behind a section $(1/2)z_n\delta t$ of smooth waveform with slope $1/t$.

Now careful study of the system of equations (25), (26) and (20) shows that their completely general solution takes the form

$$tz_n = Y_n \ , \ \mathrm{v}_n = b_n + (Y_n / 2t) \ , \ X_n = a_n + ½Y_n + b_n t \ , \tag{27}$$

where a_n, b_n and Y_n are constants satisfying

$$a_{n-1} - a_n = Y_n \ . \tag{28}$$

This solution gives a constant value Y_n to the product of the shock strength z_n with the time t. Here, as noted earlier, t is measured from an origin ($t = 0$) when the waveform was initiated, so that it later developed into a sawtooth wave with the slope $1/t$ for all smooth sections thereof.

Great interest attaches to the distance $h_n = X_{n-1} - X_n$ between two adjacent shocks. Equations (27) and (28) show that

$$h_n = (Y_{n-1} + Y_n)/2 + (b_{n-1} - b_n)t. \tag{29}$$

Now, as noted earlier, the analysis in this lecture is only concerned with propagation of random sawtooth waves once they have appeared; say, from $t = T$ when h_n takes the form

$$H_n = (Y_{n-1} + Y_n)/2 + (b_{n-1} - b_n)T. \tag{30}$$

Eliminating the unknown constant $b_{n-1} - b_n$ between (29) and (30) we obtain an equation for the evolution of h_n in the form

$$h_n = (t/T)[H_n - (Y_{n-1} + Y_n)\sigma], \tag{31}$$

where it proves useful to introduce a new time-like variable σ. The equations

$$\sigma = (t - T)/2t, \quad t = T/(1 - 2\sigma) \tag{32}$$

make σ an increasing function of t and map the unbounded interval $T < t < \infty$ into the finite interval $0 < \sigma < 1/2$. It turns out that σ is a highly convenient measure of the times at which unions of shocks occur.

As just a preliminary example of this, we note from equation (31) that the very first union of two shocks occus when

$$\sigma = \min_{1 \le n \le N} [H_n / (Y_{n+1} + Y_n)], \tag{33}$$

since it is this value of σ that first allows one of the h_n (distances between shocks) to fall to zero. Because of the periodicity assumption which was exhaustively discussed earlier, the minimum (33) must of course be attained not only for a particular value n within the period $1 \le n \le N$ but also for the corresponding value ($n \pm mN$, with m an integer) in any other period. Attention is here focussed, however, on a single period — except for the fact that when equation (28) or (33), or any other equation involving $n - 1$, is applied to (say) $n = 1$, periodicity is used to interpret $n - 1$ as N.

On the other hand, there is no need in a random sawtooth wave to consider the possibility (really an impossibility — since it would occur with zero measure in a probability space of random variables) that the same minimum value (33) might be attained for two different n in the period $1 \leq n \leq N$. It will be assumed rather that the first union occurs for a particular n and σ, as specified by equation (33). Also, because equations (22) and (27) imply that

$$\sum_{n=1}^{N} Y_n = L = \sum_{n=1}^{N} H_n \tag{34}$$

the minimum (33) is readily shown to satisfy the requirement, $\sigma < 1/2$, for it to correspond to an actual time $t > T$ for first union between shocks.

Until this time, of course, every Y_n (defined as the product of the strength of the nth shock with the time t) has remained constant. Then a first union — say, of the nth and $(n - 1)$th shocks — produces addition of their strengths z_n and z_{n-1}, with consequences noted earlier for the high-frequency form (24) of the noise spectrum.

In the meantime, there is a reduction by 1 in the number N of shocks in a single period, but the sum on the left-hand side of (34) continues to take the value L simply because the new shock, replacing those associated with the Y-values Y_n and Y_{n-1}, takes on their added Y-value $Y_n + Y_{n-1}$. Similarly equation (34) remains unchanged at all later unions — other properties of which will now be investigated.

9 FORMULAS SPECIFYING ALL LATER UNIONS OF SHOCKS

In this section, the tendency to "bunching" of shocks — that is, an increased likelihood of union between shocks that have already participated in union with other shocks — is quantified by means of a general formula for the time of union of two shocks which allows for all preceding unions. The two shocks to be considered are taken as (i) a shock originally numbered n, into which other shocks have merged from behind, and (ii) a shock originally numbered $n - 1$, which has run into (and merged with) various shocks ahead of it.

Evidently, this is the most general possible merger. Indeed, when any pair of shocks unite, it is permissible to "identify" the back one with an original shock (numbered n, say) into which others have merged from behind, and the front one with an original shock (numbered n - 1, say) that has run into others ahead of it.

Specifically, we shall suppose that the shock originally numbered n has participated in a total number m_n of unions with shocks from behind, with Y-values

$$Y_{Bm}, \text{ occurring at times } t = t_{Bm}(m = 1 \text{ to } m_n). \tag{35}$$

Similarly, we assume that the shock originally numbered n - 1 has participated in a total number m_{n-1} of unions with shocks ahead of it, with Y-values

$$Y_{Am}, \text{ occurring at times } t = t_{Am}(m = 1 \text{ to } m_{n-1}). \tag{36}$$

It is convenient to embark upon this problem by considering first the dynamics of the shock originally numbered n immediately after it has undergone just the first of the above unions, at $t = t_{B1}$. Because we "identify" this merged shock with the shock originally numbered n, we continue to use the subscript n in quantities such as (18) and (19) associated with this shock. Then the differential equations (25) and (26) continue to apply to this united shock, as does the relationship (20), so that the form (27) of this system's general solution is unaltered. However, the constants a_n, b_n and Y_n are changed to new values

$$a_n - Y_{B1}\ ,\ b_n + (Y_{B1} / 2t_{B1})\ ,\ Y_n + Y_{B1}\ ; \tag{37}$$

while (28) is still satisfied since a_{n-1} has not been changed.

Out of these new values (37) for a_n, b_n and Y_n, the last (added Y-values for uniting shocks) has been explained above. Also, we may readily verify that the new values for a_n and b_n are the only ones which satisfy two essential conditions: that union at time $t = t_{B1}$ makes no discontinuous change in the position X_n of the nth shock, while increasing v_n (the value of v behind it) by $z_{B1} = Y_{B1}/t_{B1}$ (the merging shock's jump in v).

Similar studies show that, in each subsequent union of the nth shock with shocks from behind as specified in (35), the constants receive further changes as in (37) but with Y_{Bm} and t_{Bm} replacing Y_{B1} and t_{B1}; so that, after all of them, a_n, b_n and Y_n have become

$$a_n - \sum_{m=1}^{m_n} Y_{Bm} \ , \ b_n + \sum_{m=1}^{m_n} \left(Y_{Bm} / 2t_{Bm}\right) \ , \ Y_n + \sum_{m=1}^{m_n} Y_{Bm} \ . \tag{38}$$

Other careful studies of the unions of the (n - 1)th shock with shocks ahead of it as specified in (36) demonstrate that, when all are completed, a_{n-1}, b_{n-1} and Y_{n-1} have become

$$a_{n-1} + \sum_{m=1}^{m_{n-1}} Y_{Am} \ , \ b_{n-1} - \sum_{m=1}^{m_{n-1}} \left(Y_{Am} / 2t_{Am}\right) \ , \ Y_{n-1} + \sum_{m=1}^{m_{n-1}} Y_{Am} \ . \tag{39}$$

These results allow both X_n and X_{n-1} to be obtained by use of (27), so that the separation $h_n = X_{n-1} - X_n$ between the shocks can be expressed as

$$h_n = \tfrac{1}{2}\left(Y_{n-1} + Y_n\right) + \left(b_{n-1} - b_n\right)t - \tfrac{1}{2}\sum_{m=1}^{m_{n-1}} Y_{Am}\left(t / t_{Am} - 1\right) - \tfrac{1}{2}\sum_{m=1}^{m_n} Y_{Bm}\left(t / t_{Bm} - 1\right) . \tag{40}$$

Here, equation (30) may be used to substitute for $b_{n-1} - b_n$ in terms of the value H_n of the separation h_n at the initial time $t = T$. Then, with times t, t_{Am} and t_{Bm} substituted in terms of the corresponding σ-values from (32), we obtain

$$(1 - 2\sigma)h_n = H_n - \left(Y_{n-1} + Y_n\right)\sigma - \sum_{m=1}^{m_{n-1}} Y_{Am}\left(\sigma - \sigma_{Am}\right) - \sum_{m=1}^{m_n} Y_{Bm}\left(\sigma - \sigma_{Bm}\right) \ . \tag{41}$$

Equation (41) gives an elegant and completely general expression for the value of the time-like variable σ when the nth and (n - 1)th shocks unite (so that h_n becomes zero). The form of this expression becomes particularly instructive if we use σ_n to signify the ratio

$$\sigma_n = H_n / \left(Y_{n-1} + Y_n\right) \ . \tag{42}$$

Equation (31) shows how, provided that $\sigma_n < 1/2$, this value of σ specifies the time at which the nth and (n - 1)th shocks would have united if no other unions of shocks had occurred first.

On the other hand, the actual value of σ corresponding to the time at which the shocks unite is given, by using the substitution (42) after putting $h_n = 0$ in (41), as

$$\sigma = \frac{(Y_{n-1}+Y_n)\sigma_n + \sum_{m=1}^{m_{n-1}} Y_{Am}\sigma_{Am} + \sum_{m=1}^{m_n} Y_{Bm}\sigma_{Bm}}{Y_{n-1}+Y_n+\sum_{m=1}^{m_{n-1}} Y_{Am} + \sum_{m=1}^{m_n} Y_{Bm}} . \tag{43}$$

Expression (43) may be recognized as a weighted mean of the σ-values σ_n, σ_{Am} and σ_{Bm} with weights given in terms of shock strengths according to the following simple rules.

The weight attached to σ_{Am}, or to σ_{Bm}, is Y_{Am} or Y_{Bm} respectively; in other words, it is the Y-value (strength multiplied by t) of the shock that merges when $\sigma = \sigma_{Am}$ with the $(n - 1)$th shock or when $\sigma = \sigma_{Bm}$ with the nth shock. By contrast, the weight attached to σ_n is the initial combined Y-values of the nth and $(n - 1)$th shocks which would have united at $\sigma = \sigma_n$ (provided that $\sigma_n < 1/2$, though otherwise not at all) if no preceding unions had occurred. The sum of all these weights, which constitutes the denominator of (43), represents the overall Y-value of the united shock that results from union of the nth and $(n - 1)$th shocks after they have respectively undergone all the mergers (35) and (36).

10 BUNCHING TENDENCIES AND THEIR IMPLICATIONS

Very simple properties of the weighted mean (43) tell us that, because all these mergers at the times (35) and (36) have preceded the union of the nth and $(n - 1)$th shocks, so that

$$\sigma > \sigma_{Am} \text{ and } \sigma > \sigma_{Bm} \text{ for all } m, \tag{44}$$

therefore the quantity σ_n, which is the only other among all the σ-values of which (43) represents the weighted mean, must satisfy the opposite inequality $\sigma < \sigma_n$. In words, the union of the nth and $(n - 1)$th shocks takes place earlier, in consequence of any preceding unions in which either have participated, than would have been the case if no such preceding unions had occurred.

But still more valuable than any such purely qualitative statement is the quantitative expression (43) for the reduced value of σ at the instant of union. For example, in cases when the Y-values of the nth and $(n - 1)$th shocks were initially rather small — before they merged with stronger shocks—it is of course quite possible for σ_n, as defined by the ratio (42), to be considerably greater than 1/2. This implies that the shocks are initially moving farther apart, so that their union could never have taken place if neither had merged with other shocks. Yet, in the weighted mean (43), the combined weights attached to all the σ_{Am} and σ_{Bm} may greatly exceed that attached to σ_n if the Y-values of the merging shocks amount in total to much more than the initial combined value $Y_{n-1} + Y_n$; and, evidently, this allows the value (43) for σ (associated with the time of union of the nth and $(n - 1)$th shocks) to be only a little bigger than the greatest of the σ_{Am} and σ_{Bm}.

Such considerations allow us to infer that a random sawtooth wave may be subject to a very marked "bunching" process. This is a process of "snowball" type where early local unions of shocks act to stimulate further unions with neighbouring shocks. Because this process, while leaving the sum of all the Y-values unchanged, produces an increase in the sum of their squares, it may have an important effect of enhancing (as expression (24) shows) the high-frequency part of the noise spectrum.

11 AN ALGORITHM FACILITATING NUMERICAL EXPERIMENTS ON BUNCHING

Another useful route to the quantitative study of the bunching process is through numerical experiments. In principle, these might be attempted by computing "weak" solutions of equation (17) and scrutinizing their evolution from instant to instant to identify those times when unions of shocks occur. On the other hand such identifications can be a little inconvenient and somewhat lacking in precision with practical values of the grid spacing.

A much more straightforward approach is one that treats all shocks as discontinuities and which makes use of just a single time-step between each union and the next. The basic result (33), for defining when an initial set of N shocks within a period L becomes reduced to a set of $N - 1$ shocks after the first union of two shocks has occurred, may be applied successively for this purpose in the following algorithmic treatment.

The algorithm starts from N shocks filling the period L at a given time t, with Y-values Y_n (n = 1 to N) and with spacings $h_n = X_{n-1} - X_n$ between shocks, where

$$\sum_{n=1}^{N} h_n = L \text{ and, by (22), } \sum_{n=1}^{N} Y_n = L . \tag{45}$$

The form of the algorithm is a respecification of all these variables at the instant when the first union of any pair of adjacent shocks has occurred. (This algorithm can then, systematically, be given successive applications to determine subsequent developments, including bunching.)

The most obvious change in one of the variables after that union is that N has been replaced by $N - 1$. All of the other changes, however, depend critically on those quantities which have been shown to define when a union first occurs; namely the minimum (33) and the value of n for which the minimum is attained.

In the present case we define, then,

$$\sigma_m = \underset{1 \leq n \leq N}{Min} \left[h_n / \left(Y_{n-1} + Y_n \right) \right], \text{ attained for } n = m, \tag{46}$$

and note that, according to equations (45), this minimum value cannot exceed 1/2. Then the time t needs to be respecified as the time

$$t / \left(1 - 2\sigma_m \right) \tag{47}$$

at which a union first occurs. At that instant the mth and $(m - 1)$th shocks have united, and are designated thereafter as the $(m - 1)$th, with Y-value $Y_{m-1} + Y_m$. For other values of n, the nth shock retains its Y-value Y_n, and is still numbered as the nth shock for $n < m - 1$ but now needs to be renumbered as the $(n - 1)$th shock for $n > m$. Also, equation (31) tells us that the spacing h_n is changed to

$$\left[h_n - \sigma_m \left(Y_{n-1} + Y_n \right) \right] / \left(1 - 2\sigma_m \right) \tag{48}$$

for $n \leq m - 1$. The same change occurs for $n \geq m + 1$ except that the value of n is simultaneously replaced by $n - 1$.

A special note is required to the effect that quantities involving Y_{n-1} need when $n = 1$ to be reinterpreted (by periodicity) with Y_{n-1} replaced by Y_N. Similarly, when $m = 1$, the united shock is then renumbered as the $(N - 1)$th shock, with Y-value $Y_N + Y_1$.

The overall algorithm may be summarised as follows. Given a positive integer N and a time t, and a set of intervals and of Y-values (shock strengths multiplied by the time) h_n and Y_n for $n = 1$ to N, satisfying equations (45), we determine the minimum (46) where Y_o is to be interpreted as Y_N. Then the algorithm

	replaces each of	**by the new value**
(a)	N	$N - 1$
(b)	t	$t/(1 - 2\sigma_m)$
(c)	Y_n for $n < m - 1$	Y_n
(d)	Y_{m-1} if $m > 1$	$Y_{m-1} + Y_m$
(e)	Y_{N-1} if $m = 1$	$Y_N + Y_1$
(f)	Y_n for $m \leq n \leq N - 1$ (unless $n = N - 1$ and $m = 1$)	Y_{n+1}
(g)	h_n for $n < m$	$[h_n - \sigma_m (Y_{n-1} + Y_n)]/(1 - 2\sigma_m)$
(h)	h_n for $m \leq n \leq N - 1$	$[h_{n+1} - \sigma_m(Y_n + Y_{n+1})]/(1 - 2\sigma_m)$

but it should perhaps be mentioned that, evidently, there are no values of n for which (c) can be applied if $m = 1$ or 2, and none for which (f) can be applied if $m = N$. Similarly, there are no values of n for which (g), or else (h), can be applied if $m = 1$, or $m = N$, respectively. Subject to these interpretations, the algorithm can be very readily executed, and then re-executed as many times as required to determine all subsequent developments of the random sawtooth wave.

Figure 3 shows the results of applying this algorithm to the random sawtooth wave illustrated in Figure 2. That configuration at the initial time $t = T$, with its twenty-five shocks, occupies the bottom of the diagram; the rest of which depicts the paths of the shocks, including all unions, for $T \leq t \leq 7T$.

During this period there have occurred one bunching of nine shocks, two bunchings of five shocks and one bunching of three shocks; on the other hand, three out of the original twenty-five shocks have avoided participating in any unions. All of this appears consistent with the suggested definition of bunching as a tendency "for shock waves that have already formed unions with other shock waves to acquire an increased proneness to form further unions."

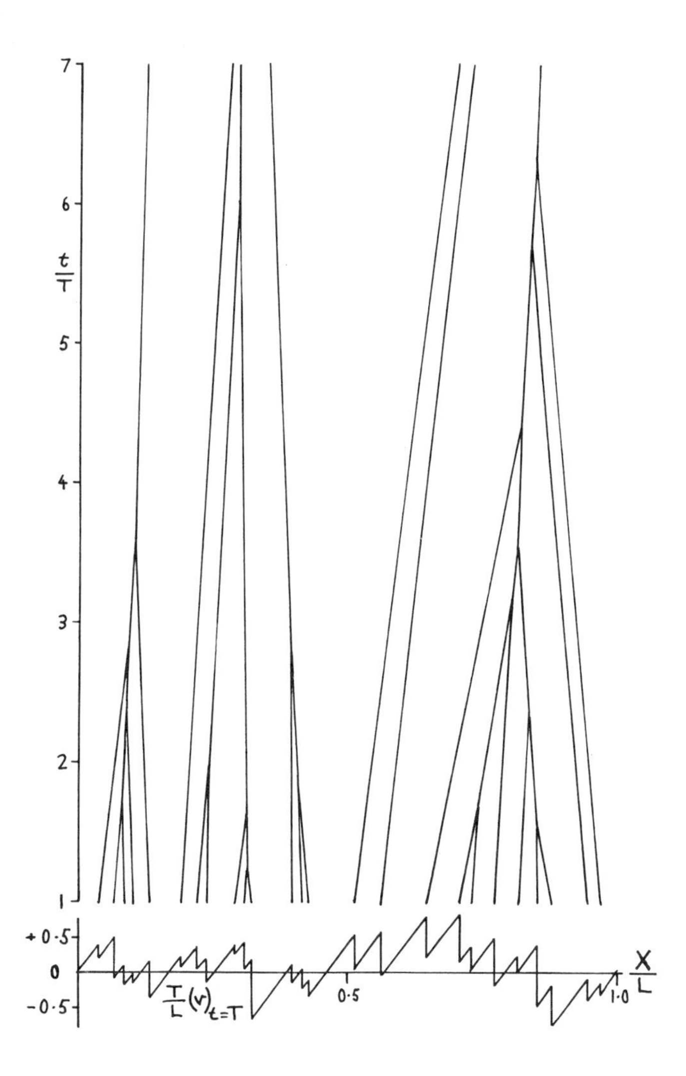

Figure 3. Results of applying the algorithm of section 11 to the initial waveform of Figure 2. This is reproduced at the bottom of the diagram; the rest of which depicts (by plots of X/L against t/T) the paths of the shocks, including all unions, for $T \leq t \leq 7T$. Reading from left to right, note two buncings of five shocks each, one of three shocks and one of nine shocks, interspersed with three shocks that have avoided participating in any unions.

It may be natural to ask whether Figure 3 represents in any sense a "selected" example; this question, however, has a negative answer. Figure 2 was produced, as explained in section 7, by means of a random number generator, and the algorithm of this section was simply applied to the very first wave so generated. I shall be quite content to leave to later investigators the pleasure of executing all those numerous "runs" of the algorithm that may be necessary to establish statistical trends!

At each and every union, as already mentioned, the sum of the Y_n remains constant but the sum of their squares increases discontinuously. Figure 4 shows this process in the case of the sawtooth waveform evolution illustrated in Figure 3. Within the interval $T \leq t \leq 7T$ quite a pronounced rise in $\Sigma Y_n{}^2$ is found, and it is intriguing that this discontinuously increasing dependence on t deviates only a little from a straight line through the origin, which would represent a simple proportionality to t.

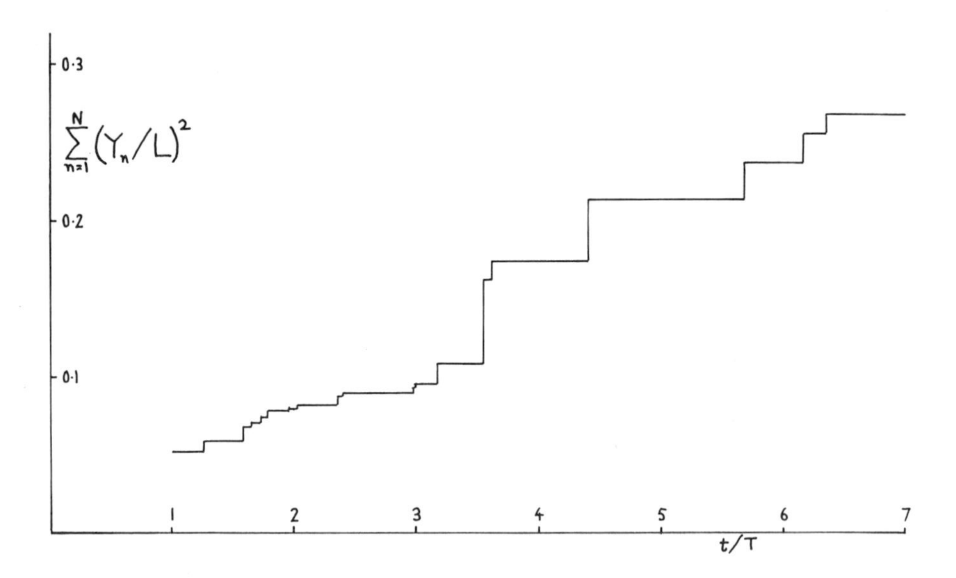

Figure 4. A plot of $\sum_{n=1}^{N}(Y_n/L)^2$ against t/T, for the sawtooth waveform evolution illustrated in Figure 3, is used to indicate how bunching enhances the high frequency part of the noise spectrum.

The implication here is that, although the inherent tendency of each shock strength $z_n = Y_n t^{-1}$ is to vary as t^{-1}, nevertheless the sum Σz_n^2 which appears in the high-frequency noise spectrum (24) shows a variation very different from t^{-2} and, indeed, one rather close to t^{-1}. The bunching process, in short, significantly enhances the high-frequency part of the noise spectrum.

12 CONCLUSION

This has been a lecture which, in the words of its title, has aimed at exploring just "some aspects" of jet noise fields. Indeed, even that single aspect which has been covered in most detail (the nonlinear propagation of the noise signal from extreme-speed jets) has been investigated only in a noise field quite close to the jet, where a rather consistently conical propagation of noise is to be found. This noise has its origin in the jet mixing region, separating an internal core flow of velocity U from the outside atmosphere, with eddies in this region convected at an approximately constant velocity V which defines the Mach direction (12) for noise emission.

On the other hand, there exists only a finite length of jet mixing region, beyond which the core flow vanishes and eddy convection velocities immediately start to fall off with distance. The beam of truly conical noise propagation, then, is only of finite width and, accordingly, can maintain its conical character for only a limited distance from the jet axis. Beyond that distance the propagation must change progressively to a propagation which, although still directional, is beginning to exhibit an essentially spherical attenuation.

Nonetheless, although conscious of all this, I have here placed emphasis on the field of conical noise propagation rather close to the jet for two main reasons. The first applies if our concern really is with just an isolated jet radiating into still air.

Then general nonlinear acoustics [8] tells us that waveform modifications due to nonlinear propagation effects are enormously greater in a conically spreading wave than in a spherically spreading wave; specifically, the square-root transformation (A4) derived in the Appendix below is replaced by a logarithmic transformation for a spherical wave which, accordingly, undergoes far less modification from nonlinear propagation. The conical field is specially important, then, as the region of occurrence of most of the spectrum-modification effects of nonlinear propagation.

But theoretical studies may have even more importance for the problem of how aero-engine installations are to be designed so as to reduce ground levels. In this context a jet noise field where propagation is directionally concentrated (say, in the Mach direction) may offer engine-installation designers a rather special opportunity to incorporate shielding devices aimed at limiting community noise. This, then, is yet another good reason why, in this Keynote Lecture at the outset of the celebrations of Bill's 80th birthday, I have presented to my distinguished audience quite a detailed analysis of how the noise spectrum in that conical field with its "random assemblage of shocks" may be influenced by the general tendency on which I have affectionately ventured to bestow the name "bunching."

APPENDIX. CONICAL-WAVE PROBLEM TRANSFORMED INTO PLANE-WAVE CASE

A classical transformation converts the problem of conically propagating waves incorporating relatively weak shocks into a plane-wave problem. The transformation is effective in any region which for linear acoustics would be described as a far field, with amplitudes varying as $r^{-1/2}$ along linearized characteristics.

The corresponding law for nonlinear acoustics, in continuous parts of the field, is that amplitudes vary as $r^{-1/2}$ along characteristics as defined for the nonlinear problem [7]. In both cases r represents propagation distance measured from a zero on the axis of the cones, but a certain lower limit for r exists beyond which this far-field representation can be applied.

In the conical-wave problem treated here, it may be convenient to use a notation t_c to signify the time, to avoid confusion with the use of t as the time for the plane-wave problem, treated above, into which the conical-propagation problem will be transformed. Moreover a suitable dependent variable (also designated by subscript c) will be the defect in reciprocal signal velocity [8],

$$v_c = c_o^{-1} - (u+c)^{-1} \quad , \tag{A1}$$

rather than the excess (16) in the signal velocity itself.

Now the condition that this dependent variable varies as $r^{-1/2}$ along characteristics satisfying

$$dt_c = (u+c)^{-1}dr = \left(c_o^{-1} - v_c\right)dr \tag{A2}$$

may be written

$$\left\{\partial / \partial r + \left(c_o^{-1} - v_c\right)\partial / \partial t_c\right\}\left(r^{1/2}v_c\right) = 0\ . \tag{A3}$$

Then a simple transformation of coordinates

$$2r^{1/2} = t\ ,\ \ c_o^{-1}r - t_c = X \tag{A4}$$

transforms the operator in braces into

$$r^{-1/2}\partial / \partial t + \partial / \partial X - \left(c_0^{-1} - v_c\right)\partial / \partial X\ ; \tag{A5}$$

so that, if the dependent variable is changed also to $v = r^{1/2}v_c$, the equation for v becomes

$$\left(\partial / \partial t + v\partial / \partial X\right)v = 0\ \ , \tag{A6}$$

exactly as in equation (17) above.

It follows that all the results for the plane-wave case can be directly applied to the conical-wave problem, with v standing for $r^{1/2}v_c$ and t for $2r^{1/2}$. The temporal evolution (as t increases) of the spatially unlimited waveform (in $-\infty < X < \infty$) for plane waves is then seen, according to (A4), as a representation of the spatial evolution (as r increases) of the temporally unlimited waveform (in $-\infty < t_c < \infty$) for conical waves.

In this representation the important quantity Y_n, which remains constant for each shock until it unites with another shock (when their Y-values add), stands for the product of t with the change Δv between the values of v ahead of and behind the shock. Its corresponding meaning for conical propagation is therefore $2r\Delta v_c$, where Δv_c is the shock strength z_n. Accordingly, in a conically propagating random sawtooth wave, the inherent tendency of shock strengths (while their Y-values remain constant) is to decay like the reciprocal r^{-1} of the propagation distance r from the axis of the cones, although this tendency is opposed wherever shocks unite (through their Y-values adding).

In the meantime the second of equations (A4) implies that the quantity h_n, which in the plane-wave case stands for the distance between adjacent shocks at a given time, represents in the conical-wave problem the interval between times of passage of adjacent shocks past a given position with a specified value of r. Also the quantity T, used above to signify the time when a plane wave of random sawtooth type can be considered established, must in the conical-wave problem be given the value $2R^{1/2}$, where R stands for the corresponding propagation distance from the axis at which such a wave has become established.

Calculations of times for unions of shocks in the plane-wave case treated above are expressed in terms of a nondimensional variable σ, whose defining equations (32) correspond for conical propagation to the equations

$$\sigma = \left(r^{1/2} - R^{1/2}\right)/\,2r^{1/2}\ ,\quad r = R\,/\,(1-2\sigma)^2\ . \tag{A7}$$

Expressions for the values of σ at which shocks unite, like (33) for the first such union or (43) for a quite general subsequent union — as given in terms of the Y-values of shocks and the initial values H_n of h_n (now the initial time intervals between shocks) — can then immediately be interpreted, by (A7), as specifying the distance r required for merger of adjacent shocks. This in turn can be used to quantify tendencies to bunching in conical-wave propagation.

Finally, if numerical experiments like that of Section 11 indicate a general tendency for $\Sigma Y_n{}^2$ to increase in approximate proportionality to $t = 2r^{1/2}$, this has implications for the quantity

$$\sum_{n=1}^{N} z_n^2 = \sum_{n=1}^{M} (Y_n\,/\,2r)^2 \tag{A8}$$

which equation (24) associates with the high-frequency part of the noise spectrum. Although the inherent tendency of z_n to decay at r^{-1} would cause this sum (A8) to vary as r^{-2}, bunching tendencies modify this decay law for high-frequency noise into a variation as $r^{-3/2}$.

ACKNOWLEDGMENT

I am warmly grateful to Professor D.G. Crighton for many helpful comments on the material in this paper.

REFERENCES

1 Lighthill, M.J.: Jet Noise, The Wright Brothers Lecture, *AIAA J.*, **1**, pp. 1507-1517, 1963.

2 Lighthill, M.J.: On Sound Generated Aerodynamically, I. General theory, *Proc. Roy. Soc. A*, **211**, pp. 564-587, 1952.

3 Lighthill, M.J.: On Sound Generated Aerodynamically, II. Turbulence as a source of sound, *Proc. Roy. Soc. A*, **222**, pp. 1-32, 1954.

4 Ffowcs Williams, J.E.: The Noise from Turbulence Convected at High Speed, *Phil. Trans. Roy. Soc. A*, **255**, pp. 469-503, 1963.

5 Ffowcs Williams, J.E. and Maidanik, G.: The Mach Wave Field Radiated by Supersonic Turbulent Shear Flows, *J. Fluid Mech.*, **21**, pp. 641-657, 1965.

6 Hussaini, M.Y. and Hardin, J.C. (eds.): *Proceedings of the ICASE/NASA La RC Workshop on Computational Aeroacoustics*, National Aeronautics and Space Administration, 1993.

7 Whitham, G.B.: On the Propagation of Weak Shock Waves, *J. Fluid Mech.*, **1**, pp. 290-318, 1956.

8 Lighthill, J.: *Waves in Fluids*, Cambridge University Press, 1978.

9 Lighthill, M.J.: Sound Generated Aerodynamically, The Bakerian Lecture, *Proc. Roy. Soc. A*, **267**, pp. 147-182, 1962.

10 Curle, N.: The Influence of Solid Boundaries on Aerodynamic Sound, *Proc. Roy. Soc. A*, **231**, pp. 505-514, 1955.

11 Ffowcs Williams, J.E. and Hawkings, D.L.: Sound Generation by Turbulence and Surfaces in Arbitrary Motion, *Phil. Trans. Roy. Soc. A*, **264**, pp. 321-342, 1969.

12 Crighton, D.G.: Basic Principles of Aerodynamic Noise Generation, *Prog. Aerospace Sci.* **16**, pp. 31-96, 1975.

13 Dowling, A.P., Ffowcs Williams, J.E. and Goldstein, M.E.: Sound Production in a Moving Stream, *Phil. Trans. Roy. Soc. A*, **288**, pp. 321-348, 1978.

Supersonic Sources Make Focused Waves

J.E. Ffowcs Williams

Today is St David's Day and in Tucson this year we celebrate Bill Sears' 80th birthday. My subject on this glorious day is the field of sources moving supersonically and the relevance thereto of a new category of very highly directed wave pulses. Starting with Brittingham [1], a rapidly expanding body of knowledge (Belanger [2] Sezginger [3] Ziolkowski [4] Wu [5] Myers, Shen, Wu and Brandt [6] and many others) has emerged in which exact solutions to straightforward linear wave equations display features that border on the scientifically improbable.

Waves spreading into three-dimensional space obey the inverse square law, at least that is the common perception of what is required on energetic grounds. The new field concerns waves whose energy density falls off more slowly than r^{-2} and because of that must dominate in the far radiation field where inverse square law spreading has taken its toll of conventional waves. The ability to sustain wave strength (without violating basic laws) rests on an unexpected degree of ray bunching and the formation of concentrated isolated pockets of wave activity, a property which has attracted a new descriptive nomenclature to the subject. Focus waves, super-collimated beams, wave missiles or bullets, wave bundling and caustic propagation feature in recent treatments of both the electromagnetic and acoustic wave equation and the creation of such bullets is a demanding challenge for the theoretician and experimentalist alike.

Intriguing implications of wave fields continued into complex space are part of that subject as is the output of electromagnetic sources moving at super-luminal speed. Aspects of these waves are counter-intuitive and their acceptance as models of realisable fields is slow to take hold. Slow also is the emergence of a consensus on the significance of highly directional wavelets that are launched from supersonically moving sources [7] and beamed pulses that are predicted to travel without decay from supersonic surface sources [8]. The two issues have distinct similarities, similarities that this paper aims to expose.

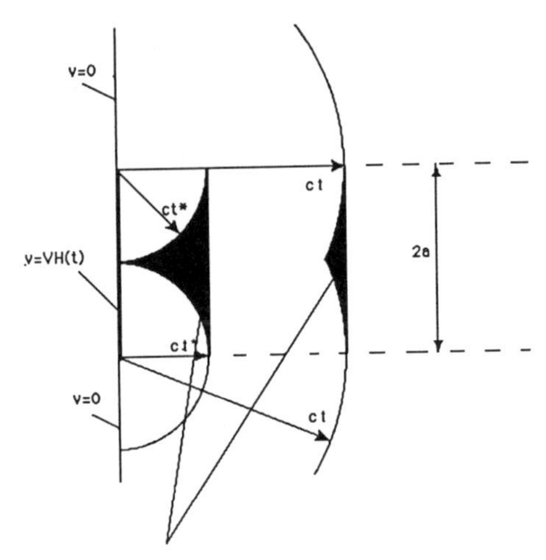

Waves from outside the source aperture cannot enter these lenticular bullets in which pressure is independent of range.

Figure 1. Impulsive motion of a source aperture.

The acoustic bullet is a wave packet whose strength falls off to infinity slower than is required by the inverse square law -- but is an exact solution of the homogeneous wave equation where sources of compact support normally impose inverse square law decay at infinity. One member of the new field is especially clear [9, 10], the wave strength not decreasing at all with range and not diffracting into an ever widening beam. The lack of both decay and spreading run counter to most intuitive notions that diffraction inevitably defocuses pulses and lowers their amplitude. In this example it is the duration of the wave rather than its strength that falls, the energy in the pulse decreasing only linearly with distance. The pulse evolves to dominate over all inverse square law elements, but the pulse volume diminishes with range. The probability of finding a pulse in unit volume of distant space falls as the cube of range - but within the pulse the field level is undiluted and remains forever as high as it was at the pulse's birth. This is illustrated in Figure 1. The field lies in the right half space and is generated by impulsive (t =0) motion within a circular aperture in the x=0 boundary, which otherwise inhibits motion. Waves move into the quiescent space at speed c but no t =0 information can reach the bullet interior from the x =0 surface *outside* the source. The important region is shaded in the Figure and marks the bullet propagating away from the source without spreading and unaware of the source aperture's finite extent. It shares the wave-form with that which would be launched by an unbounded plane surface impulsively started at $t = 0$, a step wave in which the pressure rises abruptly from 0 to ρcV at $t = \frac{r}{c}$, a value it maintains throughout the bullet volume - for ever.

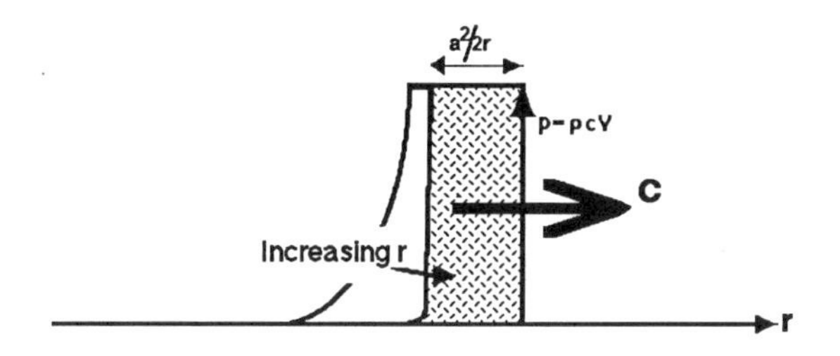

Missile Spectrum, FT of p

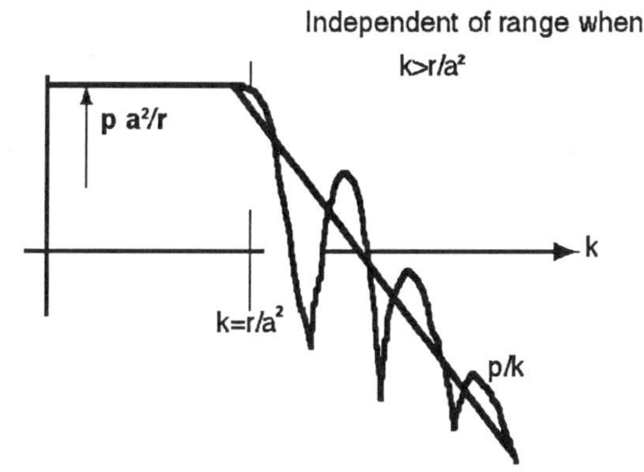

Figure 2. Missile Waveform.

The spectrum of the bullet pulse is obtained by Fourier transformation of the wave-form; its properties are sketched in Figure 2. On the axis of symmetry the constant pressure region extends over a pulse thickness $\frac{a^2}{2r}$ and then falls rapidly to zero. Its strongest spectral elements are at long wavelength, the transform being flat from wave number zero to $\frac{r}{a^2}$ and then falls off as k^{-1} to infinity.

It is the range-dependent bandwidth of the pulse that confounds the inverse square law, the low wave number spectral density of the signal conforming with it perfectly well -- but the bandwidth increases with range. The bullet volume gets ever smaller as it travels, a concentration of waves that might be expected for only as long as the bullet remains within the Rayleigh distance of its launcher -- but that is evidently not so. Those wave elements of the distant bullets that are still within their Rayleigh distance are range independent in strength, their lower frequency bound increasing with range making their contribution to the energy in the bullet also decay linearly with range. But the pulse's strongest spectral elements come from low frequencies and for them the bullet has propagated well beyond the Rayleigh distance, a distance having no bearing on the bullet's dimensions even though it is crucial to the collimation of any, and all, harmonic beams.

The field generated when a thin supersonically moving aerofoil intersects a flow defect was examined by Guo and Ffowcs Williams [11] in their study of a potentially noisy aspect of supersonic propellers. The wake of one blade might well be impacted by a following blade and they modelled that interaction according to linear theory in an example involving a straightforward geometrical arrangement. An intense pulse is formed that travels in the Mach wave direction, the pulse amplitude not attenuating at all, neither does its area change as it radiates to infinity. But the pulse duration falls with range in precisely the same way as does the bullet pulse -- and the spectral characteristics are identical. Figure 3 illustrates those features which are exact within the confines of linear theory for this precise geometry.

Sound is mostly modelled by linear theory, sound is usually weak enough a perturbation for its linear modelling to be relevant. Where linear theory indicates strong fields, there the real fields are likely to be specially important. Where linear theory predicts very strong fields non-linear real-fluid effects are crucial but how they will modify the linear prediction has to be determined by a more accurate account of the physics. Wave bullets are very strong relative to other linear far field terms and it is a most interesting matter to determine whether real-fluid finite-amplitude effects obliterate them. If they don't and the pointers of linear theory prove accurate, then the strongest wave pulses would seem to exist in isolated low volume density regions where the normally expected spherical decay features of sound are surprisingly absent. A new branch of the subject might be starting. The audibility or otherwise of such very small duration, very strong sound pulses is one of the issues about which there is bound to be strong debate, and the experimental task of capturing very big but volumetrically improbable pulses is not a straightforward extension of procedures developed for conventional continuous signals. I think bullets are probably real but only time will tell whether they are important.

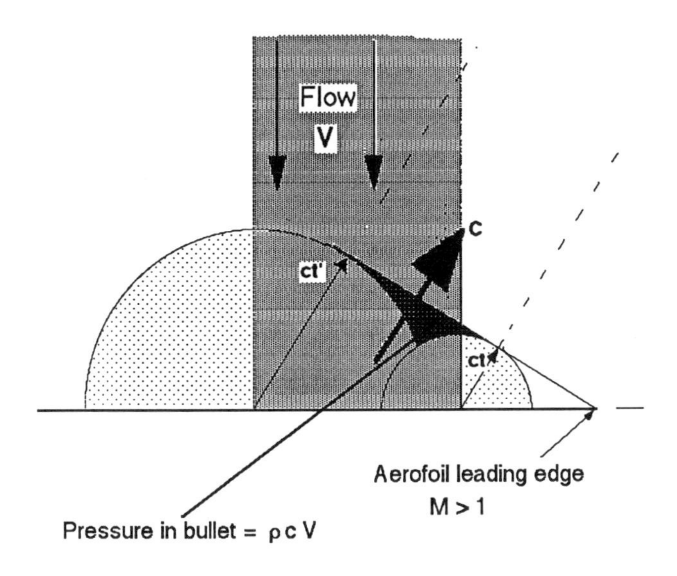

Figure 3. Aerofoil/Wake Interaction at $M>1$.

Source motion alters the frequency of sound from that of its source, the Doppler factor being the ratio of the two. When a source of frequency ω is approaching at Mach number M_r, then $\omega/(1-M_r)$ is the Doppler-shifted frequency that will be heard by the observer. If the source approaches at exactly the speed of sound only the steady element of the source, its zero frequency component, can be heard at finite frequency. That condition is special and marks the Mach wave regime and its electromagnetic equivalent, Cerenkov radiation. The frequency content of Mach waves is set by the spatial rather than temporal form of the source and many of the notions relating the source to its radiation characteristics are strange, the normal expectation being formed on the behaviour of a periodic source concentrated at a moving point. That field is actually singular at the Mach wave condition, the infinite field strength being evident from the Lienard-Wiechert potential for a moving point charge

$$\Phi = \frac{Q^*}{r^*\left(1 - M_r^*\right)} \tag{1}$$

$M_r^* = 1$ marking the Cerenkov condition.

Sonic approach contracts the acoustic wavelength, to zero were the Doppler factor $(1-M_r)$ the whole story. Short waves are easily directed, rays of wavelength λ being steerable to a focus as far away as L^2/λ from a collimating lens of size L. Sonic approach changes the influence of source frequency on sound and the formal vanishing of the Doppler factor hints at a vanishing of the wave scale also. The existence of very small distant concentrated packets of wave energy is not then so contrary to the usual criterion because the effective Rayleigh distance $\frac{L^2}{\lambda}$ might be infinite for Mach waves.

Source symmetries impose lower dimensionality on wave fields. For example if the source is independent of one axial co-ordinate, so will its sound be, and because of that will propagate in the remaining two space dimensions, spreading cylindrically rather than spherically. The motion of steady sources has the same effect. Consider for example the field generated by a source which is steady in a reference frame moving along the x axis with speed cM, the substantive derivative of both source and field quantities being zero,

$$\frac{\partial}{\partial t}+cM\frac{\partial}{\partial x}=0.$$

The symmetry reduces the inhomogeneous three-dimensional wave equation,

$$\left\{\frac{\partial^2}{\partial x^2}+\frac{\partial^2}{\partial y^2}+\frac{\partial^2}{\partial z^2}-\frac{1}{c^2}\frac{\partial^2}{\partial t^2}\right\}\Phi = q(x-cMt,y,z)$$

to the two-dimensional equation

$$\left\{\frac{\partial^2}{\partial y^2}+\frac{\partial^2}{\partial z^2}+\left(\frac{1}{M^2}-1\right)\frac{1}{c^2}\frac{\partial^2}{\partial t^2}\right\}\Phi = q(x-cMt,y,z) \tag{2}$$

The field of subsonically moving sources is evanescent but the field of a supersonically moving steady source spreads out cylindrically from its source with wave speed $c/\sqrt{1-M^{-2}}$.

For the supersonically moving point charge q is a three-dimensional delta function and its field is the Lienard-Wiechert potential, spreading cylindrically -- despite the contrary appearance afforded by the symbolism of Equation (1). The asterisks there mean that the values of r and M_r must be chosen to be appropriate to the time when the source emitted the waves currently at the observer and wherever multiple emission points arise contributions must be summed. The wave field driven by the supersonic source, the field described in Equation (1) is the Green function of Equation (2) and is depicted in its more usual diagrammatic form in Figure 4. The field is zero ahead of a conical front moving normal to itself at speed c having travelled in that direction a distance $r=ct$ since it left the point source at time $t=0$. The field at the front is singular but decays behind. The projection of the wave onto the (y,z) plane expands cylindrically at speed $c/\sqrt{1-M^{-2}}$. In fact the field strength, which is given by Equation (1) is equal to

$$\Phi = \frac{Q}{r^*\left(1-M_r^*\right)} = \frac{Q}{\sqrt{c^2t^2-r^2}}, \quad r < ct \tag{3}$$

$$= 0, \; r > ct.$$

This is a cylindrical wave whose far field is a function of $(r-ct)$ times $r^{-1/2}$ quite different from the $f(r - ct)$ times r^{-1} structure of spherical waves. (Whitham [12])

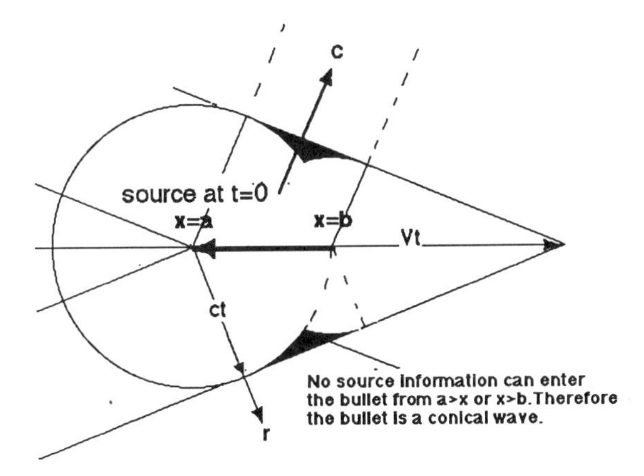

Figure 4. The conical wavefield of a supersonically moving charge.

The cylindrical form can be sustained in small regions even when the source's supersonic constant-speed journey is limited and to demonstrate this fact curves are indicated in Figure 4 delineating the boundaries containing all wavelets coming from the source prior to its arrival at $x=a$ and after its passage beyond $x=b$. Between the conical Mach cone and those two spherical boundary sections, only source activity between $x=a$ and $x=b$ can possibly reach at sound speed c. That region is shaded and marked as a bullet: the wave field there is exactly the same cylindrical wavefield as is generated by the completely steady source, because no information can reach the bullet from outside the axial region $b>x>a$. It is immaterial what the source did or where it was outside that region. The short lived supersonic source generates a field which within the bullet is identical to that of an infinitely long-lived one, so we have a cylindrically spreading packet of wave activity generated in three-dimensional space by a source of compact support. That supersonic source generates within the thinning bullet annulus a field that must be louder than any spherically spreading distant field.

Angular source motion has a similar dimension-lowering aspect and that too arouses the prospect that pulses decaying with distance slower than the inverse square law, will dominate small regions of distant space. Sources, steady in a rotating frame, are thought to model important aspects of helicopter and propeller noise, and also the electromagnetic radiation form pulsars [13, 14]. Such a source is a function of the azimuthal angle minus ωt, a symmetry it imposes on its field. The

$$\frac{\partial}{\partial t}+\omega\frac{\partial}{\partial \theta}=0$$

property allows the wave equation in cylindrical polar co-ordinates to be written

$$\frac{1}{r}\frac{\partial}{\partial r}\left(r\frac{\partial \phi}{\partial r}\right)+\frac{\partial^2 \phi}{\partial z^2}+\left(\frac{c^2}{\omega^2 r^2}-1\right)\frac{1}{c^2}\frac{\partial^2 \phi}{\partial t^2}=q(r,z,\theta-\omega t)$$

an equation that is wavebearing only in its hyperbolic region outside the sonic radius ; $r \geq c/\omega$. There is evidently an analogy between the sound of steadily rotating sources and that generated by annular sources in a radially stratified medium. The speed of sound in the analogy falls from infinity at the sonic radius to c at infinity, a refractive effect that would bend all rays emanating from sources on the sonic radius to go to infinity in planes of contstant z. That is a clue that sound is confined in the analogy to spread in lower dimension and that non-spherical spreading behaviour is expected.

Definite non-spherical behaviour is found by Ardavan [15] in the field of quadrupole sources that are steady in a rotating reference frame. Such sources are thought to represent important acoustic effects of non-linear flow over the blades of high speed rotating machinery, a much studied problem in which there is a universal expectation that spherical, inverse square law spreading controls distant field levels. In fact the asymptotic field where this is assumed to hold attracts most fundamental attention and it is quite a jolt to be warned that a non-spherical component exists and that it might well model the loudest sound, an element that has previously escaped the attention of the subject's experts. Ardavan deduces that new wave's characteristics based on an exact study of the 'spiral Green's function' [16], the field of a fixed charge rotating at constant angular velocity. He finds the unfamiliar behaviour at caustics of the distant field where sound accumulates from wavelets launched by the source at three different previous times, launched when the source was approaching the observer at exactly the speed of sound. The new waves are closely related to acoustic bullets but are distinct in that their detailed structure depends on fine details of the spiral Green's function, details which are lost in the usually adequate far field approximations that suppress wave curvature.

These new features add zest to a subject in danger of appearing so mature that few expect it to contain the possibility of novel cures for the loud noises of modern life. But I do not believe the subject's most important high speed noise problems are at all well understood. High Mach number aeroacoustics is developing nicely but as yet it is far from being able to predict the noise of unusual flows. To handle design of the radically different configurations that will be needed for quiet supersonic flight, to discover the quiet propulsive flows, that I'm perfectly sure are there to be found, aircraft minded researchers must be encouraged to probe the new and strange cracks in the subject. That is where the new principles of noise control will be found and that is where it is most fun. Aeronautical progress has always relied on the new technology borne of open minded enquiry into the subject's fundamentals and great pleasure and fellowship comes from doing that well. No one epitomises that better than Bill and Mabel Sears, whose friendship, wisdom and ever-youthful attitude we celebrate today.

REFERENCES

1 Brittingham, J. N.: *J. App. Phys.,* **54**, pp. 1179, 1983.

2 Belanger, P. A.: Packet-like Solutions of the Homogeneous Wave Equation, *J. Opt. Soc. Amer.* A, **1**(7), pp. 723-724, 1984.

3 Sezginger, A.: A General Formulation of Focus Wave Modes, *J. App. Phys.,* **57** (3), pp. 678-683, 1985.

4 Ziolkowski, R.W.: Exact Solutions of the Wave Equation with Complex Source Locations, *J. Math. Phys.,* **26**(4), pp. 861-863, 1985.

5 Wu, T.T. : Electromagnetic Missiles, *J. App. Phys.,* **57**(7), pp. 2370-2373, 1985.

6 Myers J., Shen, H.M., Wu T.T. and Brandt H.: Fun with Pulses, *Physics World,* **39**, November 1990.

7 Ffowcs Williams, J.E.: On the Development of Mach Waves Radiated by Small Disturbances, *J. Fluid Mech.,* **22**, pp. 1, 1965.

8 Ffowcs Williams, J.E. and Hawkings, D.L.: Sound Generation by Turbulence and Surfaces in Arbitrary Motion, *Phil. Trans. Roy. Soc. A,* **264**, pp. 321-342, 1969.

9 Shen H.M. and Wu ,T.T.: The Transverse Energy Pattern of an Electromagnetic Missile from a Circular Current Disk, SPIE Vol. 1061, Microwave and Particle Beam Sources and Directed Energy Concepts, 1989.

10 Ffowcs Williams, J.E.: Computing the Sources of Sound, *Computational Acoustics,* Vol. 1, Lau R.L., Lee D. and Robinson A.R. (eds.), Elsevier Science Publishers B.V., North Holland, 1992.

11 Guo, Y.P. and Ffowcs Williams, J.E.: Sound Generated by the Interruption of a Steady Flow by a Supersonically Moving Aerofoil, *J. Fluid Mech.,* **195**, pp. 113-135, 1988.

12 Whitham, G.B.: *Linear and Nonlinear Waves,* Wiley, 1974.

13 Farassat, F.: Linear Acoustic Formulas for the Calculation of Rotating Blade Noise, *AIAA J.,* **19**, pp. 1122-1130, 1981.

14 Ardavan, H. : Is the Light Cylinder the site of Emission in Pulsars? *Nature,* **287**, pp. 44-45, 1981.

15 Ardavan, H.: Asymptotic Analysis of the Radiation by Quadrupole Sources in Supersonic Rotor Acoustics. Submitted to *J. Fluid Mech.,* 1993.

16 Ardavan, H. : The Speed of Light Catastrophe, *Proc. Roy. Soc. Lond. A,* **424**, pp. 113-141, 1989.

An Introduction to Generalized Functions with Some Applications in Aerodynamics and Aeroacoustics

F. Farassat

ABSTRACT

Since the early fifties when Schwartz published his theory of distributions, generalized functions have found many applications in various fields of science and engineering. One of the most useful aspects of this theory in applications is that discontinuous functions can be handled with the same ease as continuous or differentiable functions. This provides a powerful tool in formulating and solving many problems of aerodynamics and acoustics. Furthermore, generalized function theory elucidates and unifies many ad hoc mathematical approaches used by engineers and scientists in these two fields. In this paper, we start with the definition of generalized functions as continuous linear functionals on the space of infinitely differentiable functions with compact support. The concept of generalized differentiation is introduced next. This is the most important concept in generalized function theory and the applications we present utilize mainly this concept. First, some of the results of classical analysis, such as Leibniz rule of differentiation under the integral sign and the divergence theorem, are derived using the generalized function theory. It is shown that the divergence theorem remains valid for discontinuous vector fields provided that the derivatives are all viewed as generalized derivatives. This implies that all conservation laws of fluid mechanics are valid as they stand for discontinuous fields with all derivatives treated as generalized derivatives. Once these derivatives are written as ordinary derivatives and jumps in the field parameters across discontinuities, the jump conditions can be easily found. For example, the unsteady shock jump conditions can be derived from mass and momentum conservation laws. By using generalized function theory, this derivation becomes trivial. Other applications of the generalized function theory in aerodynamics discussed in this paper are derivation of general transport theorems for deriving governing equations of fluid mechanics, the interpretation of finite part of divergent integrals, derivation of Oswatitsch integral equation of transonic flow, and analysis of velocity field discontinuities as sources of vorticity. Applications in aeroacoustics presented here include the derivation of the Kirchhoff formula for moving surfaces, the noise from moving surfaces, and shock noise source strength based on the Ffowcs Williams-Hawkings equation.

LIST OF SYMBOLS

a	a constant
$A(x)$	coefficient of the second order term of a linear ordinary differential equation
$A(\alpha)$	lower limit of the integral in Leibniz rule depending on parameter α
b	a constant
$B(\alpha)$	upper limit of the integral in Leibniz rule depending on parameter α
BC, BC_1, BC_2	boundary conditions
$B(x)$	coefficient of the first order term in a second order linear ordinary differential equation
c	a constant, also speed of sound
C, C_1, C_2	constants
$C(x)$	coefficient of zero order term (the unknown function) in a second order ordinary differential equation
D	the space of infinitely differentiable functions with bounded support (test functions)
D'	the space of generalized functions based on D
$\exp(x)$	exponential function e^x
E_1, E_2	expressions in the integrands of the Kirchhoff formula for moving surfaces
$E(\alpha)$	a fuction defined by Equation (3.70)
E_h	shift operator $E_h f(x) = f(x+h)$
E_{ij}	viscous stress tensor
$f(x), f(\bar{x})$	arbitrary ordinary functions
$f_1(x)$	arbitrary function
$f_i(\tau)$	$i = 1-3$, components of a moving compact force
$f(\bar{x}, t)$	equation of a moving surface defined as $f(\bar{x}, t) = 0$, $f > 0$ outside surface.
$\tilde{f}(\bar{x}, t)$	a moving surface defined by $\tilde{f}(\bar{x}, t) = 0$ whose intersection with $f(\bar{x}, t) = 0$ defines the edge of the open surface $f = 0$, $\tilde{f} > 0$
F	in $F[\phi]$, defines a linear functional on a test function space, a generalized function
$F(\bar{y}; \bar{x}, t)$	equals $[f(\bar{y}, \tau)]_{ret} = f\left(\bar{y}, t - \frac{r}{c}\right)$

$\tilde{F}(\bar{y};\bar{x},t)$	equals $\left[\tilde{f}(\bar{y},\tau)\right]_{ret} = \tilde{f}\left(\bar{y},t-\frac{r}{c}\right)$
$g(x,y), g(\bar{x},\bar{y})$	Green's function
g	equals $\tau - t + \frac{r}{c}$
$g_1(x,y), g_2(x,y)$	define Green's function for $x < y$ and $x > y$, respectively
$g_{(2)}$	determinant of the coefficients of the first fundamental form of a surface
$g(x), g(\bar{x})$	arbitrary functions
h	a constant
$h(x)$	Heaviside function
H	in $H[\phi]$, linear functional $\int_0^\infty \phi(x)dx$ based on the Heaviside function
H_f	local mean curvature of surface $f = 0$
$H(x,\alpha)$	function defined by Equation (3.71)
$h_\varepsilon(x)$	a function of x indexed by the continuous parameter ε
i	equals $\sqrt{-1}$
I	an interval on the real line, an expression given by an integral, an expression
k	a non-negative integer
$k(\bar{x}), k(\bar{x},t)$	equation of a shock or wake surface given by $k = 0$
K	in $K[\phi]$, defines a linear functional on a test function space, a generalized function
ℓ	in ℓu, a second order linear ordinary differential equation
L	in dL, length parameter of the edge of the Σ surface given by $F = \tilde{F} = 0$
$\bar{M}$	Mach number vector
M_r	equals $\bar{M} \cdot \bar{\hat{r}}$
M_n	equals $\bar{M} \cdot \bar{n}$
M_v	equals $\bar{M} \cdot \bar{v}$
m	index of summation of Fourier series
n	a non-negative integer
$\bar{n}$	unit outward normal
$\bar{n}'$	unit outward normal
$\bar{n}_1$	vector $(n_1, 0, 0)$ based on $\bar{n} = (n_1, n_2, n_3)$
$\bar{N}$	unit normal to $F = 0$

$\bar{\tilde{N}}$ unit normal to $\tilde{F} = 0$

o in $o(\varepsilon)$, small order of ε

p blade surface pressure

p' acoustic pressure

PV principal value

P_{ij} the compressive stress tensor

$Q(\bar{x}, t), \bar{Q}(\bar{x}, t)$

source strength of inhomogeneous term of the wave equation

r equals $|\bar{x} - \bar{y}|$

r_i $i = 1 - 3$, components of vector $\bar{r} = \bar{x} - \bar{y}$

$\hat{r}_i$ $i = 1 - 3$, components of unit radiation vector $\frac{\bar{r}}{r}$

S in dS, surface area of a given surface

S space of rapidly decreasing test functions

S' space of generalized functions based on S

S_k portion of surface $k = 0$ inside the surface $\partial\Omega$

$\bar{s}(t)$ position vector of a compact force in motion

t a variable, time variable

$\bar{t}_1$ unit vector in the direction of projection of $\bar{\hat{r}}$ onto the local tangent plane to $f(\bar{x}, t) = 0$

t_1 in $\frac{\partial}{\partial t_1}$, directional derivative in the direction of $\bar{t}$

T_{ij} Lighthill stress tensor

u_i $i = 1 - 3$, components of fluid velocity

u_n local fluid normal velocity

u^i $i = 1 - 3$, curvilinear coordinate variables

v_n local normal velocity of a surface

$v_{n'}$ local normal velocity of a surface

$\bar{x}$ observer variable

$\bar{y}$ source variable

Greek Symbols

α a constant, a parameter

α_f a constant depending on the shape of the surface $f = 0$

β a constant

$\bar{\Gamma}$ strength of vorticity

Γ	in $d\Gamma$, length parameter along curve of intersection of surfaces $f = 0$ and $g = 0$
γ	height of a cylinder
$\delta(x), \delta(\bar{x}), \delta(f)$	the Dirac delta function
$\delta[\phi]$	the linear functional representing the Dirac delta function
δ_{ij}	the Kronecker delta, $\delta_{ij} = 0$ if $i \neq j$, $\delta_{ii} = 1$
Δ	jump in a function at a discontinuity
ε	a small parameter
$\bar{\eta}$	Lagrangian variable
θ	the angle between ∇f and ∇g, the angle between the radiation direction $\bar{\hat{r}}$ and the local normal to a surface $\bar{n}$
θ_1	the angle between $\bar{\hat{r}}$ and $\bar{n}_1$
θ'	the angle between $\bar{N}$ and $\bar{\tilde{N}}$
Λ	equals to $\lvert\nabla F\rvert$, $F = [f]_{ret}$, $\lvert\nabla f\rvert = 1$
$\tilde{\Lambda}$	equals to $\lvert\nabla\tilde{F}\rvert$, $\tilde{F} = [\tilde{f}]_{ret}$, $\lvert\nabla\tilde{f}\rvert = 1$
Λ_0	equals to $\lvert\nabla F \times \nabla\tilde{F}\rvert$, F, and $\tilde{F}$ as defined above
$\bar{\nu}$	unit inward geodesic normal
ξ	variable of Fourier transform
ρ	density
ρ_o	density of undisturbed medium
ϕ	a test function, an arbitrary function
ϕ_1, ϕ_2	test functions
ϕ_n	a sequence of test functions, also the normal component of a vector field $\bar{\phi}$ to a surface
$\phi^{(k)}$	k-th derivative of ϕ
$\Phi(\bar{x}, t)$	the unknown function of an inhomogeneous wave equation
$\tilde{\phi}$	the extension of the function ϕ to the unbounded space
$\phi_{1,i}$	$i = 1 - 3$, the components of function $\bar{\phi}_1$
τ	source time
Ω	an open interval or region of space, $d\Omega$ boundary of Ω
$\Omega(t)$	the sphere $r = c(t - \tau)$, $(\bar{x}, t, \tau)$ kept fixed

Subscripts

h	in E_h, shift of a function by the amount h to the right or the left
n, n'	component of a vector field in the direction of the local normal $\bar{n}$ or $\bar{n}'$
n	index of a sequence such as ϕ_n
o	in ρ_o, indicates condition of undisturbed medium
ret	retarded time
x	in ℓ_x, indicates that the derivatives in ℓ act on the variable x in $\ell_x g(x, y)$
ε	the continuous index in a function such as $h_\varepsilon(x)$

Superscripts

k	in $\phi^{(k)}$, k-th derivative of ϕ
n	in $\phi^{(n)}$, n-th derivative of ϕ

Other Symbols

$\Box^2$	D'Alembertian, the wave operator $\frac{1}{c^2}\frac{\partial^2}{\partial t^2} - \nabla^2$
$[\ ]$	in $F[\phi]$, indicates functional evaluated for ϕ, ϕ a test function
$supp$	support of a function
$\sim$	in $\tilde{\phi}$, indicates the restriction of ϕ to the support of a delta function
$\wedge$	in $\hat{\psi}$, indicates Fourier transform
$*$	in τ^*, indicates emission time
∇	gradient operator
∇_2	surface gradient operator
∇_y	gradient operator acting on variable $\bar{y}$
$-$	over a derivative, such as $\bar{f}'(x)$, indicates generalized differentiation
∂	in $\partial\Omega$, indicates the boundary of region Ω

1 INTRODUCTION

In the early fifties, Schwartz published his theory of distributions or, what we call in this paper, generalized functions [1]. Earlier Dirac had introduced the delta "function" $\delta(x)$ by the sifting property

$$\int_{-\infty}^{\infty} \phi(x)\delta(x)dx = \phi(0) \tag{1.1}$$

He recognized that no ordinary function could have the sifting property. Nevertheless, he thought of $\delta(x)$ as a useful mathematical object in algebraic manipulations which could be viewed as the limit of a sequence of ordinary functions. The Dirac delta function is a generalized function in the theory of distributions. Schwartz established rigorously the properties of generalized functions. His theory has had an enormous impact on many of the areas of mathematics particularly partial differential equations. Generalized function theory has been used in many fields of science and engineering.

To include mathematical objects such as the Dirac delta "function" into analysis, we must somehow extend the concept of a function. The process we use to introduce new objects is familiar in mathematics. We extended natural numbers to integers, integers to rationals and rationals to real numbers. We also extended real numbers to complex numbers. In each extension, new objects were introduced in the number system while most properties of the old number system were retained. Furthermore, for each extension, we had to think of the new number system in a different way from the old system. For example, in going from integers to rationals, we view numbers as ordered pairs of integers (a, b) where $b \neq 0$. We identify ordered pairs $(a, 1)$ with integer a. The new number system (the rationals) includes the old number system (the integers). We must now think of numbers as ordered pairs (a, b), which we usually write $\frac{a}{b}$, instead of a single number a in thinking about integers. Similarly, to extend the concept of function to include the Dirac delta "function", we must think of functions differently. We will explain in Section 2 how to think of functions as functionals, i.e., the mapping of a suitable function space into scalars. In this way the Dirac delta function can naturally be included in this extended space of functions which we call distributions or generalized functions. The usefulness of this theory stems from the powerful operational properties of generalized functions. In addition, solutions with discontinuities can be handled easily in the differential equation or in the Green's function approach. Many of the ad hoc mathematical methods used by engineers and scientists are unified and elucidated by generalized function theory. In fluid dynamics, the derivations of transport theorems, conservation laws and jump conditions are facilitated by this theory. Geometric identities for curves, surfaces and volumes, particularly when in motion and deformation, can be derived with ease using generalized function theory.

In Section 2 we define generalized functions as continuous linear functionals on some space of test functions. Some operations on generalized functions are defined in this section. In a subsection, various approaches that can be used to introduce generalized functions in mathematics are discussed. In Section 3 we present some definitions and results on generalized functions. In particular, we give some important results on generalized derivatives, multidimensional delta functions and the finite part of divergent integrals. In Section 4 various applications in aerodynamic applications include derivation of two transport theorems, the interpretation of velocity discontinuity as vortex sheet and the derivation Oswatitsch integral equation of transonic flow. The aeroacoustic applications include the derivation of the wave equation with various inhomogeneous source terms, the Kirchhoff equation for moving surfaces, the Ffowcs Williams-Hawkings equations and shock noise source strength. All these applications depend on the concept of generalized differentiation. Concluding remarks and references are in Sections 5 and 6 respectively.

There are many articles and books on generalized function theory. Most of these works are inaccessible to engineers and scientists because of the abstract nature of the presentation. In particular, multidimensional generalized functions, which are most useful in applications, are often treated cursorily in applied mathematics and physics books. There are, of course, some exceptions [2-7]. It is relatively easy to learn and use multidimensional generalized functions if the theory is stripped of some abstraction. To work with multidimensional generalized functions, some knowledge of differential geometry and tensor analysis is required. There are accessible books in these areas also [8-9]. In this paper, we present the rudiments of generalized function theory with emphasis on applications in aerodynamics and aeroacoustics. The presentation is expository. It is the intention of the author to make the readers interested in the subject and to impress them with the power of the generalized function theory. Some illustrative examples are given in this paper to help in the understanding of the concepts involved.

It is a great honor for me to be an invited speaker in this symposium. I became interested in generalized function theory when I was involved in my Ph.D. work on helicopter rotor noise under the supervision of Professor Sears. My Ph.D. problem was given to me in the first week of the fall semester of 1970 when I joined the Graduate School of Aerospace Engineering at Cornell. I remember vividly that Professor Sears wrote down the wave equation with a delta function as the inhomogeneous source term. I had then some rudimentary knowledge of the delta function. I immediately saw the connection of my Ph.D. problem with the now classical paper of Ffowcs Williams and Hawkings which had been published a year earlier in 1969. I had great difficulty in following this paper because of the use of advanced generalized function theory with which I was not familiar. Fortunately in the spring semester of 1971 in a course on mathematical methods of physics, generalized function theory in one variable was taught from Lighthill's excellent little book. Once I learned the foundations, I progressed to the multidimensional generalized functions with the help of Gel'fand and Shilov. I was then able to understand fully the work of Ffowcs Williams and Hawkings. Throughout this period of learning, I had many interesting and lively discussions at the blackboard with Professor Sears. One of his policies was that he encouraged his students to take advanced courses in mathematics. He recognized the importance of advanced mathematics in engineering early in his career. My scientific association with him has been a delightful and rewarding experience. His enthusiasm in scientific and technical discovery is contagious as are his research style and high standards. It is with utmost respect and affection that I dedicate this paper to Bill Sears on the occasion of his eightieth birthday.

2 WHAT ARE GENERALIZED FUNCTIONS?

It can be shown from classical Lebesgue integration theory that the Dirac delta "function" cannot be an ordinary function. By an ordinary function we mean a locally Lebesgue integrable function, i.e. one that has a finite integral over any bounded region. To include the Dirac delta "function" in mathematics we must change the way we think of an ordinary function $f(x)$. Conventionally, we think of this function as a table of ordered pairs $(x, f(x))$. This is, of course, often a table with an uncountably infinite number of ordered pairs. We show this table as a curve representing the function in a plane. In generalized function theory, we also describe a function $f(x)$ by a table of numbers. These numbers are produced by the relation

$$F[\phi] = \int_{-\infty}^{\infty} f(x)\phi(x)dx \tag{2.1}$$

where the function $\phi(x)$ comes from a given space of functions called the test function space. For a fixed function $f(x)$, Equation (2.1) is a mapping of the test function space into reals or complex numbers. Such a mapping is called a functional. We will use square brackets to denote functionals, e.g., $F[\phi]$ and $\delta[\phi]$. Therefore, a function $f(x)$ is now described by a table of its functional values over a given space of test functions. We must first, however, specify the test function space.

The test function space that we use here is the space D of all infinitely differentiable functions with bounded support. The support of a function $\phi(x)$ is the closure of the set on which $\phi(x) \neq 0$. For an ordinary function $f(x)$, the functional $F[\phi]$ is linear in the following sense. If ϕ_1 and ϕ_2 are in D and α and β are two constants, then

$$F[\alpha\phi_1 + \beta\phi_2] = \alpha F[\phi_1] + \beta F[\phi_2] \tag{2.2}$$

The functional $F[\phi]$ is also continuous in the following sense. Take a sequence of functions $\{\phi_n\}$ in D and let this sequence have the following two properties:

i) there is a bounded interval I such that for all n, *supp* $\phi_n \subset I$,

ii) $\lim_{n\to\infty} \phi_n^{(k)}(x) = 0$ uniformly for all $k = 0, 1, 2, \ldots.$

Such a sequence is said to go to zero in D and written $\phi_n \xrightarrow{D} 0$. Here *supp* ϕ_n stands for support of ϕ_n. We then say that the functional $F[\phi]$ is continuous if $F[\phi_n] \to 0$ for $\phi_n \xrightarrow{D} 0$. We will have more to say about the space D and why we require the two conditions above in the definition of $\phi_n \xrightarrow{D} 0$.

We have not yet given an example of a function $\phi(x)$ in D. Here is an important example. For a given finite $a > 0$, we define

$$\phi(x; a) = \begin{cases} \exp\left(\dfrac{a^2}{x^2 - a^2}\right) & |x| < a \\ 0 & |x| \geq a \end{cases} \tag{2.3}$$

we note that *supp* $\phi(x; a) = [-a, a]$ i.e., bounded. We can show that $\phi(x; a)$ is infinitely differentiable. Therefore, $\phi(x; a) \in D$. The proof of infinite differentiability at $x = \pm a$ is somewhat messy and algebraically complicated. We will not belabor this point here because it is irrelevant to the aims of this paper. We can show that from any continuous function $g(x)$, we can construct another function $\psi(x)$ in D from the relation

$$\psi(x) = \int_b^c g(t)\phi(x - t; a)dt \tag{2.4}$$

where the interval $[b, c]$ is finite. The support of $\psi(x)$ is $[b - a, c + a]$ which is bounded. The infinite differentiability of $\psi(x)$ follows from infinite differentiability of $\phi(x; a)$. Therefore, $\psi(x) \in D$. There is an uncountably infinite number of continuous functions (All polynomials with integer coefficients are constructed from subsets of integers. The number of these subsets has the cardinality of the continuum, therefore uncountable). It follows from the above argument that there is an uncountably infinite number of functions in space D, so our table constructed from $F[\phi]$ by Equation (2.1) representing the ordinary function $f(x)$ has an uncountably infinite number of elements. This fact has an important consequence. Two ordinary functions f and g which are not equal in the Lebesgue sense, i.e. two functions which are not equal on a set with nonzero measure, generate tables by Equation (2.1) which differ in some entries. This means that space D is large enough so that functionals on D generated by Equation (2.1) can distinguish different ordinary functions.

We now give an example of a sequence $\{\phi_n\}$ in D such that $\phi_n \overset{D}{\to} 0$. Using the function $\phi(x; a)$ in Equation (2.3), we define

$$\phi_n(x) = \frac{1}{n}\,\phi(x; a) \tag{2.5}$$

It can easily be shown that this sequence satisfies the two conditions required for $\phi_n \overset{D}{\to} 0$. We note in particular that *supp* $\phi_n = [-a, a]$ for all n.

Now with all the above preliminaries we come to the definition of generalized functions. First we note that for an ordinary function $f(x)$, i.e., a locally Lebesgue integrable function, the functional $F[\phi]$ given by Equation (2.1) is linear and continuous. The proof of linearity is obvious. The proof of continuity only requires that $\phi_n \to 0$ uniformly which already follows from $\phi_n \overset{D}{\to} 0$. Remembering that we are now looking at functions by their table of functional values over the space D and this functional is linear and continuous, we ask the following question. Are all continuous linear functionals on space D generated by ordinary functions through the relation given in Equation (2.1)? The answer is no! There are continuous linear functionals on space D which are not generated by ordinary functions. Here is an example:

$$\delta[\phi] = \phi(0) \tag{2.6}$$

Proof of linearity is obvious. Continuity follows again from $\phi_n \to 0$ uniformly. But this functional has the sifting property that the Dirac delta "function" requires. As we said earlier, no ordinary function has the sifting property. Therefore, here is a way to introduce the delta "function" into mathematics rigorously. We define generalized functions as continuous linear functionals on space D. The space of generalized functions on D is denoted as D'. Figure 1 shows schematically how we have extended the space of ordinary functions to generalized functions. We call ordinary functions regular generalized functions while all other generalized functions such as the Dirac delta function are called singular generalized functions.

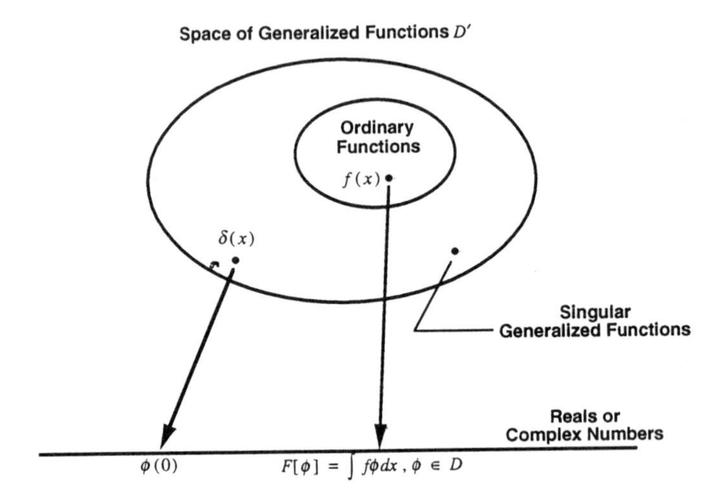

Figure 1. Generalized functions are continuous linear functionals on the space D of test functions.

For the purpose of algebraic manipulations, it is convenient that we retain the notation of ordinary functions for generalized functions. We symbolically introduce the notation $\delta(x)$ for the Dirac delta function by the relation

$$\delta[\phi] = \phi(0) = \int \delta(x)\phi(x)dx \tag{2.7}$$

Note that the integral on the right side of Equation (2.7) does not stand for conventional integration of a function. Rather it is a symbol standing for $\phi(0)$. We can now use $\delta(x)$ in mathematical expressions as if it is an ordinary function. We must, however, remember that singular generalized functions are not, in general, defined pointwise and that they define a functional (i.e. a function from our new point of view) whenever they are multiplied by a test function and then appear under an integral sign. Thus, whenever a singular generalized function appears in an expression, it is always in an intermediate stage of solving a real physical problem.

We close this section by giving some more facts about space D in multidimensions, convergence to zero in D and the concept of continuity of a functional. The multidimensional test function space D is defined as the space of infinitely differentiable functions with bounded support. An example is, for $a > 0$

$$\phi(\bar{x};a) = \begin{cases} \exp\left(\dfrac{a^2}{|\bar{x}|^2 - a^2}\right) & |\bar{x}| < a \\ 0 & |\bar{x}| \geq a \end{cases} \tag{2.8}$$

where $|\bar{x}| = \left(\sum_{i=1}^{n} x_i^{\,2}\right)^{\frac{1}{2}}$ is the Euclidean norm. Other functions in this space can be constructed by using any continuous function $g(\bar{x})$ and the convolution relation

$$\psi(\bar{x}) = \int_{\Omega} g(\bar{t})\phi(\bar{x} - \bar{t};a)d\bar{t} \tag{2.9}$$

where Ω is a bounded region. The multidimensional generalized function space D' is defined as the space of continuous linear functionals on the space D. In the multidimensional case, a number of important singular generalized functions of the delta function type appear in applications. In one dimension, the support of $\delta(x)$ consists of one point, $x = 0$. We will define the support of a generalized function later. In the multidimensional case in addition to $\delta(\bar{x})$ which has the support $\bar{x} = 0$, there is also $\delta(f)$ with support on the surface $f(\bar{x}) = 0$. We will have more to say about $\delta(f)$ in the next section.

We now discuss the definition of continuity of linear functionals on the space D. Continuity is a topological property. Space D is a linear or vector space. It is made into a topological vector space by defining the neighborhood of $\phi(x) = 0$ by a sequence of seminorms. The two conditions required above in the definition of $\phi_n \xrightarrow{D} 0$ follow from the conditions used in the definition of neighborhood of $\phi(x) = 0$ [10,11]. The definition of continuity of linear functionals on space D can be based on the weak or strong topologies of space D' [11]. It so happens that the definitions of continuity based on these topologies are equivalent to what we have defined earlier, i.e., $F[\phi_n] \to 0$ if and only if $\phi_n \xrightarrow{D} 0$. We note that since D is a linear space, we can define $\phi_n \xrightarrow{D} \phi$ if $\phi_n - \phi \xrightarrow{D} 0$. Because of the linearity of $F[\phi]$, we can also say $F[\phi]$ is continuous if whenever $\phi_n \xrightarrow{D} \phi$, we have $F[\phi_n] \to F[\phi]$.

We mention one more important fact here. Although in this paper we will confine ourselves to the test function space D, in many applications we should use a different test function space. For example, to define Fourier transformation, we should use a test function space S of infinitely differentiable functions which go to zero at infinity faster than $|x|^{-n}$ for any $n > 0$ (the space of rapidly decreasing functions). Other test function spaces are defined in Bremermann [12,13] and Carmichael and Mitrovic [10]. Generalized functions on these spaces are defined as continuous linear functionals after a suitable definition of convergence to zero in the test function space is given to get a topological vector space. One should remember that in all these spaces of generalized functions, the important singular generalized functions, such as the Dirac delta function, are retained with properties essentially similar to what we study below in space D'. Note that if $A \subset B$, and A and B are two test function spaces used to define generalized function spaces A' and B', respectively, then we have $A \subset B \subset B' \subset A'$, i.e. the space of generalized functions A' is larger than B'. In particular, $D \subset S \subset S' \subset D'$ where S is the space of rapidly decreasing functions defined above.

2.1 How Generalized Functions Can Be Introduced In Mathematics

Although L. Schwartz developed the theory of distributions, like many great ideas in mathematics and science, the subject has a long history. We quote here a statement from Synowiec [14] to this effect: "Evolution of the concepts of distribution theory followed a familiar pattern in mathematics of multiple and simultaneous discoveries, because the appropriate ideas were "in the air"". There are several good sources of history of theory of distributions [15,16]. We will not, therefore, present a detailed history here. There are now many different approaches in mathematics to introduce and develop systematically generalized function theory. We mention several of them here.

i) The Functional Approach

In this approach, generalized functions are defined as continuous linear functionals. We use this approach in this paper. This approach was originally used by Schwartz [1] and is the most popular and direct method of studying generalized functions [3,4,7]. The operations on ordinary functions such as differentiation and Fourier transformation are extended by first writing these operations in the language of functionals for ordinary functions and then using them to define the operations for all generalized functions. Once the rules of these operations are obtained, one uses the usual notation of ordinary functions for all generalized functions. It is relatively easy to develop a working knowledge with this notation without confusion. Some elementary knowledge of functional analysis is needed in this approach.

ii) The Sequential Approach

This approach is essentially based on the original idea of Dirac in defining a delta function as the limit of a sequence of ordinary functions. The approach originates from a theorem in distribution theory that the space of generalized functions is complete. Therefore, singular generalized functions such as the delta function, can be defined as the limit of ordinary (i.e., regular) functions. This is much like defining irrational numbers as limits of a Cauchy sequence of rational numbers. There are many good books on this subject [2,6,17]. To define a generalized function, one is required to construct and work with a sequence of ordinary functions which are usually infinitely differentiable. Although one uses mathematics of the level of advanced calculus, the algebraic manipulations are technical and laborious. An extension to multidimensional case appears to be more difficult than with the functional approach.

iii) Bremermann's Approach

In this approach, generalized functions of real variables are viewed as the boundary value of analytic functions on the real axis. This approach was originally developed by Bremermann [12,13] and has its basis in earlier works on Fourier transformation in the complex plane to define Fourier transforms of polynomials. This approach employs some of the powerful results of analytic function theory and is particularly useful in Fourier analysis and partial differential equations. A recent book on the subject is by Carmichael and Mitrovic [10].

iv) Mikusinski's Approach

This approach is based on ideas from abstract algebra. A commutative ring is constructed from functions with support on a semi-infinite axis by defining the operations of addition and multiplication as ordinary addition and convolution of functions, respectively. This commutative ring has no zero divisors by a theorem of Titchmarsh. It can, therefore, be extended to a field by the addition of a multiplicative identity and the multiplicative inverses of all functions. This multiplicative identity turns out to be the Dirac delta function. Mikusinski's approach explains Heaviside's operational calculus rigorously and solves other problems such as the solution of recursion relations. One of the limitations of this approach is that the supports of the functions are confined to semi-axis or half-space in the multidimensional case. A good source of this approach is a two volume book by Mikusinski [18] and an excellent expository book is by Erdelyi [19].

v) Other Approaches

There are also other important approaches to introduce generalized functions in mathematics. One is based on the nonstandard analysis of Robinson [20]. Nonstandard analysis uses formal logic theory to extend the real line by including infinitesimals of Leibniz rigorously. Many interesting applications of this theory, particularly in dynamical systems, are now available. Another more recent approach is by Colombeau [21,22] and Rosinger [23]. This approach uses advanced algebraic and topological concepts to develop a theory of generalized functions where multiplication of arbitrary functions is allowed. This approach is gaining popularity at present. Applications to nonlinear partial differential equations are given by Rosinger [23] and Obeguggenberger [24].

3 SOME DEFINITIONS AND RESULTS

In this section, some important definitions and results will be presented. In the subsections, generalized derivative, multidimensional delta functions and the finite part of divergent integrals are discussed. Since this paper is application oriented, we are very selective about the material presented here. Note that we freely refer to a generalized function by a symbolic or a functional notation.

i) Multiplication of a generalized function by an infinitely differentiable function.

Let $f(x)$ be a generalized function in D' defined by the functional $F[\phi]$ and let $a(x)$ be an infinitely differentiable function, then $a(x)f(x)$ is defined by the following rule

$$aF[\phi] = F[a\phi]. \tag{3.1}$$

Note that aF stands for the functional defining af. Also since ϕ is in D, so is $a\phi$. We can use this definition to define $a(x)\delta(x)$. Let $\delta[\phi]$ be the Dirac function given by Equation (2.6), then

$$a\delta[\phi] = \delta[a\phi] = a(0)\phi(0). \tag{3.2}$$

Symbolically this is interpreted as

$$a(x)\delta(x) = a(0)\delta(x). \tag{3.3}$$

In the space D', multiplication of two arbitrary generalized functions is not defined. This statement needs clarification. Obviously, since ordinary functions are also generalized functions and since any two ordinary functions can be multiplied, they can also be multiplied in the sense of distributions. However, multiplication of a regular and a singular generalized function or two singular generalized functions may not always be defined. As an example, the multiplication of $\delta(x)$ by itself, i.e., $\delta^2(x)$ is not defined, neither is $f(x)\delta(x)$ where $f(x)$ has a jump discontinuity or a singularity at $x = 0$. In applications, one usually knows from experience or inconsistencies in the results that some multiplication of two generalized functions is not allowed. This occurs occasionally in applications. Sometimes this problem can be removed by rewriting the expression in such a way that the troublesome multiplication is avoided. For example, in finding shock jump conditions (Subsection 4.1), if we use the mass continuity and momentum equations in nonconservative forms, difficulties with multiplication of distributions appear. These difficulties can be removed by using these equations in conservative forms. To overcome the problem of multiplication of distributions in space D', new spaces of generalized functions have been defined [22-26]. Colombeau [26, chapters 1-3] gives an intuitive description of the problem of multiplication of distributions and how to remedy this problem.

ii) Shift operator.

Let $f(x)$ be an ordinary function and define the shift operator as $E_h f(x) = f(x+h)$. Then if $F[\phi]$ is the functional representing $f(x)$ by Equation (2.1) and if the shifted function $E_h f(x)$ is represented by $E_h F[\phi]$, we have

$$\begin{aligned} E_h F[\phi] &= \int f(x+h)\phi(x)dx \\ &= \int f(x)\phi(x-h)dx \\ &= F[E_{-h}\phi]\,. \end{aligned} \tag{3.4}$$

This rule can now be used for all generalized functions in D' since $E_{-h}\phi$ is in D. For example, $E_h\delta(x) = \delta(x+h)$ has the following property:

$$\begin{aligned} E_h\delta[\phi] &= \int \delta(x+h)\phi(x)dx \\ &= \delta[\phi(x-h)] \\ &= \phi(-h)\,. \end{aligned} \tag{3.5}$$

Note that the integral in Equation (3.5) is meaningless and stands for the functional $E_h\delta[\phi]$ which in turn is given by $\delta[E_{-h}\phi]$.

iii) Equality of two generalized functions $f(x)$ and $g(x)$ on an open set Ω.

Two generalized functions f and g in D' given by functionals $F[\phi]$ and $G[\phi]$ on D, respectively, are equal on an open set Ω if $F[\phi] = G[\phi]$ for all ϕ such that *supp* $\phi \subset \Omega$. For example, $\delta(x) = 0$ on open sets $(0,\infty)$ and $(-\infty, 0)$. We mention here that generalized functions are only compared on open intervals.

iv) Support of a generalized function.

The support of a generalized function $f(x)$ is the complement with respect to the real line of the open set on which $f(x) = 0$. For example, the support of $\delta(x)$ is the set $\{0\}$, that is, the point $x = 0$.

v) Sequence of generalized functions.

A sequence of generalized functions $F_n[\phi]$ is convergent if the sequence of numbers $\{F_n[\phi]\}$ is convergent for all ϕ in D. For example, let

$$\delta_n(x) = \begin{cases} n^2\left(\dfrac{1}{n} - |x|\right) & |x| \le \dfrac{1}{n} \\ 0 & |x| > \dfrac{1}{n} \end{cases} . \tag{3.6}$$

This function is shown in Figure 2. It is, of course, an ordinary function. It can be shown that

$$\lim_{n\to\infty} \delta_n(x) = \delta(x) \tag{3.7}$$

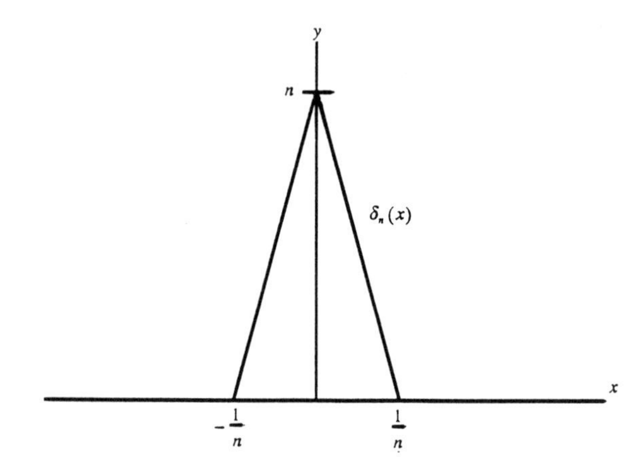

Figure 2. An example of a δ-sequence.

This means that, for the functional $\delta_n[\phi]$ representing $\delta_n(x)$ as follows

$$\delta_n[\phi] = \int \delta_n(x)\phi(x)dx \tag{3.8}$$

where ϕ is in D, we have

$$\lim_{n\to\infty} \delta_n[\phi] = \phi(0) = \delta[\phi] \tag{3.9}$$

The following important theorem characterizes D' and has significant applications [7].

Theorem: The space D' is complete.

This theorem means that a convergent sequence of generalized functions in D' always converges to a generalized function in D'. The index in the definition of convergence can be a continuous variable. For example, $F_\varepsilon[\phi]$ is convergent as $\varepsilon \to 0$ if $\lim_{\varepsilon\to 0} F_\varepsilon[\phi]$ exists for all ϕ in D. We will use this theorem later in this section when we discuss the finite part of divergent integrals.

vi) Odd and even generalized functions.

A generalized function $F[\phi]$ is even if $F[\phi(-x)] = F[\phi(x)]$ and odd if $F[\phi(-x)] = -F[\phi(x)]$. For example $\delta(x)$ is even while x is odd.

vii) Derivative of a generalized function.

This is the most important operation used in this paper. Let $f(x)$ be an ordinary function with a continuous first derivative, i.e., f is a C^1 function. If $f(x)$ is represented by the functional $F[\phi]$ in Equation (2.1), then it is natural that we identify its derivative $f'(x)$ with $F'[\phi]$ given by the functional

$$F'[\phi] = \int f'\phi \, dx . \tag{3.10}$$

Now integrate by parts and use the fact that ϕ has compact support to get

$$F'[\phi] = -\int f\phi' \, dx = -F[\phi'] . \tag{3.11}$$

Note that since $\phi \in D$, it follows that $\phi' \in D$. Thus $F[\phi']$ is a functional on D. We now use Equation (3.11) to define the derivative of all generalized functions in D'. We can keep taking higher order derivatives and obtain the following result

$$F^{(n)}[\phi] = (-1)^n F[\phi^{(n)}] \qquad (n = 1, 2, \dots) . \tag{3.12}$$

We have thus arrived at the following important theorem:

Theorem: Generalized functions have derivatives of all orders.

What we have obtained is a very surprising result. Even locally Lebesgue integrable functions that are discontinuous are infinitely differentiable in generalized function sense What does this result mean in applications? We will say much more about generalized derivatives in the next subsection. Let us first see some examples.

Example 1. The derivative of the delta function, $\delta'(x)$, has the following property

$$\begin{aligned} \delta'[\phi] &= -\delta[\phi'] \\ &= -\phi'(0) . \end{aligned} \tag{3.13}$$

Symbolically, we can write

$$\int \delta'(x)\phi(x)dx = -\phi'(0) . \tag{3.14}$$

Note that $\delta'(x)$ is an odd generalized function.

Example 2. The Heaviside function is defined as

$$h(x) = \begin{cases} 1 & x > 0 \\ 0 & x < 0 \end{cases} \tag{3.15}$$

or in functional notation

$$H[\phi] = \int_0^\infty \phi(x)dx . \tag{3.16}$$

This function is discontinuous at $x = 0$. What is its generalized derivative? We use Equation (3.11) as follows:

$$\begin{aligned} H'[\phi] &= -H[\phi'] \\ &= -\int_0^\infty \phi' dx \\ &= \phi(0) \\ &= \delta[\phi] . \end{aligned} \tag{3.17}$$

Or symbolically, we write

$$\overline{h'}(x) = \delta(x) . \tag{3.18}$$

Note the use of the bar over h' to signify generalized differentiation since $h'(x) = 0$ where now h' stands for the ordinary derivative.

We give one more important characterization of space D' [7] known as the structure theorem of distribution theory:
Theorem: Generalized functions in D' are generalized derivatives of a finite order of continuous functions.
As an example, we note that the Dirac delta function is the second generalized derivative of the continuous function

$$f(x) = \begin{cases} x & x \geq 0 \\ 0 & x < 0 \end{cases}. \tag{3.19}$$

viii) Fourier transforms of generalized functions.

We now work with the space of rapidly decreasing test functions S. In this space the Fourier transform of each test function is again in S. We define the Fourier transform of an ordinary function $\psi(x)$ as follows

$$\hat{\psi}(\xi) = \int_{-\infty}^{\infty} \psi(x) e^{2\pi i x \xi} dx . \tag{3.20}$$

Let $f(x)$ be an ordinary function which has Fourier transform $\hat{f}(\xi)$, e.g., let f be square integrable on $(-\infty, \infty)$. Then for $\psi(x)$ in S, the Parseval's relation gives

$$\int_{-\infty}^{\infty} \hat{f}(x)\psi(x)dx = \int_{-\infty}^{\infty} f(x)\hat{\psi}(x)dx . \tag{3.21}$$

If now $F[\psi]$ is identified with $f(x)$, then we should identify $\hat{F}[\psi]$ with $\hat{f}(\xi)$. However, Equation (3.21) is actually the following relation

$$\hat{F}[\psi] = F[\hat{\psi}] . \tag{3.22}$$

We use this as the definition of the Fourier transform of generalized functions in space S'. As an example, we have

$$\hat{\delta}[\psi] = \delta[\hat{\psi}] = \hat{\psi}(0) = \int_{-\infty}^{\infty} \psi(x)dx . \tag{3.23}$$

The last integral is the functional generated by function 1 so that

$$\hat{\delta}(\xi) = 1. \tag{3.24}$$

This means that the Fourier transform of the Dirac delta function is the constant function one.

We will not say more about this subject here since we will not use Fourier transforms in this paper. We note, however, that if ψ is in D, then $\hat{\psi}$ is not necessarily in D so that Equation (3.22) is meaningless in D'. That is why we have to change the test function space from D to S. Another method of fixing this problem is that we use the Fourier transforms of functions in space D as a new test function space $\hat{D}$. The Fourier transformation of functions in D' are now continuous linear functionals on space $\hat{D}$. These generalized functions are called ultradistributions [27].

ix) <u>Exchange of limit processes.</u>

One of the most powerful results in generalized function theory is that the limit processes can be exchanged. For example, the following exchanges are all permissible

$$\frac{\bar{d}}{dx}\int_{\Omega}\cdots = \int_{\Omega}\frac{\bar{d}}{dx}\cdots \tag{3.25-a}$$

$$\frac{\bar{d}}{dx}\sum_{n}\cdots = \sum_{n}\frac{\bar{d}}{dx}\cdots \tag{3.25-b}$$

$$\sum_{n}\int_{\Omega}\cdots = \int_{\Omega}\sum_{n}\cdots \tag{3.25-c}$$

$$\lim_{n\to\infty}\int_{\Omega}\cdots = \int_{\Omega}\lim_{n\to\infty}\cdots \tag{3.25-d}$$

$$\frac{\bar{d}}{dx}\lim_{n\to\infty}\cdots = \lim_{n\to\infty}\frac{\bar{d}}{dx}\cdots \tag{3.25-e}$$

$$\lim_{n\to\infty}\sum_{m}\cdots = \sum_{m}\lim_{n\to\infty}\cdots \tag{3.25-f}$$

$$\frac{\bar{\partial}^2}{\partial x_i \partial x_j}\cdots = \frac{\bar{\partial}^2}{\partial x_j \partial x_i}\cdots . \tag{3.25-g}$$

Here, as before, we have used a bar over the derivative to indicate generalized differentiation. As an example, let us consider the Fourier series of the following simple periodic function

$$f(x) = \begin{cases} 1 & 0 < x < \pi \\ -1 & -\pi < x < 0 \end{cases} \tag{3.26}$$

which is

$$f(x) = \sum_{m=0}^{\infty} \frac{4}{(2m+1)\pi} \sin(2m+1)x \,. \tag{3.27}$$

This function is shown in Figure 3. The function $f(x)$ has a jump of $2(-1)^n$ at $x = n\pi$ for $n = 0, \pm 1, \pm 2$. By a result given in the next subsection (Equation (3.43))

$$\frac{\bar{d}f}{dx} = 2 \sum_{n=-\infty}^{\infty} (-1)^n \delta(x - n\pi) \tag{3.28}$$

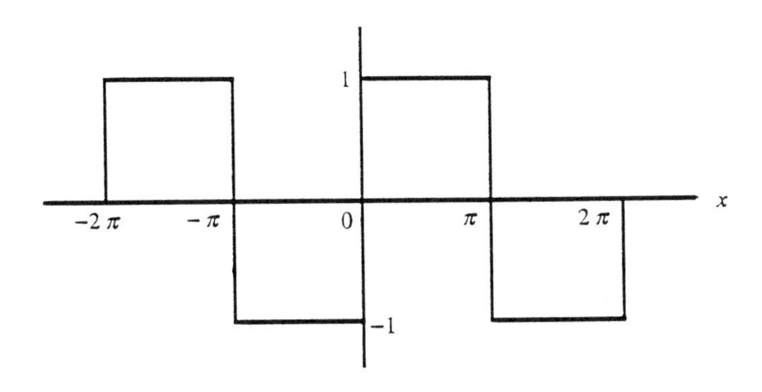

Figure 3. A periodic function with jump discontinuity of $2(-1)^n$ at $x = n\pi, n = 0, \pm 1, \pm 2, \ldots$

Also by Equation (3.25-b), we have

$$\begin{aligned} \frac{\bar{d}f}{dx} &= \frac{\bar{d}}{dx} \sum_{m=0}^{\infty} \frac{4}{(2m+1)\pi} \sin(2m+1)x \\ &= \sum_{m=0}^{\infty} \frac{\bar{d}}{dx} \frac{4}{(2m+1)\pi} \sin(2m+1)x \\ &= \sum_{m=0}^{\infty} \frac{4}{\pi} \cos(2m+1)x \,. \end{aligned} \tag{3.29}$$

From Equations (3.28) and (3.29), we conclude that

$$\frac{2}{\pi}\sum_{m=0}^{\infty}\cos(2m+1)x = \sum_{n=-\infty}^{\infty}(-1)^n\,\delta(x-n\pi) \tag{3.30}$$

The series on the left is divergent in the classical sense. Nevertheless, such a result is often useful in signal analysis. Another important application of exchange of limit processes is in obtaining the finite part of a divergent integral (see Subsection 3.3, below).

x) Integration of generalized functions.

We say that $G[\phi]$ is an integral of $F[\phi]$ if

$$G'[\phi] = F[\phi]. \tag{3.31}$$

For example, we can easily show that the Heaviside function is an integral of the Dirac delta function since

$$\begin{aligned} H'[\phi] &= -H[\phi'] \\ &= \phi(0) \\ &= \delta[\phi] \end{aligned} \tag{3.32}$$

Let $K[\phi]$ be a generalized function such that

$$K'[\phi] = 0 \tag{3.33}$$

for all $\phi \in D$. Then, if $G[\phi]$ is an integral of $F[\phi]$, it follows that $(G+K)[\phi] = G[\phi] + K[\phi]$ is also an integral of $F[\phi]$. It can be shown [7,28] that the only solution of Equation (3.33) in D' is

$$K[\phi] = \int c\,\phi(x)dx \tag{3.34}$$

where c is an arbitrary constant, i.e., $K[\phi]$ is a constant distribution. This result corresponds to the classical indefinite integration of a function

$$\int f(x)dx = g(x) + c. \tag{3.35}$$

We use the same notation symbolically for all generalized functions. For example, we write

$$\int \delta(x)dx = h(x) + c \tag{3.36}$$

where $h(x)$ is the Heaviside function. Note that the integral on the left of Equation (3.36) is meaningless in terms of the classical integration theories.

3.1 Generalized Derivative

In view of the importance of this concept in generalized function theory, we will devote this subsection to it and give some very useful results for applications here. Indeed, it is the results themselves, rather than the mathematical rigor used in deriving them, that are of interest to us in this paper. We remind the readers that we use a bar over the differentiation symbol to denote generalized derivatives if there is an ambiguity in interpretation. For example, we use $\frac{\bar{d}f}{dx}$, $\bar{f}'(x)$, $\frac{\bar{\partial} f}{\partial x_i}$,and $\frac{\bar{\partial}^2 f}{\partial x_i \partial x_j}$ to denote generalized derivatives of ordinary functions but we do not use a bar over $\delta'(x)$ and $\frac{\partial}{\partial x_i}\delta(f)$ because it is obvious that these derivatives can only be generalized derivatives.

i) Functions with discontinuities in one dimension.

Let $f(x)$ be a piecewise smooth function with one discontinuity at x_0 with a jump at this point defined by the relation

$$\Delta f = f(x_{0+}) - f(x_{0-}). \tag{3.37}$$

We want to find the generalized derivative of $f(x)$. Let ϕ be in D and let x_0 be in the support of $\phi(x)$. Then if $F[\phi]$ is the functional representing $f(x)$ by Equation (2.1), we have for *supp* $\phi = [a, b]$, the following result:

$$\begin{aligned} F'[\phi] &= -F[\phi'] \\ &= -\int_a^b f(x)\phi'(x)dx \\ &= -\left[\int_a^{x_{0-}} + \int_{x_{0+}}^b f(x)\phi'(x)dx\right] \\ &= \int_a^b f'(x)\phi(x)dx + \left[f(x_{0+}) - f(x_{0-})\right]\phi(x_0) \\ &= \int_a^b f'(x)\phi(x)dx + \Delta f\phi(x_0). \end{aligned} \tag{3.38}$$

We have performed an integration by parts to get to the last step. We have also used the fact that $\phi(a) = \phi(b) = 0$ in the integration by parts. Noting that $\phi(x_0) = \delta[\phi(x + x_0)] = E_{-x_0}\delta[\phi(x)]$, where E_{-x_0} is the shift operator, we write Equation (3.38) symbolically as

$$\bar{f}'(x) = f'(x) + \Delta f \delta(x - x_0) \tag{3.39}$$

One question that comes to the mind is what the use of $\bar{f}'(x)$ is as compared to the ordinary derivative $f'(x)$. Let us study eq. (3.38). The functional $F'[\phi]$ corresponding to $\bar{f}'(x)$ indeed has retained the memory of the jump Δf on the right side of this equation. Symbolically, $\bar{f}'(x)$ can be integrated over $[c, x]$ where $c < x_0 < x$ to give the result

$$f(x) = \int_c^x f'(x)dx + f(c) + \Delta f \tag{3.40}$$

This means that we have recovered the original discontinuous function. We note, however, that

$$f(x) \neq \int_c^x f'(x)dx + f(c) \tag{3.41}$$

because the memory of the jump Δf is not retained in $f'(x)$ but is retained in $\bar{f}'(x)$. If a function $f(x)$ has n discontinuities at x_i, $i = 1 - n$ with the jump Δf_i at x_i defined by

$$\Delta f_i = f(x_{i+}) - f(x_{i-}), \tag{3.42}$$

then

$$\bar{f}'(x) = f'(x) + \sum_{i=1}^{n} \Delta f_i \delta(x - x_i) \tag{3.43}$$

This is the first indication that when we work with discontinuous functions in applications, the proper setting for the problem is in the space of generalized functions. In particular if an integral method, such as the Green's function approach, is used to find the solution, essentially no significant changes to algebraic manipulations are needed in finding discontinuous solutions provided we stay in the space of generalized functions. Again, we will have more to say about this later.

ii) Functions with discontinuities in multidimensions.

Let us now consider the function $f(\bar{x})$ which is discontinuous across the surface $g(\bar{x}) = 0$. Let us define the jump Δf across $g = 0$ by the relation

$$\Delta f = f(g = 0+) - f(g = 0-) \tag{3.44}$$

Note that $g = 0+$ is on the side of the surface $g = 0$ into which ∇g points. We would like to find $\frac{\bar{\partial} f}{\partial x_i}$. To do this we will use the result of (*i*) above as follows. Let us put a surface coordinate system $\left(u^1, u^2\right)$ on $g = 0$ and extend these coordinates to the space in the vicinity of this surface along normals. Let $u^3 = g$ be the third coordinate variable which is well defined by the function g in the vicinity of this surface. We note that f in variables u^1 and u^2 is continuous but it is discontinuous in variable u^3. Therefore, we have

$$\frac{\bar{\partial} f}{\partial u^i} = \frac{\partial f}{\partial u^i} \qquad i = 1, 2 \tag{3.45-a}$$

$$\frac{\bar{\partial} f}{\partial u^3} = \frac{\partial f}{\partial u^3} + \Delta f \delta(u^3) \tag{3.45-b}$$

In Equation (3.45-b), we have used Equation (3.39). Thus, we get, using the summation convention

$$\begin{aligned} \frac{\bar{\partial} f}{\partial x^i} &= \frac{\bar{\partial} f}{\partial u^j} \frac{\partial u^j}{\partial x_i} \\ &= \frac{\partial f}{\partial u^j} \frac{\partial u^j}{\partial x_i} + \Delta f \frac{\partial u^3}{\partial x_i} \delta(u^3) \\ &= \frac{\partial f}{\partial x_i} + \Delta f \frac{\partial g}{\partial x_i} \delta(g) \end{aligned} \tag{3.46}$$

We can write this in vector notation as

$$\bar{\nabla} f = \nabla f + \Delta f \nabla g \delta(g) \tag{3.47}$$

In the next subsection, we will discuss how to interpret $\delta(g)$ when $g = 0$ is a surface. We can similarly define generalized divergence and curl as follows

$$\bar{\nabla} \cdot \bar{f} = \nabla \cdot \bar{f} + \nabla g \cdot \Delta \bar{f} \delta(g) \tag{3.48-a}$$

$$\bar{\nabla} \times \bar{f} = \nabla \times \bar{f} + \nabla g \times \Delta \bar{f} \delta(g) \tag{3.48-b}$$

The rigorous derivation of both these results requires some knowledge of the invariant definition of divergence and curl in general curvilinear coordinate systems [8,9]. We can combine the above three results by using $*$ for the three operations

$$\overline{\nabla} * \bar{f} = \nabla * \bar{f} + \nabla g * \Delta \bar{f} \delta(g) \tag{3.49}$$

iii) Ordinary differential equations and Green's function.

We give a few simple results here. One important question which was discussed in connection with integrals of generalized functions is the solution of

$$\bar{f}'(x) = 0 \tag{3.50}$$

in D'. It can be shown easily that the only solution of this equation is the classical solution [7,28], that is

$$f(x) = C \text{ (a constant)} \tag{3.51}$$

However, the solution of the equation

$$xf(x) = 0 \tag{3.52}$$

which is not a differential equation, is

$$f(x) = C\delta(x) \tag{3.53}$$

where C is a constant. To get this solution, some simple results from the generalized Fourier transform are used [28]. Taking the Fourier transform of both sides of eq. (3.52), we get

$$\frac{\bar{d}}{d\xi} \hat{f}(\xi) = 0. \tag{3.54}$$

Therefore, we have after integration of Equation (3.54).

$$\hat{f}(\xi) = C \tag{3.55}$$

By taking the inverse Fourier transform of both sides of eq. (3.55), we get eq. (3.53).

From this result, the solution of

$$x\bar{f}'(x) = 1 \tag{3.56}$$

is found as

$$f(x) = \ln|x| + C_1 + C_2 h(x) \tag{3.57}$$

where C_1 and C_2 are constants and $h(x)$ is the Heaviside function. The solution $C_2 h(x)$ comes from the fact that the solution of the homogeneous equation

$$x\bar{f}'(x) = 0 \tag{3.58}$$

is, from Equation (3.53)

$$\bar{f}'(x) = C_2 \delta(x) \tag{3.59}$$

and thus the solution of the homogeneous Equation (3.58) is

$$f(x) = C_2 h(x) + C_1 \tag{3.60}$$

Let us now consider a linear second order ordinary differential equation with two linear and homogeneous boundary conditions (BC) as follows

$$\begin{cases} \ell u = A(x)u'' + B(x)u' + C(x) = f(x) \qquad x \in [0,1] \\ \mathrm{BC}_1[u] = 0 \\ \mathrm{BC}_2[u] = 0 \end{cases} \tag{3.61}$$

Let us also assume that we know u is a C^1 function and u'' is Lebesgue integrable so that $\overline{u''} = u''$ and $\overline{u'} = u'$. Suppose there exists a function $g(x, y)$, the Green's function such that

$$u(x) = \int_0^1 f(y)g(x, y)dy\,. \tag{3.62}$$

Since $u \in C^1$, then $\bar{\ell}u = \ell u$ by continuity of u and u'. We know we can take $\bar{\ell}$ into the integral in Equation (3.62) but not ℓ since $g(x, y)$ may not belong to C^1. Therefore, using ℓ_x to indicate that derivatives in ℓ are with respect to x, we get

$$\begin{aligned} \ell u &= \bar{\ell}u \\ &= \bar{\ell}_x \int_0^1 f(y)g(x, y)dy \\ &= \int_0^1 f(y)\bar{\ell}_x g(x, y)dy \\ &= f(x) \qquad \text{(from Equation (3.61)).} \end{aligned} \tag{3.63}$$

We see that $\bar{\ell}_x g(x, y)$ must have the sifting property

$$\bar{\ell}_x g(x, y) = \delta(x - y) \tag{3.64}$$

Because of the linearity of the BC's, we also have

$$\mathrm{BC}[u] = \int_0^1 f(y)\mathrm{BC}_x[g(x, y)]dy \tag{3.65}$$

Therefore, other conditions on $g(x, y)$ are

$$\mathrm{BC}_{1x}[g(x, y)] = 0 \tag{3.66-a}$$
$$\mathrm{BC}_{2x}[g(x, y)] = 0 \tag{3.66-b}$$

where the x in the subscripts of the BC's indicates that $g(x, y)$ in variable x satisfies the two boundary conditions.

From Equation (3.64), we conclude that, since $\bar{\ell}_x$ is a second order ordinary differential equation, $g(x, y)$ must be continuous at $x = y$ and $\frac{\partial g}{\partial x}$ must have a jump discontinuity at $x = y$. The reason is that if $g(x, y)$ has a discontinuity at $x = y$, the first generalized derivative with respect to x will give a $\delta(x - y)$ by Equation (3.39). A second generalized derivative would give $\delta'(x - y)$ in the result. But since $\delta'(x - y)$ is missing on the right of Equation (3.64), $g(x, y)$ cannot be discontinuous at $x = y$. Assuming that $g(x, y)$ is defined by

$$g(x, y) = \begin{cases} g_1(x, y) & 0 \le x < y \\ g_2(x, y) & y < x \le 1 \end{cases} \tag{3.67}$$

Equation (3.64) means

$$\ell_x g_1(x, y) = \ell_x g_2(x, y) = 0 \tag{3.68-a}$$
$$g_1(y, y) = g_2(y, y) \tag{3.68-b}$$
$$\frac{\partial g_2}{\partial x}(y, y) - \frac{\partial g_1}{\partial x}(y, y) = \frac{1}{A(y)} \tag{3.68-c}$$

Equation (3.68-a) means that g_1 and g_2 in variable x are solutions of the homogeneous equation $\ell u = 0$. Equation (3.68-b) expresses continuity of g at $x = y$ while Equation (3.68-c) gives the jump of $\frac{\partial g}{\partial x}$ at $x = y$. To get Equation (3.68-c), we note that

$$\begin{aligned}
\bar{\ell}_x g &= \ell_x g + A(y)\left[\frac{\partial g_2}{\partial x}(y,y) - \frac{\partial g_1}{\partial x}(y,y)\right]\delta(x-y) \\
&= A(y)\left[\frac{\partial g_2}{\partial x}(y,y) - \frac{\partial g_1}{\partial x}(y,y)\right]\delta(x-y) \\
&= \delta(x-y) \qquad \text{(from Equation (3.64))}
\end{aligned} \tag{3.69}$$

Equation (3.68-c) follows from the fact that the coefficient of $\delta(x-y)$ in the expression after the second equality sign must be equal to 1. Note that Equation (3.68-a) is the same as $\ell_x g = 0$ which is used above. The Green's function is now determined from Equations (3.66) and (3.68).

iv) Leibniz rule of differentiation under the integral sign.

We want to find the result of taking the derivative with respect to variable α in the following expression in which A, B and f are continuous functions and $B(\alpha) > A(\alpha)$ for $\alpha \in [a,b]$:

$$E(\alpha) = \frac{d}{d\alpha}\int_{A(\alpha)}^{B(\alpha)} f(x,\alpha)dx\,. \tag{3.70}$$

Let us define the function $H(x,\alpha)$ as follows

$$H(x,\alpha) = h[x - A(\alpha)]h[B(\alpha) - x] \tag{3.71}$$

where $h(x)$ is the Heaviside function. The function $H(x,\alpha)$ is equal to one when $A(\alpha) < x < B(\alpha)$ and to zero, otherwise. Using $H(x,\alpha)$, we can write $E(\alpha)$ as follows

$$\begin{aligned}
E(\alpha) &= \frac{\bar{d}}{d\alpha}\int_{-\infty}^{\infty} H(x,\alpha)f(x,\alpha)dx \\
&= \int_{-\infty}^{\infty}\left(\frac{\bar{\partial} H}{\partial\alpha} f + H\frac{\bar{\partial} f}{\partial\alpha}\right)dx\,.
\end{aligned} \tag{3.72}$$

We have

$$\begin{aligned}
\frac{\bar{\partial} H}{\partial\alpha}(x,\alpha) &= -A'(\alpha)h[B(\alpha) - x]\delta[x - A(\alpha)] \\
&\quad + B'(\alpha)h[x - A(\alpha)]\delta[B(\alpha) - x] \\
&= -A'(\alpha)\delta[x - A(\alpha)] + B'(\alpha)\delta[B(\alpha) - x]\,.
\end{aligned} \tag{3.73}$$

Note that we have used

$$h[B(\alpha) - x]\delta[x - A(\alpha)] = h[B(\alpha) - A(\alpha)]\delta[x - A(\alpha)] = \delta[x - A(\alpha)] \tag{3.74}$$

This is because $B(\alpha) - A(\alpha) > 0$ and thus the Heaviside function is one. Similarly, we do the same as in Equation (3.74) for the second product of the Heaviside and the delta functions in Equation (3.73). Using Equation (3.73) in Equation (3.72) and integrating with respect to x, we get the Leibniz rule of differentiation under the integral sign:

$$E(\alpha) = \int_{A(\alpha)}^{B(\alpha)} \frac{\partial f}{\partial \alpha}(x, \alpha)dx + B'(\alpha)f[B(\alpha), \alpha] - A'(\alpha)f[A(\alpha), \alpha] . \tag{3.75}$$

3.2 Multidimensional Delta Functions

In multidimensions, $\delta(\bar{x})$ has a simple interpretation given by

$$\int \phi(\bar{x})\delta(\bar{x})d\bar{x} = \phi(0) . \tag{3.76}$$

One easily sees that

$$\delta(\bar{x}) = \delta(x_1)\delta(x_2)\ldots\delta(x_n) \tag{3.77}$$

where $\bar{x} = (x_1, x_2, \ldots, x_n)$. In this section, we will confine ourselves to three dimensional space. Of interest to us in applications are $\delta(f)$ and $\delta'(f)$ where $f = 0$ is a surface in three dimensional space. We can always assume that f is defined so that $|\nabla f| = 1$ at every point on $f = 0$. If f does not have this property, then $f_1 = \dfrac{f}{|\nabla f|}$ has this property. Thus, redefine the surface.

i) <u>Interpretation of $\delta(f)$.</u>

Consider the integral

$$I = \int \phi(\bar{x})\delta(f)d\bar{x} . \tag{3.78}$$

Assume that we define a curvilinear coordinate system (u^1, u^2) on the surface $f = 0$ and extend these variables locally to the space near this surface along local normals. Let $u^3 = f$ which, because $|\nabla f| = \dfrac{df}{du^3} = 1$, u^3 is the local distance from the surface. Thus, $f = u^3 =$

constant $\neq 0$ is a surface parallel to $f = 0$. Of course, we assume u^3 is small. From differential geometry [8,9], we have

$$d\bar{x} = \sqrt{g_{(2)}(u^1, u^2, u^3)}du^1du^2du^3 \tag{3.79}$$

where $g_{(2)}\left(u^1, u^2, u^3\right)$ is the determinant of the first fundamental form of the surface $f = u^3 =$ constant. Using Equation (3.79) in Equation (3.78) and integrating with respect to u^3 gives

$$\begin{aligned} I &= \int \phi[\bar{x}(u^1, u^2, u^3)]\delta(u^3)\sqrt{g_{(2)}(u^1, u^2, u^3)}du^1du^2du^3 \\ &= \int \phi[\bar{x}(u^1, u^2, 0)]\sqrt{g_{(2)}(u^1, u^2, 0)}du^1du^2 \\ &= \int_{f=0} \phi(\bar{x})dS\,. \end{aligned} \tag{3.80}$$

That is, I is the surface integral of ϕ over the surface $f = 0$.

ii) Interpretation of $\delta'(f)$.

We want to interpret

$$\begin{aligned} I &= \int \phi(\bar{x})\delta'(f)d\bar{x} \\ &= \int \phi[\bar{x}(u^1, u^2, u^3)]\delta'(u^3)\sqrt{g_{(2)}(u^1, u^2, u^3)}du^1du^2du^3 \end{aligned} \tag{3.81}$$

Here we have used the coordinate system $\left(u^1, u^2, u^3\right)$ defined above. Integrating the above equation with respect to u^3 gives

$$I = -\int \frac{\partial}{\partial u^3}\left[\phi(\bar{x})\sqrt{g_{(2)}(u^1, u^2, u^3)}\right]_{u^3=0} du^1du^2\,. \tag{3.82}$$

Again, from differential geometry, we have

$$\frac{\partial}{\partial u^3}\sqrt{g_{(2)}(u^1, u^2, u^3)} = -2H_f(u^1, u^2, u^3)\sqrt{g_{(2)}(u^1, u^2, u^3)} \tag{3.83}$$

where H_f stands for the local mean curvature of the surface $f = u^3 =$ constant. Taking the derivative of the integrand of Equation (3.82) and using the result of (3.83), we obtain

$$
\begin{aligned}
I &= -\int \frac{\partial \phi}{\partial u^3}[\bar{x}(u^1,u^2,0)]\sqrt{g_{(2)}(u^1,u^2,0)}du^1du^2 \\
&\quad +\int 2H_f(u^1,u^2,0)\phi[\bar{x}(u^1,u^2,0)]\sqrt{g_{(2)}(u^1,u^2,0)}du^1du^2 \\
&= \int_{f=0} [-\frac{\partial \phi}{\partial n} + 2H_f(\bar{x})\phi(\bar{x})]dS
\end{aligned}
\tag{3.84}
$$

where $\frac{\partial \phi}{\partial n}$ is the usual normal derivative of ϕ. Intuitively, the appearance of the term $2H_f\phi$ in the integrand is not at all obvious. This is a clear indication of the importance of differential geometry in multidimensional generalized function theory.

iii) A simple trick.

We have already shown that $\phi(x)\delta(x) = \phi(0)\delta(x)$. Taking derivatives of both sides of this equation, we get

$$
\phi'(x)\delta(x) + \phi(x)\delta'(x) = \phi(0)\delta'(x)\,. \tag{3.85}
$$

It is obvious that the right side is simpler than the left side. Let us consider the following expression

$$
E = \phi(\bar{x})\delta(f) = \phi[\bar{x}(u^1,u^2,u^3)]\delta(u^3) \tag{3.86}
$$

where again we have used the coordinate system (u^1,u^2,u^3) defined above. We know that

$$
\phi[\bar{x}(u^1,u^2,u^3)]\delta(u^3) = \phi[\bar{x}(u^1,u^2,0)]\delta(u^3) \tag{3.87}
$$

We use the notation $\underset{\sim}{\phi}(\bar{x})$ for $\phi\left[\bar{x}\left(u^1,u^2,0\right)\right]$, that is, $\underset{\sim}{\phi}(\bar{x})$ is the restriction of $\phi(\bar{x})$ to the support of the delta function which is the surface $f = 0$. We note that $\frac{\partial \underset{\sim}{\phi}}{\partial n} = \frac{\partial}{\partial u^3}\phi\left[\bar{x}\left(u^1,u^2,0\right)\right] = 0$. Using $\underset{\sim}{\phi}(\bar{x})$, we can write E in two forms

$$
\begin{aligned}
E &= \phi(\bar{x})\delta(f) && \text{(first form)} \\
&= \underset{\sim}{\phi}(\bar{x})\delta(f) && \text{(second form).}
\end{aligned}
\tag{3.88}
$$

Is there any advantage to the second form as compared to the first form? The answer is yes! Let us take the gradient of E for the two forms in Equation (3.88):

$$\bar{\nabla}E = \nabla\phi\delta(f) + \phi(\bar{x})\nabla f\delta'(f) \qquad \text{(first form)}$$
$$= \nabla_2 \underset{\sim}{\phi}\, \delta(f) + \underset{\sim}{\phi}(\bar{x})\nabla f\delta'(f) \qquad \text{(second form).} \tag{3.89}$$

Here, $\nabla_2 \underset{\sim}{\phi}$ is the surface gradient of $\underset{\sim}{\phi}(\bar{x})$ on $f = 0$. From Equation (3.84), we note that in the integration of $\delta'(f)$ in the first form, there is the term $\frac{\partial\phi}{\partial n}$ which cancels a similar term in the integration of $\nabla\phi\delta(f)$. In the second form, since $\frac{\partial \underset{\sim}{\phi}}{\partial n} = 0$, $\frac{\partial \underset{\sim}{\phi}}{\partial n}$ does not appear in the integration of $\delta'(f)$ and obviously it is also absent in the integration of $\nabla_2 \underset{\sim}{\phi}\, \delta(f)$. Therefore, algebraic manipulations are reduced. It is, thus, expedient to restrict functions multiplying the Dirac delta function to the support of the delta function. Note carefully that this is not true for $\delta'(x)$, i.e. $\phi(x)\delta'(x) \neq \phi(0)\delta'(x)$.

iv) The divergence theorem revisited.

Let Ω be a finite volume in space and let $\bar{\phi}(\bar{x})$ be a C^1 vector field. Let us define the discontinuous vector field $\bar{\phi}_1(\bar{x})$ as follows

$$\bar{\phi}_1(\bar{x}) = \begin{cases} \bar{\phi}(\bar{x}) & \bar{x} \in \Omega \\ 0 & \bar{x} \notin \Omega \end{cases}. \tag{3.90}$$

Let the surface $f = 0$ denote the boundary $\partial\Omega$ of region Ω in such a way that $\bar{n} = \nabla f$ points to the outside of $\partial\Omega$ and $|\nabla f| = 1$ on $f = 0$. We have

$$\bar{\nabla} \cdot \bar{\phi}_1 = \nabla \cdot \bar{\phi}_1 + \Delta\bar{\phi}_1 \cdot \bar{n}\delta(f)$$
$$= \nabla \cdot \bar{\phi}_1 - \bar{\phi}(\bar{x}) \cdot \bar{n}\delta(f)\,. \tag{3.91}$$

We note that $\Delta\bar{\phi}_1 = \bar{\phi}_1(f = 0+) - \bar{\phi}_1(f = 0-) = -\phi(f = 0)$. We have

$$\int\int_{-\infty}^{\infty}\int \frac{\bar{\partial}\phi_{1,1}}{\partial x_1}\, dx_1 dx_2 dx_3 = \iint \phi_1 \Big|_{-\infty}^{\infty} dx_2 dx_3 = 0 \tag{3.92}$$

similarly, we get zero for integrals of $\frac{\bar{\partial}\phi_{1,2}}{\partial x_2}$ and $\frac{\bar{\partial}\phi_{1,3}}{\partial x_3}$, where $\phi_{1,i}$ is the *i-th* component of $\bar{\phi}_1$.

Therefore,

$$\int \bar{\nabla} \cdot \bar{\phi}_1 d\bar{x} = 0 \tag{3.93}$$

Now, the integration of the right side of Equation (3.91) using Equation (3.80) gives

$$\int_{\Omega} \nabla \cdot \bar{\phi} d\bar{x} - \int_{\partial\Omega} \phi_n dS = 0 \tag{3.94}$$

Here we have used the fact that, from Equation (3.90)

$$\nabla \cdot \bar{\phi}_1 = \begin{cases} \nabla \cdot \bar{\phi} & \bar{x} \in \Omega \\ 0 & \bar{x} \notin \Omega \end{cases} \tag{3.95}$$

Also, we define $\phi_n = \bar{\phi} \cdot \bar{n}$. Equation (97) is the divergence theorem.

We note that Equation (3.93) is valid if $\bar{\phi}_1$ has a discontinuity across the surface $k = 0$ within Ω as shown in Figure 4. Equation (3.94) is therefore valid if $\nabla \cdot \bar{\phi}$ in the volume integral is replaced by $\overline{\nabla} \cdot \bar{\phi}$ where the only jump of $\bar{\phi}$ in the generalized divergence comes from the discontinuity on $k = 0$. That is we write

$$\overline{\nabla} \cdot \bar{\phi} = \nabla \cdot \bar{\phi} + \Delta\bar{\phi} \cdot \bar{n}'\delta(k) \tag{3.96}$$

where $\bar{n}' = \nabla k$ is the unit normal to $k = 0$. Equation (3.94) can now be written

$$\int_{\Omega} \overline{\nabla} \cdot \bar{\phi} \, d\bar{x} = \int_{\partial\Omega} \phi_n dS \tag{3.97}$$

which, by using Equation (3.96), can also be written as

$$\int_{\Omega} \nabla \cdot \bar{\phi} \, d\bar{x} = \int_{\partial\Omega} \phi_n dS - \int_{S_k} \Delta\phi_{n'} dS \tag{3.98}$$

where $\Delta\phi_{n'} = \Delta\bar{\phi} \cdot \bar{n}'$ and S_k is the part of the surface $k = 0$ enclosed in region Ω (see Figure 4).

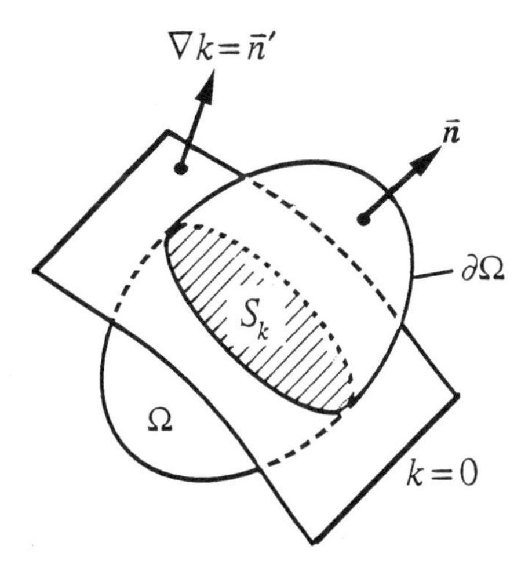

Figure 4. Control volume Ω intersecting a surface of discontinuity of vector field $\bar{\phi}$. Used for deriving the generalized divergence theorem

The divergence theorem is used in deriving conservation laws in fluid mechanics and physics in differential form. The fact that it remains valid for discontinuous vector fields, as shown in Equation (3.97), implies that such conservation laws are valid when all the derivatives are interpreted as generalized derivatives. What this really means is that the jump conditions across the surface of discontinuities are inherent in these conservation laws as will be shown in the next section. This interpretation of conservation laws eliminates the need for the pill-box analysis of jump conditions.

v) Product of two delta functions.

We have said earlier that, in general, the product of two arbitrary generalized functions may not be defined. Here we give the interpretation of the product of two multidimensional generalized functions for which multiplication is possible. Let $f = 0$ and $g = 0$ be two surfaces intersecting along a curve Γ as shown in Figure 5. Assume $\nabla f = \bar{n}$ and $\nabla g = \bar{n}'$ where $|\bar{n}| = |\bar{n}'| = 1$. We want to interpret

$$I = \int \phi(\bar{x})\delta(f)\delta(g)\, d\bar{x} . \tag{3.99}$$

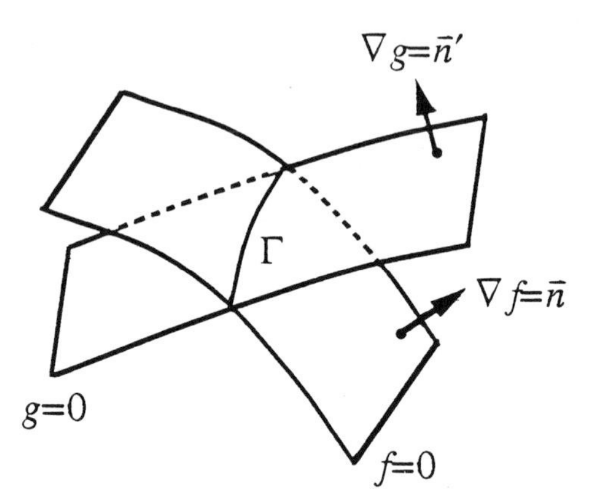

Figure 5. Diagram for the integration of $\delta(f)\delta(g)$ for two intersecting surfaces $f = 0$ and $g = 0$.

On the local plane normal to the Γ-curve, define $u^1 = f$ and $u^2 = g$ and $u^3 = \Gamma$, where Γ is the distance along the Γ-curve. Extend u^1 and u^2 to the space in the vicinity of this plane along a local normal to this plane. We have

$$d\bar{x} = \frac{du^1 du^2 du^3}{\sin\theta} \tag{3.100}$$

where $\sin\theta = |\bar{n} \times \bar{n}'|$. Using equation(3.100) in Equation (3.99) and integrating the resulting integral with respect to u^1 and u^2, we get,

$$I = \int \frac{\phi(\bar{x})}{\sin\theta}\,\delta(u^1)\delta(u^2)\,du^1du^2du^3 = \int_{\substack{f=0\\ g=0}} \frac{\phi(\bar{x})}{\sin\theta}\,d\Gamma\,. \tag{3.101}$$

This result is useful in applications.

3.3 Finite Part Of Divergent Integrals

We come here to a subject which is important in aerodynamics. The classical procedure to find the finite part of divergent integrals appears ad hoc and leads to some questions about validity of the procedure. First, one wonders if the appearance of divergent integrals in applications may not be as the result of errors in modeling the physics of the problem. Second, one would like to know if the method will lead to a unique analytic expression or whether there are different analytic expressions leading to equivalent numerical results. We will see that generalized function theory clearly answers these questions.

Let us first examine the function $f(x) = \ln|x|$ which is locally integable. The ordinary derivative of this function is

$$\frac{d}{dx}\ln|x| = \frac{1}{x} \tag{3.102}$$

which is not locally integrable over any interval which includes $x = 0$. We know, however, that as a generalized function, $\ln|x|$ has generalized derivatives of all orders. The question is what the relation of the generalized derivative of $\ln|x|$ is to the ordinary derivative $f'(x) = \frac{1}{x}$. Let us work with $F[\phi]$ as follows representing $\ln|x|$:

$$F[\phi] = \int \ln|x|\phi(x)dx,\ \phi \in D\,. \tag{3.103}$$

We have, using the definition of generalized derivative,

$$F'[\phi] = -\int \ln|x|\phi'(x)dx\,. \tag{3.104}$$

We need some integration by parts to get the term $\frac{1}{x}$ in the integrand of Equation (3.104). This cannot be performed since $\frac{1}{x}$ is not locally integrable. We solve this problem by using a new functional depending on ε whose limit is $F'[\phi]$ as follows. Let $h_\varepsilon(x)$ be a function defined below for some constant $\alpha > 0$ and a parameter $\varepsilon > 0$.

$$h_\varepsilon(x) = \begin{cases} 0 & -\varepsilon < x < \alpha\varepsilon \\ 1 & \text{otherwise} \end{cases}\,. \tag{3.105}$$

This function is shown in Figure 6. Then it is obvious that $\ln|x|$ can be written as the limit of an indexed generalized function as follows:

$$\lim_{\varepsilon\to 0} h_\varepsilon(x)\ln|x| = \ln|x|\,. \tag{3.106}$$

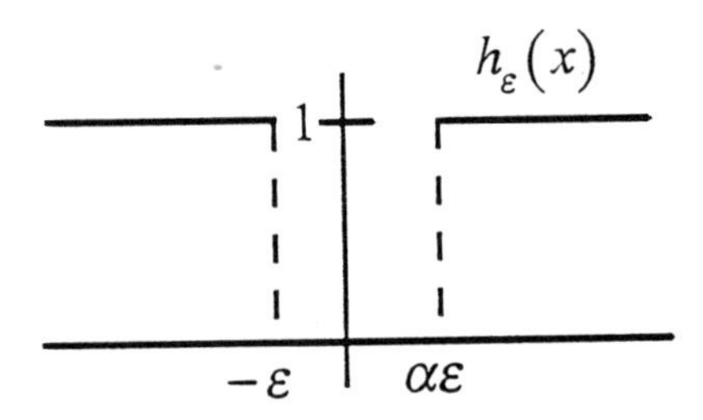

Figure 6. Function $h_\varepsilon(x)$. Used in defining the finite part of the divergent integral $\int \frac{\phi(x)}{x}\,dx$.

Note that if we define $F'_\varepsilon[\phi]$ as

$$F'_\varepsilon[\phi] = -\int h_\varepsilon(x) \ln|x| \phi'(x) dx \ , \tag{3.107}$$

then, we have from completeness theorem of D' that

$$\lim_{\varepsilon \to 0} F'_\varepsilon[\phi] = F'[\phi] \, . \tag{3.108}$$

The function $h_\varepsilon(x) \ln|x|$ has two jump discontinuities at $x = -\varepsilon$ and $x = \alpha\varepsilon$. We can either apply the classical integration by parts to Equation (3.107) by breaking the real line into two intervals or using the generalized derivative as follows

$$F'_\varepsilon[\phi] = \int \frac{\overline{d}}{dx} [h_\varepsilon(x) \ln|x|] \phi(x) dx \, . \tag{3.109}$$

Here we are integrating over *supp* ϕ and we do not worry about the terms coming from the limits of the integral in the integration by parts since $\phi = 0$ at the limit points.

We now take the derivative of the term in square brackets in Equation (3.109):

$$\frac{\bar{d}}{dx}[h_\varepsilon(x)\ln|x|] = \frac{h_\varepsilon(x)}{x} - \ln\varepsilon\,\delta(x+\varepsilon) + \ln(\alpha\varepsilon)\delta(x-\alpha\varepsilon) \tag{3.110}$$

Here, and below, we have used the result that $\phi(x)\delta(x-x_0) = \phi(x_0)\delta(x-x_0)$. We, thus, have, after using Equation (3.110) in Equation (3.109) and integrating with respect to x

$$\begin{aligned} F'_\varepsilon[\phi] &= -\ln\varepsilon\,\phi(-\varepsilon) + \ln(\alpha\varepsilon)\phi(\alpha\varepsilon) + \int \frac{h_\varepsilon(x)}{x}\phi(x)dx \\ &= \phi(0)\ln\alpha + \int \frac{h_\varepsilon(x)}{x}\phi(x)dx + o(\varepsilon) \end{aligned} \tag{3.111}$$

where $o(\varepsilon)$ stands for terms of order ε and higher. Now from Equation (3.108), we have

$$\begin{aligned} F'[\phi] &= \lim_{\varepsilon\to 0} F'_\varepsilon[\phi] \\ &= \phi(0)\ln\alpha + \lim_{\varepsilon\to 0}\int \frac{h_\varepsilon(x)}{x}\phi(x)dx \\ &= \phi(0)\ln\alpha + \lim_{\varepsilon\to 0}\left[\int_{-\infty}^{-\varepsilon} + \int_{\alpha\varepsilon}^{\infty}\frac{\phi(x)}{x}dx\right]. \end{aligned} \tag{3.112}$$

We can show that the limit of the integral on the right of Equation (3.112) exists. If now $\alpha = 1$, then $\ln\alpha = 0$ and

$$F'[\phi] = \lim_{\varepsilon\to 0}\left[\int_{-\infty}^{-\varepsilon} + \int_{\varepsilon}^{\infty}\frac{\phi(x)}{x}dx\right] \tag{3.113}$$

which is known as the Cauchy principal value (*PV*) of the integral. But one does not have to take $\alpha = 1$ and Equation (3.112) is numerically the same as Equation (3.113). The above limit procedure is called taking the finite part of a divergent integral.

What have we really achieved? First note that over any open interval that does not include $x = 0$, we have

$$\frac{\bar{d}}{dx}\ln|x| = \frac{1}{x}. \tag{3.114}$$

But whenever $x = 0$ belongs to the open interval, then the classically divergent integral must be interpreted in such a way that the functional $F'[\phi]$ corresponding to $\frac{\bar{d}}{dx}\ln|x|$ is recovered. As the above simple function demonstrates, there may be more than one different analytic expression for the procedure to find the finite part of a divergent integral. But they are all numerically equivalent. We define the principal value of $\frac{1}{x}$ as follows

$$PV\left(\frac{1}{x}\right) = \frac{\bar{d}}{dx} \ln|x| \tag{3.115}$$

This means that when $x = 0$ is in the interval of integration of $\frac{1}{x}$, the finite part of the divergent integral must by taken to get the numerical value of $F'[\phi]$ where $F[\phi]$ is given by Equation (3.103). Note that the term regularizing a divergent integral is also used in mathematics. The procedure given here corresponds to the canonical regularization of Gel'fand and Shilov [7].

What is the use of this procedure in applications? Suppose we have reduced the solution of a problem to the evaluation of the expression

$$u(x) = \frac{d}{dx} \int_{\Omega} \phi(y) \ln|x - y| dy \tag{3.116}$$

where $x \in \Omega$. Let us assume that we know that the integral is continuous as a function of x so that $\frac{d}{dx}$ can be replaced by $\frac{\bar{d}}{dx}$ and taken inside the integral. We get

$$\begin{aligned} u(x) &= \int_{\Omega} \phi(y) \frac{\bar{d}}{dx} \ln|x - y| dy \\ &= \int_{\Omega} \phi(y) PV\left(\frac{1}{x - y}\right) dy \end{aligned} \tag{3.117}$$

which is interpreted as the finite part of the divergent integral by the procedure defined earlier. We remind the readers that the procedure will result in exactly what Equation (3.116) would give had we been able to perform the integration analytically. Also, assuming that $\phi = 0$ at the boundaries of Ω, an integration by parts of the first integral in Equation (3.117) would give

$$u(x) = -\int_{\Omega} \phi'(y) \ln|x - y| dy \tag{3.118}$$

which is also a legitimate result if this integral exists. The problem is that often in applications, Equation (3.116) is an integral equation for the unknown function $\phi(x)$ which itself has integrable singularities at the boundaries of the interval Ω. This means that the above integration by parts is invalid and in any case, the integral Equation (3.118) is divergent.

Therefore, one is left with the integral equation with the principal value of $\frac{1}{x - y}$ which is a well-known kernel in the theory of singular integral equations.

We now give an advanced example in three dimensions with a surprising implication in the numerical solution of an integral equation of transonic flow which we will discuss in the next section. Let us consider the following integral

$$I(\bar{x}) = \frac{\partial^2}{\partial x_1^2} \int_\Omega \frac{\phi(\bar{y})}{r} \, d\bar{y} \tag{3.119}$$

$$r^2 = (x_1 - y_1)^2 + (x_2 - y_2)^2 + (x_3 - y_3)^2 \tag{3.120}$$

where Ω is a region in space and $\bar{x} \in \Omega$. In this problem $\phi(\bar{x})$ is a C^1 function which is the unknown of the aerodynamic problem. Assuming that the integral is a C^1 function in $\bar{x}$, we can replace $\frac{\partial^2}{\partial x_1^2}$ with $\frac{\bar{\partial}^2}{\partial x_1^2}$ and take the derivatives inside the integral

$$\begin{aligned} I(\bar{x}) &= \int_\Omega \phi(\bar{y}) \frac{\bar{\partial}^2}{\partial x_1^2} \left(\frac{1}{r} \right) d\bar{y} \\ &= \int_\Omega \phi(\bar{y}) \frac{\bar{\partial}^2}{\partial y_1^2} \left(\frac{1}{r} \right) d\bar{y} \,. \end{aligned} \tag{3.121}$$

The reason that we use generalized differentiation rather than ordinary differentiation is that the latter will result in a divergent integral. Note that

$$\frac{\partial}{\partial y_1} \left(\frac{1}{r} \right) = \frac{r_1}{r^3} \tag{3.122-a}$$

$$\frac{\partial^2}{\partial y_1^2} \left(\frac{1}{r} \right) = \frac{3r_1^2 - r^2}{r^5} \tag{3.122-b}$$

where $r_1 = x_1 - y_1$. Since $\frac{r_1}{r^3}$ is integrable, we write

$$\frac{\bar{\partial}^2}{\partial y_1^2} \left(\frac{1}{r} \right) = \frac{\bar{\partial}}{\partial y_1} \left(\frac{r_1}{r^3} \right) \tag{3.123}$$

and we proceed to find the finite part of the divergent integral in Equation (3.122).

Let $f(\bar{y}, \bar{x}, \varepsilon) = g(r_1, r_2, r_3) - \varepsilon = 0$ be a piecewise smooth surface enclosing the point $\bar{y} = \bar{x}$ where $r_i = x_i - y_i$, $i = 1 - 3$ and g is a homogeneous function of order 1, i.e., $g(\alpha r_1, \alpha r_2, \alpha r_3) = \alpha g(r_1, r_2, r_3)$. This condition assures that the surface $g(\alpha r_1, \alpha r_2, \alpha r_3) - \varepsilon = 0$ corresponds to $g(r_1, r_2, r_3) - \dfrac{\varepsilon}{\alpha} = 0$ for $\alpha \neq 0$. Thus all the surfaces $g - \varepsilon = 0$ corresponding to various values of α are similar in shape. From homogeneity of g, it follows that $f(\bar{y}, \bar{x}, 0) = g(r_1, r_2, r_3) = 0$ consists of a single point $\bar{y} = \bar{x}$. For example, for a sphere with center at $\bar{y} = \bar{x}$ and radius ε, we have

$$f(\bar{y}, \bar{x}, \varepsilon) = \sqrt{r_1^2 + r_2^2 + r_3^2} - \varepsilon = 0 . \tag{3.124}$$

In addition, we assume $\nabla_y f = \bar{n}$, where $\bar{n}$ is the local unit outward normal to the surface. Let $f > 0$ outside and $f < 0$ inside this surface, respectively. We introduce the function $h_\varepsilon(\bar{y})$ by the relation

$$h_\varepsilon(\bar{y}) = \begin{cases} 1 & f > 0 \\ 0 & f < 0 \end{cases} \tag{3.125}$$

Now, we define the required generalized derivative in Equation (3.123) by the relation

$$\begin{aligned} \frac{\bar{\partial}^2}{\partial y_1^2}\left(\frac{1}{r}\right) &= \lim_{\varepsilon \to 0} \frac{\bar{\partial}}{\partial y_1}\left[\frac{h_\varepsilon(\bar{y}) r_1}{r^3}\right] \\ &= \lim_{\varepsilon \to 0}\left[\frac{r_1 n_1}{r^3}\delta(f) + \frac{3r_1^2 - r^2}{r^5} h_\varepsilon(\bar{y})\right] \end{aligned} \tag{3.126}$$

where n_1 is the component of $\bar{n}$ along the y_1 -axis. Therefore, $I(x)$ can be written as

$$\begin{aligned} I(\bar{x}) = {} & \lim_{\varepsilon \to 0} \int_{f=0} \frac{r_1 n_1}{r^3} \phi(\bar{y}) dS \\ & + \lim_{\varepsilon \to 0} \int_{\Omega} \frac{3r_1^2 - r^2}{r^5} h_\varepsilon(\bar{y}) \phi(\bar{y}) d\bar{y} \end{aligned} \tag{3.127}$$

where we have used Equation (3.80) to integrate $\delta(f)$ in Equation (3.126).

Using a Taylor series expansion of $\phi(\bar{y})$ at $\bar{y} = \bar{x}$, we find that

$$\lim_{\varepsilon \to 0} \int_{f=0} \frac{r_1 n_1}{r^3} \phi(\bar{y}) dS = \alpha_f \phi(\bar{x}) \tag{3.128}$$

where α_f is a constant depending on the shape of the surface $f = 0$. For example, for the sphere given by Equation (3.124), we have

$$\alpha_f = \frac{4\pi}{3} \tag{3.129}$$

If we take the surface $f = 0$ to be a circular cylinder with its axis parallel to the y_1 -axis such that the base radius is ε and its height is γ, $\frac{\varepsilon}{\gamma} >> 1$, then

$$\alpha_f = 4\pi . \tag{3.130}$$

Equation (128) is thus written as

$$I(\bar{x}) = \alpha_f \phi(\bar{x}) + \lim_{\varepsilon \to 0} \int_{\Omega} \frac{3r_1^2 - r^2}{r^5} h_\varepsilon(\bar{y}) \phi(\bar{y}) d\bar{y} \tag{3.131}$$

Numerically, $I(\bar{x})$ is the same regardless of the shape of $f = 0$. Since $\frac{3r_1^2 - r^2}{r^5}$ near $\bar{y} = \bar{x}$ takes both positive and negative values, the shape of $f = 0$ as $\varepsilon \to 0$ affects the value of the integral in the summation process. This is similar to a well-known result for conditionally convergent series. Such a series can be made to converge to any value by rearranging the terms of the series. The term $\alpha_f \phi(\bar{x})$ in Equation (3.131) compensates for the change in the value of the volume integral when $f = 0$ is changed so that $I(\bar{x})$ is numerically the same.

What is the implication of the above result in applications? In practice, the volume integration is performed numerically. The volume integral has a hole enclosing $\bar{y} = \bar{x}$ whose surface is given by $f = 0$. The value of α_f must, therefore, correspond to the grid system used in volume integration. If the shape of the hole is rectangular, which is often the case, then neither of the above two α_f's in Equation (3.129) and (3.130) are appropriate for the problem.

We have left one question unanswered. When does the appearance of a divergent integral not imply the breakdown of the physical modeling? The answer is when we have wrongly taken an ordinary derivative inside an improper integral. Such a step can make the integral divergent and it is because of the wrong mathematics, not physics, that such an integral appeared. One should, thus, always check the cause of the appearance of divergent integrals in applications. In classical aerodynamics, it is almost always the inappropriate mathematics which causes the appearance of divergent integrals.

4 APPLICATIONS

In this section we give some applications in aerodynamics and aeroacoustics. One of the purposes of the presentation is to show the power and the beauty of generalized function theory. We use the results of the previous sections here. There are many areas of aerodynamics and aeroacoustics where generalized function theory can be used. Although in some instances one can use other methods that do not use this theory, the use of generalized function theory is almost always more direct and simpler to get results. In addition, for many problems involving partial differential equations, there does not seem to be an alternate method for finding a solution. We give a partial list of applications of generalized function theory in aerodynamics (or fluid mechanics) and aeroacoustics:

i) Fluid mechanics and aerodynamics:

- Derivation of transport theorems
- Derivation of governing conservation laws (such as for two phase flows)
- Derivation of jump conditions across flow discontinuities, velocity discontinuity as a vortex sheet
- Derivation of the governing equation for boundary element or field panel methods
- Subsonic, transonic, and supersonic aerodynamic theory

ii) Aeroacoustics:

- Sound from moving singularities
- Derivation of the governing equation for the boundary element method
- Derivation of the Kirchhoff formula for moving surfaces
- Study of noise from moving surfaces using the acoustic analogy
- Identification of new noise generation mechanisms and their source strength (such as shock noise)

In addition, in both aerodynamics and aeroacoustics, generalized function theory can help in the derivation of geometric identities involving curves, surfaces, and volumes, particularly when deforming and in motion.

4.1 Aerodynamic Applications

We give here four applications which have been previously derived by other classical methods. The method based on generalized function theory, as expected, is much shorter and more elegant. Some other examples in aerodynamics are presented in De Jager's book [5].

i) Two transport theorems.

We will give two results here which are used in the derivation of conservation laws. We want to take the time derivative inside the following integral

$$I = \frac{d}{dt}\int_{\Omega(t)} Q(\bar{x},t)\,d\bar{x} \tag{4.1}$$

where $\Omega(t)$ is a time dependent region of space and $Q(\bar{x},t)$ is a C^1 function. Let us assume the boundary $\partial\Omega(t)$ of Ω is piecewise smooth and is given by the surface $f = 0$ in such a way that $f > 0$ in Ω. Assume also that $\nabla f = \bar{n}'$ where $\bar{n}'$ is the unit inward normal. Suppose we can ascertain that the integral in Equation (4.1) is continuous in time. Then, we can replace $\frac{d}{dt}$ by $\frac{\bar{d}}{dt}$ and bring the derivative inside the integral. We write

$$\begin{aligned} I &= \frac{\bar{d}}{dt}\int h(f)Q(\bar{x},t)\,d\bar{x} \\ &= \int\left[\frac{\partial f}{\partial t}\,\delta(f)Q(\bar{x},t) + h(f)\frac{\partial Q}{\partial t}\right]d\bar{x} \\ &= \int_{\partial\Omega(t)} \frac{\partial f}{\partial t}\,Q(\bar{x},t)\,dS + \int_{\Omega(t)} \frac{\partial Q}{\partial t}\,d\bar{x} \end{aligned} \tag{4.2}$$

where $h(f)$ is the Heaviside function. Here we have used Equation (3.80) to integrate $\delta(f)$ in the second step above. We can show that

$$\frac{\partial f}{\partial t} = -v_{n'} = v_n \tag{4.3}$$

where $v_{n'}$ and v_n are the local normal velocities in the direction of inward and outward normals, respectively. We thus have

$$I = \int_{\partial\Omega(t)} v_n Q(\bar{x},t)\,dS + \int_{\Omega(t)} \frac{\partial Q}{\partial t}\,d\bar{x}\;. \tag{4.4}$$

This equation is the generalization of the Leibniz rule of differentiation of integrals in one dimension.

We now give a second result. We want to take the time derivative inside the following integral assuming again that the integral is continuous in time and Q is a C^1 function:

$$I = \frac{d}{dt} \int_{\partial\Omega(t)} Q(\bar{x}, t) dS \ . \tag{4.5}$$

We first convert the surface integral into a volume integral

$$I = \frac{\bar{d}}{dt} \int \delta(f) \, \underset{\sim}{Q}(\bar{x}, t) d\bar{x} \ . \tag{4.6}$$

Here $f = 0$ describes $\partial\Omega(t)$ and $\nabla f = \bar{n}$, where $\bar{n}$ is the unit outward normal. Also note that $\underset{\sim}{Q}$ is the restriction of Q to $f = 0$ as explained in Subsection 3.2. We have therefore,

$$I = \int \left[\frac{\partial f}{\partial t} \delta'(f) \, \underset{\sim}{Q}(\bar{x}, t) + \delta(f) \frac{\partial \underset{\sim}{Q}}{\partial t} \right] d\bar{x} \ . \tag{4.7}$$

We now must use the results of Subsection 3.2 to integrate $\delta'(f)$ and $\delta(f)$. We note however that $\frac{\partial f}{\partial t} = -\underset{\sim}{v}_n$ and this function is restricted already to $f = 0$. This means that,

$$\frac{\partial}{\partial n} [\underset{\sim}{v}_n \underset{\sim}{Q}(\bar{x}, t)] = 0 \tag{4.8}$$

Using Equations (3.80) and (3.84), we get

$$I = \int_{\partial\Omega(t)} \left[\frac{\partial \underset{\sim}{Q}}{\partial t} - 2 v_n H_f Q(\bar{x}, t) \right] dS \tag{4.9}$$

where H_f is the local mean curvature of $\partial\Omega(t)$. What is $\frac{\partial \underset{\sim}{Q}}{\partial t}$? We have the following result, assuming Q is non-impulsive,

$$\frac{\partial \underset{\sim}{Q}}{\partial t} = \frac{\partial Q}{\partial t} + v_n \frac{\partial Q}{\partial n} \ . \tag{4.10}$$

Derivation of Equation (4.9) by other methods is not a trivial matter.

ii) Unsteady shock jump conditions.

These conditions are usually obtained by the pill-box analysis. We give another method here based on generalized function theory. We have said that the conservation laws such as the mass continuity and the momentum equations are valid as they stand if we replace all ordinary derivatives with generalized derivatives. We derive here the jump conditions from these two conservation laws. Let $k(\bar{x}, t) = 0$ describe an unsteady shock surface. Let $\nabla k = \bar{n}$, where $\bar{n}$ is the unit normal pointing in the downstream direction. We denote this downstream region as region 2 and the upstream region as region 1. We define the jump ΔQ in any parameter by

$$\Delta Q = [Q]_2 - [Q]_1 \tag{4.11}$$

where the subscripts 1 and 2 refer to the upstream and downstream regions, respectively. Applying the rules of generalized differentiation to the mass continuity equation, we have

$$\begin{aligned} \frac{\bar{\partial} \rho}{\partial t} + \bar{\nabla} \cdot (\rho \bar{u}) = & \frac{\partial \rho}{\partial t} + \nabla \cdot (\rho \bar{u}) \\ & + [\Delta \rho \frac{\partial k}{\partial t} + \Delta(\rho \bar{u}) \cdot \bar{n}] \delta(k) = 0 \end{aligned} \tag{4.12}$$

where ρ is the density and $\bar{u}$ is the fluid velocity. The sum of the first two terms on the right of the first equality sign is the ordinary mass continuity equation and is zero. The coefficient of $\delta(k)$ must thus be zero. This gives

$$-\Delta \rho v_n + \Delta(\rho u_n) = \Delta[\rho(u_n - v_n)] = 0 \tag{4.13}$$

where $v_n = -\frac{\partial k}{\partial t}$ is the local shock normal velocity and $u_n = \bar{u} \cdot \bar{n}$ is the local fluid normal velocity. This is the first shock jump condition.

The momentum equation in tensor notation using the summation convention gives

$$\begin{aligned} \frac{\bar{\partial}}{\partial t}(\rho u_i) + \frac{\bar{\partial}}{\partial x_j}(\rho u_i u_j) + \frac{\bar{\partial} p}{\partial x_i} = & \frac{\partial}{\partial t}(\rho u_i) + \frac{\partial}{\partial x_j}(\rho u_i u_j) + \frac{\partial p}{\partial x_i} \\ & + \left[\Delta(\rho u_i) \frac{\partial k}{\partial t} + \Delta(\rho u_i u_j) n_j + \Delta p n_i \right] \delta(k) = 0 \end{aligned} \tag{4.14}$$

where p is the pressure. The sum of the three terms after the first equality sign is the ordinary momentum equation. The coefficient of $\delta(k)$ must, therefore, be zero:

$$\Delta\left[\rho u_i(u_n - v_n)\right] + \Delta p n_i = 0 . \tag{4.15}$$

This is the second shock jump condition. We can derive a similar result from the energy equation.

Before we leave this subject, we mention that had we used the mass continuity and momentum equations in nonconservative form, we would have been faced with ambiguities of multiplication of generalized functions. This problem is discussed in detail by Colombeau [26]. In this reference, the remedy for the removal of these ambiguities is discussed both from the intuitive and mathematically rigorous aspects.

iii) Velocity discontinuity as a vortex sheet.

Let us consider a thin lifting wing in forward flight in an incompressible fluid as shown in Figure 7. It can be shown that the velocity field can be idealized as irrotational, i.e. $\nabla \times \bar{u} = 0$ where $\bar{u}$ is the fluid velocity. There is, however, a velocity discontinuity on the wing and on the wake. In this case $\overline{\nabla} \times \bar{u} \neq 0$ and the velocity discontinuity over the wing and the wake gives a vorticity distribution as follows

$$\begin{aligned} \overline{\nabla} \times \bar{u} &= \nabla \times \bar{u} + \bar{n} \times \Delta\bar{u}\delta(k) \\ &= \bar{n} \times \Delta\bar{u}\delta(k) \\ &= \bar{\Gamma}\delta(k) \end{aligned} \tag{4.16}$$

where $k(\bar{x}, t) = 0$ describes the wing and wake surfaces and $\nabla k = \bar{n}$, the local unit normal to these surfaces. Here we define the vorticity distribution $\bar{\Gamma} = \bar{n} \times \Delta\bar{u}$. Note that again we define $\Delta\bar{u} = [\bar{u}]_2 - [\bar{u}]_1$, where $\bar{n}$ points into region 2. The Biot-Savart law gives the velocity field:

$$\bar{u}(\bar{x}) = \int \frac{\bar{\Gamma} \times \bar{\hat{r}}}{r^2}\,\delta(k)d\bar{y} = \int_{k=0} \frac{\bar{\Gamma} \times \bar{\hat{r}}}{r^2}\,dS \tag{4.17}$$

where $\bar{\hat{r}} = \dfrac{\bar{x} - \bar{y}}{r}$.

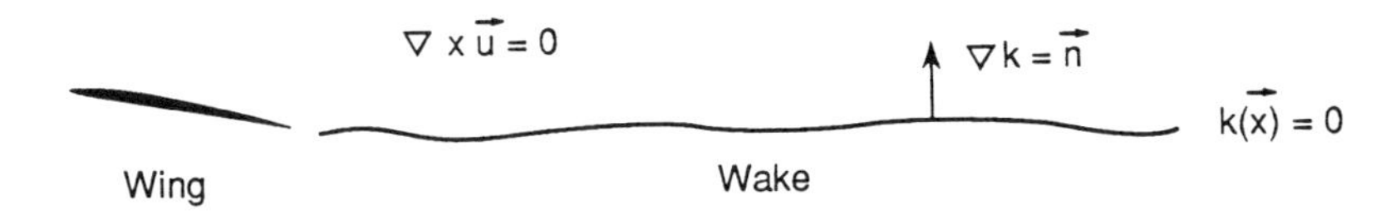

Figure 7. A thin wing in incompressible, irrotational flow with a wake.

iv) An integral equation of transonic flow.

We derive the Oswatitsch integral equation of transonic flow here [29,30]. Consider a thin wing with shock waves in transonic flow moving with uniform speed along the x_1-axis as shown in Figure 8. Let u be the perturbation velocity along the x_1-axis. The governing equation for this flow parameter in nondimensional form is

$$\nabla^2 u - \frac{1}{2}\frac{\partial^2 u^2}{\partial x_1^2} = 0 \tag{4.18}$$

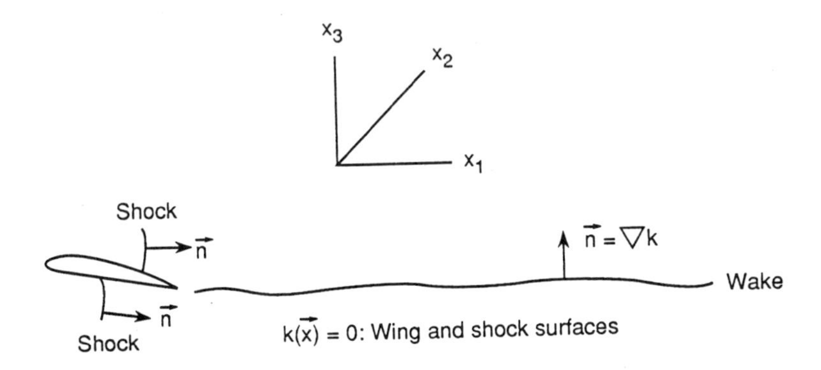

Figure 8. Diagram used in deriving Oswatitsch integral equation of transonic flow.

For simplicity, we assume that the airfoil, the shock surfaces, and the wake surface are all specified by $k(\bar{x}) = 0$. We will set up this problem in generalized function space by converting the derivatives in Equation (4.18) to generalized derivatives. We again define a jump in u or u^2 by $\Delta(\bullet) = [\bullet]_2 - [\bullet]_1$, where $\bar{n} = \nabla k$ points into region 2. For the airfoil itself, we define $[u]_1 = \left[u^2\right]_1 = 0$ since the airfoil is a closed surface. We thus get

$$\bar{\nabla}^2 u - \frac{1}{2}\frac{\bar{\partial}^2 u^2}{\partial x_1^2} = \Delta\left[\frac{\partial u}{\partial n} - \frac{\partial}{\partial n_1}\left(\frac{u^2}{2}\right)\right]\delta(k) + \nabla\cdot\left\{\Delta\left[u\bar{n} - \frac{u^2}{2}\bar{n}_1\right]\delta(k)\right\} \tag{4.19}$$

where $\bar{n} = (n_1, n_2, n_3)$ is the unit normal to the surface $k = 0$ and $\bar{n}_1 = (n_1, 0, 0)$. Note that on the right side of Equation (4.19) we have dropped the sum of two terms which by Equation (4.18) is zero.

To get an integral equation, we will use the Green's function of the Laplace equation which is $-\dfrac{1}{4\pi r}$ and treat $\dfrac{\bar{\partial}^2 u^2}{\partial x_1^2}$ as a source term:

$$4\pi u(\bar{x}) = -\frac{1}{2}\frac{\bar{\partial}^2}{\partial x_1^2}\int \frac{1}{r}u^2(\bar{y})d\bar{y}$$
$$-\int_{k=0} \frac{1}{r}\left[\frac{\partial u}{\partial n} - \frac{\partial}{\partial n_1}(\frac{u^2}{2})\right]dS$$
$$-\nabla\cdot\int_{k=0} \frac{1}{r}\left[u\bar{n} - \frac{u^2}{2}\bar{n}_1\right]dS\,. \tag{4.20}$$

Now if we bring the derivatives inside the first volume integral, which is over the unbounded space, we must use the finite part of the divergent integral introduced in Subsection 3.3, Equation (3.119). Taking $Q(\bar{y}) = u^2(\bar{y})$ in that equation, from Equation (3.131), we have

$$\frac{\bar{\partial}^2}{\partial x_1^2}\int \frac{1}{r}u^2(\bar{y})d\bar{y} = \alpha_f u^2(\bar{x}) + \lim_{\varepsilon\to 0}\int \frac{3r_1^2 - r^2}{r^5}h_\varepsilon(\bar{y})u^2(\bar{y})d\bar{y}\,. \tag{4.21}$$

The last integral in Equation (152) is

$$\nabla\cdot\int_{k=0} \frac{1}{r}\left[u\bar{n} - \frac{u^2}{2}\bar{n}_1\right]dS$$
$$= -\int_{k=0} \frac{1}{r^2}\left[u\cos\theta - \frac{u^2}{2}n_1\cos\theta_1\right]dS \tag{4.22}$$

where $\cos\theta = \bar{\hat{r}}\cdot\bar{n}$, $\cos\theta_1 = \frac{1}{n_1}\bar{\hat{r}}\cdot\bar{n}_1$, and $\bar{\hat{r}} = \frac{(\bar{x}_1 - \bar{y}_1)}{r}$. Our job is finished. The integral on the right side is convergent. Substitute Equations (4.21) and (4.22) in Equation (4.20). The result is the Oswatitsch integral equation of transonic flow. Further approximation is possible but we stop at this point. This derivation is much shorter and more direct than the original derivation [29,30].

4.2 Aeroacoustic Applications

In this subsection, we will give four examples which are all for the linear wave equation. Even for this equation, the use of generalized function theory leads to very important and useful results. Before we present the examples, we give some standard forms of the inhomogeneous source terms appearing in aeroacoustic problems. These are as follows:

$$\Box^2\Phi = Q(\bar{x},t) \tag{4.23-a}$$
$$\Box^2\Phi = Q(\bar{x},t)\delta(f) \tag{4.23-b}$$

$$\Box^2\Phi = \frac{\partial}{\partial t}\left[Q(\bar{x},t)\delta(f)\right] \tag{4.23-c}$$

$$\Box^2\Phi = \nabla\cdot\left[\bar{Q}(\bar{x},t)\delta(f)\right] \tag{4.23-d}$$

$$\Box^2\Phi = \underset{\sim}{Q}(\bar{x},t)h\left(\tilde{f}\right)\delta'(f) \tag{4.23-e}$$

$$\Box^2\Phi = Q(\bar{x},t)\delta(f)\delta\left(\tilde{f}\right). \tag{4.23-f}$$

In these equations, $f(\bar{x},t) = 0$ is a moving surface, usually assumed a closed surface. An open surface, such as a panel on a rotor blade, is described by $f = 0$ and $\tilde{f}(\bar{x},t) > 0$, where $f(\bar{x},t) = \tilde{f}(\bar{x},t) = 0$ describes the edge of the open surface, see Figure 9. Also we denote the Heaviside function as $H\left(\tilde{f}\right)$. In Equation (4.23-e), note that $\underset{\sim}{Q}$ is the restriction of Q to $f = 0$. The solution of the above equations have been given in many publications of the author and coworkers [31-34]. We will only give a brief summary here.

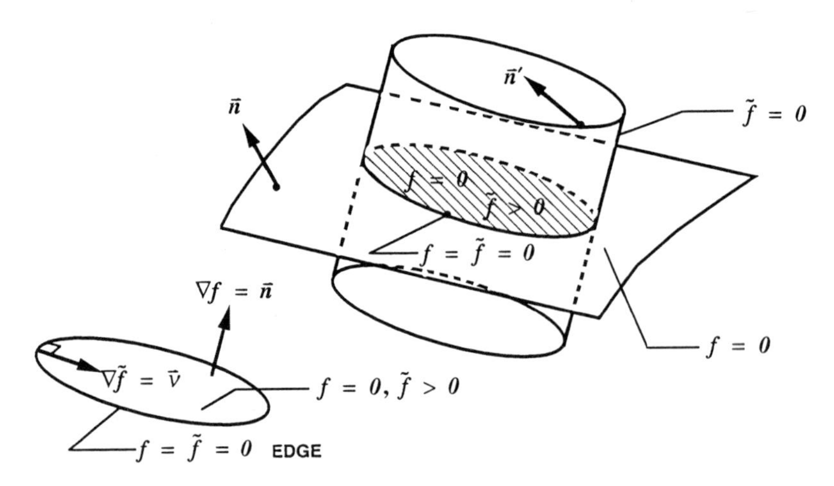

Figure 9. Definition of an open surface by the relations $f = 0, \tilde{f} = 0$. Note that the edge is defined by $f = \tilde{f} = 0$ and $\bar{\nu}$ is the unit inward geodesic normal.

The Green's function of the wave equation is

$$G(\bar{y},\tau;\bar{x},t) = \begin{cases} \dfrac{\delta(g)}{4\pi r} & \tau \le t \\ 0 & \tau > t \end{cases} \tag{4.24}$$

where

$$g = \tau - t + \frac{r}{c}. \tag{4.25}$$

In this equation, $(\bar{x}, t)$ and $(\bar{y}, \tau)$ are the observer and the field (source) space-time variables, respectively. The sound speed is denoted by c and $r = |\bar{x} - \bar{y}|$. The two forms of the solution of Equation (4.23-a) are

$$4\pi\Phi(\bar{x}, t) = \int \frac{1}{r}[Q]_{ret}\, d\bar{y} \tag{4.26}$$

and

$$4\pi\Phi(\bar{x}, t) = \int_{-\infty}^{t} \frac{d\tau}{t - \tau} \int_{\Omega(t)} Q(\bar{y}, \tau)\, d\Omega \tag{4.27}$$

where the subscript *ret* stands for retarded time $t - r / c$. The surface $\Omega(\tau)$ is the sphere $r = c(t - \tau)$, i.e. the sphere with center at the observer $\bar{x}$ and radius $c(t - \tau)$ with the element of the surface denoted by $d\Omega$. The two forms of the solution of Equation (4.23-a) are known as the retarded time and the collapsing sphere forms of the solution, respectively.

The solution of Equation (4.23-b) can also be written in several forms [31,32]. We give two forms here. For a rigid surface $f(\bar{x}, t) = 0$, let $M_r = \vec{M} \cdot \hat{\bar{r}}$ be the local Mach number in the radiation direction. Then

$$4\pi\Phi(\bar{x}, t) = \int_{f=0} \left[\frac{Q(\bar{y}, \tau)}{r\left|1 - M_r\right|} \right]_{ret} dS\,. \tag{4.28}$$

Note that to get this equation, the formal Green's function solution, which is

$$4\pi\Phi(\bar{x}, t) = \int \frac{1}{r} Q(\bar{y}, \tau)\delta(f)\delta(g)\, d\bar{y}\, d\tau \tag{4.29}$$

is integrated as follows. First introduce a Lagrangian variable $\bar{\eta}$ on and in the vicinity of the surface $f = 0$ such that the Jacobian of the transformation is unity. Note that we have $\bar{y} = \bar{y}(\bar{\eta}, \tau)$ and

$$r = |\bar{x} - y(\bar{\eta}, \tau)|\,. \tag{4.30}$$

Next let $\tau \rightarrow g$ which gives $\dfrac{\partial g}{\partial \tau} = 1 - M_r$. Integrate Equation (4.29) next with respect to g and finally integrate $\delta(f)$ by the method of Subsection 3.2 to get Equation (4.28).

A more interesting method of integrating the delta functions in Equation (4.29) is to let $\tau \rightarrow g$ and integrate with respect to g. this gives

$$4\pi\Phi(\bar{x}, t) = \int \frac{1}{r}[Q(\bar{y}, \tau)]_{ret}\, \delta(F) d\bar{y} \tag{4.31}$$

where $F(\bar{y}; \bar{x}, t) = [f(\bar{y}, \tau)]_{ret} = f\left(\bar{y}, t - \dfrac{r}{c}\right)$. Note, however, that even if $|\nabla f| = 1$ by definition, we have $|\nabla F| \neq 1$ in Equation (4.31). We will, therefore, give a slight modification of Equation (3.80). In the following integral, assume $|\nabla f| \neq 1$, then

$$I = \int \phi(\bar{x})\delta(f) d\bar{x} = \int_{f=0} \frac{\phi(\bar{x})}{|\nabla f|} dS . \tag{4.32}$$

This result applies to Equation (4.31). It is easily shown by differentiation that

$$|\nabla F| = (1 + M_n^2 - 2M_n \cos\theta)^{1/2} \equiv \Lambda \tag{4.33}$$

where $M_n = \dfrac{v_n}{c}$, $v_n = -\dfrac{\partial f}{\partial t}$ is the local normal velocity on $f = 0$ and $\cos\theta = \bar{n} \cdot \bar{\hat{r}}$ is the cosine of the angle between the local normal to $f = 0$ and the radiation direction $\bar{\hat{r}} = \dfrac{(\bar{x} - \bar{y})}{r}$. Using Equations (4.32) and (4.33) in Equation (4.31), we get

$$4\pi\Phi(\bar{x}, t) = \int_{F=0} \frac{1}{r}\left[\frac{Q(\bar{y}, \tau)}{\Lambda}\right]_{ret} d\Sigma \tag{4.34}$$

where $d\Sigma$ is the element of the surface area at $F = 0$. Note that for supersonic surfaces, the condition $M_r = 1$ produces a singularity in Equation (4.28). The use of Equation (4.34) removes this singularity in most cases.

The surface Σ: $F = 0$ is visualized as follows. Let the surface $f = 0$ move in space. Construct the intersection of the collapsing sphere $r = c(t - \tau)$ for a fixed $(\bar{x}, t)$ with the surface $f = 0$. The surface in space which is the locus of these curves of intersection is the Σ-surface or the influence surface for $(\bar{x}, t)$. Note that given $(\bar{x}, t)$, this surface is unique since the sphere $r = c(t - \tau)$ has center at $\bar{x}$ and $r = 0$ at $\tau = t$. Given $\tau \leq t$, since $(\bar{x}, t)$ is fixed, the sphere is specified and $f(\bar{y}, \tau) = 0$ is also specified. Therefore, the intersecting curve, if it exists, is specified. Figure 10 shows the Σ-surface for a rotating propeller blade. This figure indicates that the Σ-surface is dependent on the motion and the geometry of the surface $f = 0$. We note that there is also a possibility of a singularity in Equation (4.34) when $\Lambda = 0$. We will not address this problem here. Such a possibility can occur for supersonic propeller blades with blunt leading edges which should be avoided because of excessive drag problems.

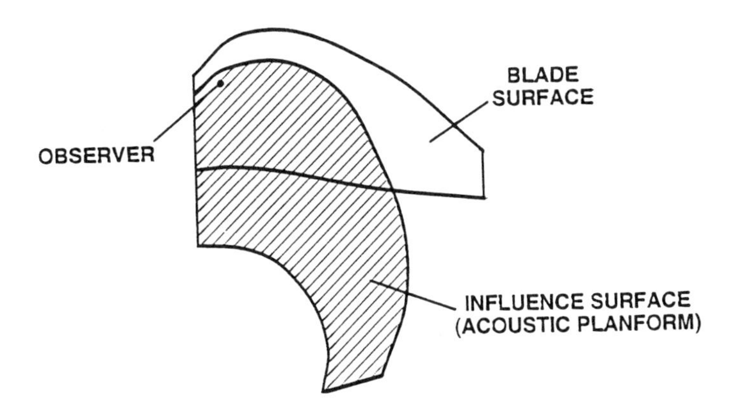

Figure 10. The influence surface for an observer on a propeller surface. Forward Mach Number = 0.334, Helical Mach Number = 0.880 (from M. H. Dunn, Ph. D. Dissertation, Old Dominion University, 1991).

The solutions of Equations (4.23-c) and (4.23-d) can be related to that of (4.23-b). For example, the solution of Equation (4.23-c) is

$$4\pi\Phi(\bar{x}, t) = \frac{\partial}{\partial t} \int_{f=0} \left[\frac{Q(\bar{y}, \tau)}{r|1 - M_r|} \right]_{ret} dS \,. \tag{4.35}$$

Now $\dfrac{\partial}{\partial t}$ can be brought inside the integral using the relation

$$\frac{\partial}{\partial t} = \frac{1}{1 - M_r} \frac{\partial}{\partial \tau} \,. \tag{4.36}$$

Note, however, that $r = |\bar{x} - \bar{y}(\bar{\eta}, \tau)|$ so that $\dfrac{\partial r}{\partial \tau} \neq 0$. See references by the author [31] and Brentner [32]. Similar manipulations can be performed for the form of solution of Equation (4.23-c) based on Equation (4.34) but it is better to work with the source terms of Equation (4.23-c) before using the Green's function approach. See the third example of this subsection.

The solution of Equation (4.23-e) is by far the most difficult of the problems considered here. We first simplify the algebraic manipulations by defining $\tilde{f}$ such that $\nabla\tilde{f} = \bar{\nu}$ where $\bar{\nu}$ is the unit outward geodesic normal to the edge. The geodesic normal is tangent to the surface $f = 0$, $\tilde{f} > 0$ and orthogonal to edge $f = \tilde{f} = 0$, see Figure 9. The formal solution of Equation (4.23-e) is

$$4\pi\Phi(\bar{x}, t) = \int \frac{1}{r} \underset{\sim}{Q}(\bar{y}, \tau) h(\tilde{f}) \delta'(f) \delta(g) d\bar{y} d\tau$$
$$= \int \frac{1}{r} [\underset{\sim}{Q}(\bar{y}, \tau)]_{ret} h(\tilde{F}) \delta'(F) d\bar{y} \tag{4.37}$$

where, as before, $F = [f]_{ret}$ and we define $\tilde{F}(\bar{y}; \bar{x}, t) = \left[\tilde{f}(\bar{y}, \tau)\right]_{ret}$. Let $\bar{N}$ be the unit normal to the surface $F = 0$. We can show that

$$\bar{N} = \frac{\bar{n} - M_n \bar{\hat{r}}}{\Lambda}. \tag{4.38}$$

Equation (4.37) is now of the form of Equation (3.81). However, again since $|\nabla F| = \Lambda \neq 1$, we must give the modification to Equation (3.81) here. In this case, we have for $|\nabla f| \neq 1$, the following result

$$\int \phi(\bar{x}) \delta'(f) d\bar{y} = \int_{f=0} \left\{ -\frac{1}{|\nabla f|} \frac{\partial}{\partial n} \left[\frac{\phi}{|\nabla f|} \right] - \frac{2 H_f \phi}{|\nabla f|^2} \right\} dS. \tag{4.39}$$

Now using F in place of f here, we get from Equation (168)

$$4\pi\Phi(\bar{x}, t) = \int_{F=0} \left\{ -\frac{1}{\Lambda} \frac{\partial}{\partial N} \left(\frac{\left[\underset{\sim}{Q}\right]_{ret}}{r\Lambda} h(\tilde{F}) \right) + \frac{2 H_F \left[\underset{\sim}{Q}\right]_{ret} h(\tilde{F})}{r\Lambda^2} \right\} d\Sigma \tag{4.40}$$

where H_F is the local mean curvature of the Σ-surface given by $F = 0$. Note that

$$\frac{\partial}{\partial N} h(\tilde{F}) = \bar{N} \cdot \nabla \tilde{F} \delta(\tilde{F}) \tag{4.41}$$

so that we must integrate this delta function in Equation (4.40). Using a curvilinear coordinate system on the Σ-surface, we can show that

$$\int_{F=0} \phi(\bar{x}) \left| \nabla \tilde{F} \right| \delta\left(\tilde{F} \right) d\Sigma = \int_{\substack{F=0 \\ \tilde{F}=0}} \frac{\phi}{\sin \theta'} dL \tag{4.42}$$

where θ' is the angle between $\bar{N}$ and $\bar{\tilde{N}} = \dfrac{\nabla \tilde{F}}{\left| \nabla \tilde{F} \right|}$. Also dL is the element of length of the edge of the Σ-surface given by $F = \tilde{F} = 0$. The final result of manipulations of Equation (4.40) using Equation (4.42) is

$$4\pi\Phi(\bar{x}, t) = \int_{\substack{F=0 \\ \tilde{F}>0}} \left\{ -\frac{1}{\Lambda} \frac{\partial}{\partial N} \left(\frac{\left[\underset{\sim}{Q} \right]_{ret}}{r\Lambda} \right) + \frac{2 H_F \left[\underset{\sim}{Q} \right]_{ret}}{r\Lambda^2} \right\} d\Sigma$$

$$- \int_{\substack{F=0 \\ \tilde{F}=0}} \frac{\left[\underset{\sim}{Q} \right]_{ret} \cot \theta'}{r\Lambda^2} dL. \tag{4.43}$$

Note that, we have defined $\bar{\tilde{N}}$ as the unit normal to $\tilde{F} = 0$ and we have

$$\bar{\tilde{N}} = \frac{\bar{v} - M_v \hat{\bar{r}}}{\tilde{\Lambda}} \tag{4.44-a}$$

$$\tilde{\Lambda} = \left| \nabla \tilde{F} \right| \tag{4.44-b}$$

$$\cos \theta' = \bar{N} \cdot \bar{\tilde{N}} . \tag{4.44-c}$$

Note that since $\underset{\sim}{Q}$ is the restriction of Q to $f = 0$, we have $\dfrac{\partial \underset{\sim}{Q}}{\partial n} = 0$. In Equation (4.43), we find the normal derivative of $\left[\underset{\sim}{Q} \right]_{ret}$ first:

$$\frac{\partial}{\partial N} \left[\underset{\sim}{Q} \right]_{ret} = \left[\frac{\partial \underset{\sim}{Q}}{\partial N} \right]_{ret} + \frac{1}{c} \left[\frac{\partial \underset{\sim}{Q}}{\partial \tau} \bar{N} \cdot \hat{\bar{r}} \right]_{ret} \tag{4.45}$$

Using Equation (4.38), we get

$$\bar{N} = \frac{1}{\Lambda}[(1 - M_n \cos\theta)\bar{n} - M_n \sin\theta\, \bar{t}_1] \tag{4.46}$$

where $\bar{t}_1$ is the unit vector along the projection of $\bar{\hat{r}}$ on the local tangent plane to $f = 0$. Therefore, we have, after using $\frac{\partial \underset{\sim}{Q}}{\partial n} = 0$,

$$\frac{\partial \underset{\sim}{Q}}{\partial N} = -\frac{M_n \sin\theta}{\Lambda}\frac{\partial \underset{\sim}{Q}}{\partial t_1} \tag{4.47}$$

where $\frac{\partial}{\partial t_1}$ is the directional derivative of $\underset{\sim}{Q}$ along $\bar{t}_1$. In this case, we no longer need restriction of Q to $f = 0$ since $\frac{\partial \underset{\sim}{Q}}{\partial t_1} = \frac{\partial Q}{\partial t_1}$. Also note that for $\frac{\partial \underset{\sim}{Q}}{\partial \tau}$ in Equation (4.45), we must use a relation similar to Equation (4.10). We mention here that the curve $F = \tilde{F} = 0$ in Equation (4.43) is generated by the intersection of the collapsing sphere $g = 0$ and the edge curve $f = \tilde{f} = 0$ of the open surface $f = 0$, $\tilde{f} > 0$.

We next consider Equation (4.23-f). Again the formal solution is

$$\begin{aligned} 4\pi\Phi(\bar{x}, t) &= \int \frac{1}{r} Q(\bar{y}, \tau)\delta(f)\delta(\tilde{f})\delta(g)d\bar{y}d\tau \\ &= \int \frac{1}{r}[Q]_{ret}\,\delta(F)\delta(\tilde{F})d\bar{y}\ . \end{aligned} \tag{4.48}$$

This is similar to Equation (3.99), except that $|\nabla F| \neq 1$ and $|\nabla \tilde{F}| \neq 1$. In this case, $\sin\theta$ in Equation (3.100) is replaced by $|\nabla F \times \nabla \tilde{F}|$ which by definition is Λ_0. Therefore, Equation (4.48) gives

$$4\pi\Phi(\bar{x}, t) = \int_{\substack{F=0 \\ \tilde{F}=0}} \frac{1}{r}\left[\frac{Q}{\Lambda_0}\right]_{ret} dL\ . \tag{4.49}$$

We now give four applications.

i) Lowson's formula for a dipole in motion.

A dipole is an idealization of a point force. A point force in motion is described by the wave equation

$$\Box^2 p' = -\frac{\partial}{\partial x_i}\{f_i(t)\delta[\bar{x} - \bar{s}(t)]\} \tag{4.50}$$

where p' is the acoustic pressure, f_i is the component of the point force and $\bar{s}(t)$ is the position of the force at time t. The formal solution of the above equation is

$$4\pi p'(\bar{x}, t) = -\frac{\partial}{\partial x_i}\int \frac{f_i(\tau)}{r}\delta[\bar{y} - \bar{s}(\tau)]\delta(g)d\bar{y}d\tau\ . \tag{4.51}$$

Let us integrate the above integral with respect to $\bar{y}$. We get

$$4\pi p'(\bar{x}, t) = -\frac{\partial}{\partial x_i}\int \frac{f_i(\tau)}{r^*}\delta(g^*)d\tau \tag{4.52}$$

where

$$r^* = |\bar{x} - \bar{s}(\tau)| \tag{4.53-a}$$

$$g^* = \tau - t + \frac{r^*}{c} \tag{4.53-b}$$

Now let $\tau \rightarrow g^*$ and note that

$$\frac{\partial g^*}{\partial \tau} = 1 - M_r \tag{4.54}$$

where $M_r = \dot{\bar{s}} \cdot \frac{\hat{\bar{r}}}{c}$ is the Mach number of the point force in the radiation direction. Integrate the resulting equation with respect to g^* to get

$$4\pi p'(\bar{x}, t) = -\frac{\partial}{\partial x_i}\left[\frac{f_i(\tau)}{r|1 - M_r|}\right]_{\tau^*} \tag{4.55}$$

where τ^* is the emission time. It is the solution of $g^* = 0$ which, assuming that the point force is in subsonic motion, has only one root. The derivative in Equation (4.55) can now be taken inside the square brackets. The resulting equation is a formula given by Lowson [35]. It is useful in noise prediction of rotating blades where the dipole sources can by assumed compact.

ii) The Kirchhoff formula for moving surfaces.

In the 1930's, Morgans published a paper in which he derived the Kirchhoff formula for moving surfaces [36]. Morgans' derivation of this formula was based on classical analysis and was very lengthy. In 1988, the author and M. K. Myers gave a modern derivation of this result based on generalized function theory [37,38]. The derivation is quite short and avoids the use of four dimensional Green's identity and the associated difficulties of dealing with surfaces and volumes in four dimensions. We present the basic idea behind this derivation here and refer the readers to the author's paper [37].

Assume that the surface in motion on which conditions on $\phi(\bar{x}, t)$ are specified is given by $f(\bar{x}, t)$. This surface can be deformable. Assume that ϕ satisfies the wave equation in the exterior of the surface $f = 0$ which is the region defined by $f > 0$. Now extend ϕ to the entire unbounded space as follows

$$\tilde{\phi}(\bar{x}, t) = \begin{cases} \phi(\bar{x}, t) & f > 0 \\ 0 & f < 0 \end{cases} \tag{4.56}$$

It is clear that $\tilde{\phi}$ satisfies the wave equation in the unbounded space. However $\tilde{\phi}$ has discontinuities across $f = 0$ which appear as source terms of the wave equation. Note that the jumps in $\tilde{\phi}$ and its derivatives depend on corresponding values for ϕ since $\Delta\tilde{\phi} = \phi(f = 0_+)$ and $\Delta \dfrac{\partial \tilde{\phi}}{\partial t} = \dfrac{\partial \phi}{\partial t}(f = 0_+)$.

Applying the rules of generalized differentiation to $\tilde{\phi}$, we get

$$\begin{aligned} \bar{\Box}^2 \tilde{\phi} = & -\left(\phi_n + \frac{1}{c} M_n \phi_t \right) \delta(f) \\ & - \frac{1}{c} \frac{\partial}{\partial t} \left[M_n \phi \delta(f) \right] \\ & - \nabla \cdot \left[\phi \bar{n} \delta(f) \right] \end{aligned} \tag{4.57}$$

where $M_n = \dfrac{v_n}{c}$ and $v_n = -\dfrac{\partial f}{\partial t}$ is the local normal velocity of the surface $f = 0$. As before, we have assumed $\nabla f = \bar{n}$, the local outward unit normal to $f = 0$. The three types of source terms on the right of Equation (4.57) are of the standard types given in Equations (4.23-a-f). The solution for a deformable surface is given by

$$4\pi\tilde{\phi}(\bar{x},t) = \int_{D(S)} \left[\frac{E_1\sqrt{g_{(2)}}}{r(1-M_r)}\right]_{\tau^*} du^1 du^2 + \int_{D(S)} \left[\frac{\phi E_2\sqrt{g_{(2)}}}{r(1-M_r)}\right]_{\tau^*} du^1 du^2 \tag{4.58}$$

where $D(S)$ is a time-independent region in u^1u^2-space onto which the surface $f = 0$ is mapped. The determinant of the coefficient of the first fundamental form is denoted as $g_{(2)}$. In this equation τ^* is the emission time of the point (u^1, u^2) on the surface $f = 0$. The expression E_1 depends on ϕ, ϕ_n, $\nabla_2\phi$ (surface gradient of ϕ) and the kinematic and geometric parameters of the surface $f = 0$. The second expression E_2 depends only on the kinematic and geometric parameters of the surface $f = 0$ [37].

iii) Noise from moving surfaces.

Let an impenetrable surface $f = 0$ be in motion such that $f > 0$ outside the body and $\nabla f = \bar{n}$, the unit outward normal. Let us assume that the fluid is extended inside this surface with conditions of undisturbed medium, i.e. density ρ_0 and speed of sound c. We know that the mass continuity and momentum equations are valid when the derivatives are written as generalized derivatives. Let us only extract the contribution of discontinuities across $f = 0$ and leave the effect of all other discontinuities, such as those across shock waves, in these equations. The mass continuity equation gives

$$\begin{aligned}\frac{\bar{\partial}\rho}{\partial t} + \bar{\nabla}\cdot(\rho\bar{u}) &= -(\rho-\rho_0)v_n\delta(f) + \rho u_n\delta(f) \\ &= \rho_0 v_n\delta(f)\end{aligned} \tag{4.59}$$

where $v_n = -\dfrac{\partial f}{\partial t}$ is the local normal velocity of $f = 0$ and we have used the impenetrability condition on this surface which is $u_n = v_n$. The momentum equation gives

$$\frac{\bar{\partial}}{\partial t}(\rho u_i) + \frac{\bar{\partial}}{\partial x_j}(\rho u_i u_j + P_{ij}) = P_{ij}n_j\delta(f) \tag{4.60}$$

where $P_{ij} = E_{ij} + (p - p_0)\delta_{ij}$ is the compressive stress tensor and E_{ij} is the viscous stress tensor. Now taking the generalized derivative of both sides of Equation (4.59) and $\dfrac{\bar{\partial}}{\partial x_i}$ of both sides of Equation (4.60), subtracting the latter from the former, and finally subtracting $c^2\dfrac{\bar{\partial}^2\rho}{\partial x_i^2}$ from both sides, we get

$$\square^2 p' = \frac{\bar{\partial}^2}{\partial x_i \partial x_j}\left[T_{ij} h(f)\right] - \frac{\partial}{\partial x_i}\left[P_{ij} n_j \delta(f)\right] + \frac{\partial}{\partial t}\left[\rho_0 v_n \delta(f)\right] \tag{4.61}$$

where $p' = c^2(\rho - \rho_0)$. Here T_{ij} is the Lighthill stress tensor. Now we have added $h(f)$, the Heaviside function, on the right side to indicate that $T_{ij} \neq 0$ outside the surface $f = 0$. This is the Ffowcs Williams - Hawkings (FW-H) equation [39]. Note that in the far field, p' is the acoustic pressure.

The source terms of Equation (4.61) are of the standard types in Equations (4.23-a-f). For a surface in subsonic motion, the solution for surface sources involving the Doppler factor is most appropriate for numerical work [31,32]. For supersonic surfaces, such as an advanced propeller blade, on which $|M_n| < 1$ everywhere, a different solution based on the Σ-surface must be used. We show here briefly how this can be done. In applications, we need to calculate the sound from an open surface such as a panel on a blade. We, therefore, define such an open surface by $f = 0$, $\tilde{f} > 0$ with the edge defined by $f = \tilde{f} = 0$ as before. The assumptions concerning the gradients of f and $\tilde{f}$ at the beginning of this subsection hold here. We are interested in the solution of equations of the types

$$\square^2 p' = \frac{\partial}{\partial t}\left[\rho_0 v_n h(\tilde{f})\delta(f)\right] \tag{4.62-a}$$

$$\square^2 p' = -\frac{\partial}{\partial x_i}\left[p n_i h(\tilde{f})\delta(f)\right] \tag{4.62-b}$$

where $h(\tilde{f})$ is the Heaviside function. Note that we have approximated P_{ij} in Equation (4.61) by $p\delta_{ij}$ where p is the surface (gage) pressure. To derive solutions for Equations (4.62-a,b) suitable for supersonic panel motion, we write

$$\begin{aligned}\frac{\partial}{\partial t}\left[\rho_0 \underset{\sim}{v}_n h(\tilde{f})\delta(f)\right] = {} & \rho_0 \frac{\partial \underset{\sim}{v}_n}{\partial t} h(\tilde{f})\delta(f) \\ & -\rho_0 \underset{\sim}{v}_n^2 h(\tilde{f})\delta'(f) \\ & -\rho_0 v_n v_\nu \delta(\tilde{f})\delta(f)\end{aligned} \tag{4.63}$$

where v_ν is the velocity of the edge in the direction of the geodesic normal. A similar operation can be performed on the right of Equation (192-b)

$$-\frac{\partial}{\partial x_i}\left[pn_i h(\tilde{f})\delta(f)\right] = -\nabla \cdot \left[\underset{\sim}{p}\,\bar{n}h(\tilde{f})\delta(f)\right]$$
$$= -\underset{\sim}{p}\,h(\tilde{f})\delta'(f) + 2pH_f h(\tilde{f})\delta(f) \qquad (4.64)$$

where H_f is the local mean curvature of the surface $f = 0$. The source terms of the right of Equations (4.63) and (4.64) are all of standard types in Equations (4.23-a-f). See the author's papers on this subject [33,34].

iv) Identification of shock noise source strength.

The first term on the right of the FW-H Equation (4.61) is known as the quadrupole source. As mentioned in the derivations, the discontinuities in the region $f > 0$, i.e. outside the body, contribute source terms after generalized differentiation is performed. Let us assume that a shock wave described by the equation $k(\bar{x}, t) = 0$ exists on a rotating blade. Then, the quadrupole term gives surface sources on the shock whose strengths are determined as follows. Let us take the generalized second derivative of T_{ij}:

$$\frac{\bar{\partial} T_{ij}}{\partial x_i} = \frac{\partial T_{ij}}{\partial x_i} + \Delta T_{ij} n'_j \delta(k)$$
$$\frac{\bar{\partial}^2 T_{ij}}{\partial x_i \partial x_j} = \frac{\partial^2 T_{ij}}{\partial x_i \partial x_j} + \Delta \frac{\partial T_{ij}}{\partial x_i} n'_j \delta(k) + \frac{\partial}{\partial x_j}\left[\Delta T_{ij} n'_j \delta(k)\right] \qquad (4.65)$$

where $\bar{n}' = \nabla k$ is the unit normal to the shock surface pointing to the downstream region. The last two terms on the right of this equation are shock surface terms which are of monopole and dipole types, respectively. The first term on the right of Equation (4.65) is a volume term of the type familiar in Lighthill's jet noise theory. In the rotating blade problem, the major contribution of this term is due to nonlinearities rather than turbulence. It was conjectured by the author that the shock surface terms contributed relatively more than the volume term in Equation (4.65) [40]. Some preliminary calculations have supported this conjecture [41].

What is interesting in the above result is that the shock source strength is obtained purely by mathematics only. Without the use of the operational properties of generalized functions, the identification of shocks as sources of sound and the determination of the source strength would be difficult. Other mechanisms of noise generation can also be identified by this method [42].

5 CONCLUDING REMARKS

In this paper, we have given the rudiments of generalized function theory and some applications in aerodynamics and aeroacoustics. These applications depend on the concept of generalized differentiation and the Green's function approach. We have briefly discussed the generalized Fourier transformation. Many more examples could be given. The power of this theory stems from its operational properties. In addition to the exchange of limit processes which leads to many useful results, one can easily obtain discontinuous solutions of linear equations using Green's function by posing the problem in generalized function space. As seen in the example of the Oswatitsch integral equation of transonic flow, a nonlinear partial differential equation with a discontinuous solution can be cast into an integral equation using the fundamental solution of the linear part of the differential equation. Generalized function theory of Schwartz has unified many ad hoc methods in mathematics and has answered some fundamental questions about linear partial differential equations. The new nonlinear theory under development, where multiplication of generalized functions is allowed, can be even more useful in applications. Engineers and scientists should recognize that generalized function theory is an extension of classical analysis and thus it gives the users added power in applications. This is much like the complex analysis which extends real analysis and is very important in applied mathematics. Finally it should be noted that multidimensional generalized functions, particularly the delta function and its derivatives, are very useful in many applications.

Acknowledgments

The author would like to thank Drs. M. K. Myers, J. C. Hardin, and M. H. Dunn for their comments on this paper resulting in improvements in the presentation of the subject. The help of Mr. P. L. Spence and Dr. M. H. Dunn in preparation of the paper is gratefully acknowledged.

6 REFERENCES

1 Schwartz, L.: *Theorie Des Distributions*, Vol. I, Vol. II, Hermann, Paris, 1950, 1951.

2 Jones, D. S.: *The Theory of Generalized Functions*, 2nd edition, Cambridge University Press, Cambridge, 1982.

3 Stakgold, I.: *Boundary Value Problems of Mathematical Physics*, Vol. 2, The MacMillan Company, London, 1967.

4 Kanwal, R. P.: *Generalized Functions - Theory and Technique*, Academic Press, New York, 1983.

5 De Jager, E. M.: *Applications of Distributions in Mathematical Physics*, 2nd Edition, Mathematical Centre, Amsterdam, 1969.

6 Lighthill, M. J.: *Introduction to Fourier Analysis and Generalized Functions*, Cambridge University Press, Cambridge, 1958.

7 Gel'fand, I. M., and Shilov, G. E.: *Generalized Functions*, Vol. 1, Academic Press, New York, 1964.

8 Aris, R.: *Vectors, Tensors, and the Basic Equations of Fluid Mechanics*, Prentice - Hall, Englewood Cliffs, N. J., 1962.

9 McConnell, A. J.: *Applications of Tensor Analysis*, Dover, New York, 1957.

10 Carmichael, R. D., and Mitrovic, D.: *Distributions and Analytic Functions*, Longman Scientific and Technical, Essex, 1989.

11 Robertson, A. P., and Robertson, W.: *Topological Vector Spaces*, Cambridge University Press, Cambridge, 1964.

12 Bremermann, H. J., and Durand, L.: On Analytic Continuation, Multiplication, and Fourier Transformation of Schwartz Distributions, *J. Math. Phys.*, **2**(2), pp. 240-258, 1961.

13 Bremermann, H. J.: *Distributions, Complex Variables and Fourier Transforms*, Addison-Wesley, Reading, MA, 1965.

14 Synowiec, J.: *Distributions: The Evolution of a Mathematical Theory*, Historia Mathematica, **10**, pp. 149-183, 1983.

15 Lutzen, J.: *The Prehistory of the Theory of Distributions*, Springer Verlag, New York, 1982.

16 Dieudonne, J.: *History of Functional Analysis*, North Holland, Amsterdam, 1981.

17 Antosik, P.; Mikusinski, I., and Sikorski, R.: *Theory of Distributions - The Sequential Approach*, Elsevier Scientific Publishing Company, Amsterdam, 1973.

18 Mikusinski, I.: *Operational Calculus*, 2nd Edition, Two Volumes, Pergamon Press, Oxford, 1983.

19 Erdelyi, A.: *Operational Calculus and Generalized Functions*, Holt, Rinehart and Winston, New York, 1962.

20 Robinson, A.: *Non-standard Analysis*, North-Holland Publishing Company, Amsterdam, 1966.

21 Colombeau, J. F.: *New Generalized Functions and Multiplication of Distributions,* North-Holland Mathematics Studies, Vol. 84, Elsevier Science Publishers, Amsterdam, 1984.

22 Colombeau, J. F.: *Elementary Introduction to New Generalized Functions*, North-Holland Mathematics Studies, Vol. 113, Elsevier Science Publishers, Amsterdam, 1985.

23 Rosinger, E. E.: *Non-linear Partial Differential Equations - An Algebraic View of Generalized Solutions*, North-Holland Mathematics Studies, Vol. 164, Elsevier Science Publishers, Amsterdam, 1990.

24 Oberguggenberger, M.: *Multiplication of Distributions and Applications to Partial Differential Equations*, Longman Scientific and Technical, Essex, 1992.

25 Rosinger, E. E. : *Generalized Solutions of Nonlinear Partial Differential Equations*, North-Holland Mathematics Studies, Vol. 146, Elsevier Science Publishers, Amsterdam, 1987.

26 Colombeau, J. F.: *Multiplication of Distributions - A Tool in Mathematics, Numerical Engineering and Theoretical Physics*, Lecture Notes in Mathematics, No. 1532, Springer-Verlag, Berlin, 1992.

27 Zemanian, A. H.: *Distribution Theory and Transform Analysis: An Introduction to Generalized Functions With Applications*, Mc-Graw Hill, New York (also reprinted as a Dover book), 1965.

28 Richards, J. I., and Youn, H, K.: *Theory of Distributions - A Non-technical Introduction*, Cambridge University Press, Cambridge, 1990.

29 Oswatitsch, K.: Die Geschwindigkeitsverteilung Bei Lokalen Uberschallgebieten An Flachen Profilen, *Zeit. fur Angew. Math. und Mech.*, **30**, pp. 17-24, 1950.

30 Spreiter, J. E., and Alksne, A.: Theoretical Prediction of Pressure Distributions on Non-lifting Airfoils at High Subsonic Speeds, NACA Report 1217, 1955.

31 Farassat, F.: Linear Acoustic Formulas For Calculation of Rotating Blade Noise, *AIAA J.*, **19**(9), pp. 1121-1130, 1981.

32 Brentner, K. S.: Prediction of Helicopter Rotor Discrete Frequency Noise, NASA Tech. Memo. 87721, 1986.

33 Farassat, F., and Myers, M. K.: The Moving Boundary Problem for the Wave Equation: Theory and Application, *Computational Acoustics - Algorithms and Applications*, Vol. 2, D. Lee, R. L. Sternberg and M. H. Schultz (Eds.), Proc. of the 1st IMACS Symposium on Computational Acoustics, New Haven, CT, August, 1986, Elsevier Science Publishers (North-Holland), Amsterdam, pp. 21-44, 1988.

34 Farassat, F., and Myers, M. K.: Aeroacoustics of High Speed Rotating Blades: The Mathematical Aspect, *Computational Acoustics - Acoustic Propagation*, Vol. 2, D. Lee, R. Vichnevetsky and A. R. Robinson (Eds.), Proc. of the 3rd IMACS Symposium on Computational Acoustics, Cambridge, MA, June 1991, Elsevier Science Publishers (North-Holland), Amsterdam, pp. 117-148, 1993.

35 Lowson, M. V.: The Sound Field of Singularities in Motion, Proc. Roy. Soc. Lond., A286, pp. 559-572, 1965.

36 Morgans, W. R.: The Kirchhoff Formula Extended to a Moving Surface, Philosophical Magazine, **9**, pp. 141-161, 1930.

37 Farassat, F., and Myers, M. K.: Extension of Kirchhoff's Formula to Radiation From Moving Surfaces, *Sound and Vib.*, **123**(3), pp. 451-460, 1988.

38 Hawkings, D. L.: Comments on "Extension of Kirchhoff's Formula to Radiation From Moving Surfaces", Letters to the Editor, *Sound and Vib.*, **133**(1), pp. 189, (Response by F. Farassat and M. K. Myers, same journal **133**(1), pp. 190), 1989.

39 Ffowcs Williams, J. E., and Hawkings, D. L.: Sound Generated by Turbulence and Surfaces in Arbitrary Motion, Phil. Trans. Roy. Soc., A264, pp. 321-342, 1969.

40 Farassat F., and Tadghighi, H.: Can Shock Waves on Helicopter Rotors Generate Noise? - A Study of Quadrupole Source, 46th Annual Forum of the American Helicopter Society, 1990.

41 Tadghighi, H., Holtz, R., Farassat F., and Lee, Y-J: Development of Shock Noise Prediction Code For High Speed Helicopters - The Subsonically Moving Shock, 47th Annual Forum of the American Helicopter Society, 1991.

42 Farassat, F., and Myers, M. K.: An Analysis of the Quadrupole Noise Source of High Speed Rotating Blades, *Computational Acoustics - Scattering, Gaussian Beams and Aeroacoustics*, Vol. 2, D. Lee, A. Cakmak, and R. Vichnevetsky (Eds.), Proc. of the 2nd IMACS Symposium on Computational Acoustics, Princeton, N. J., March, 1989, Elsevier Science Publishers (North-Holland), Amsterdam, pp. 227-240, 1990.

Section II

AERODYNAMICS & WING DESIGN

Unsteady Aerodynamics of Vortical Flows: Early and Recent Developments

H. M. Atassi

ABSTRACT

The development of aerodynamic theories of streaming motions around bodies with unsteady vortical and entropic disturbances is reviewed. The basic concepts associated with such motions, their interaction with solid boundaries and their noise generating mechanisms are described. The theory was first developed in the approximation wherein the unsteady flow is linearized about a uniform mean flow. This approach has been extensively developed and used in aeroelastic and aeroacoustic calculations. The theory was recently extended to account for the effect of distortion of the incident disturbances by the nonuniform mean flow around the body. This effect is found to have a significant influence on the unsteady aerodynamic force along the body surface and the sound radiated in the far field. Finally, the nonlinear characteristics of unsteady transonic flows are reviewed and recent results of linear and nonlinear computations are presented.

1 INTRODUCTION

Aerodynamics has achieved impressive results by the simplification of considering uniform streaming motion relative to a moving body. From the early days of aeronautics until now, this approximation, which leads to potential flow theory with some viscous correction near the solid surfaces, has been the basic engineering tool for designing lifting surfaces such as aircraft wings, turbomachine and propeller blades, and helicopter rotors.

Most natural and technological flows have some steady or unsteady nonuniformity and thus exhibit irregular flow patterns. These are caused by a variety of phenomena such as atmospheric turbulence, interaction with natural obstacles or structural components, momentum defects due to viscous boundary layers and wakes, secondary flows, installation effects, inlet distortion, etc. When a body moves in such irregular flow patterns, the unsteady fluid motion produces fluctuating forces along the body surface and radiates sound in the far field. This is, for example, what happens in the case of a wing moving into a gust or that of a turbomachine blade row rotating in the wake of another row. The unsteady aerodynamics of moving objects therefore has direct application to aeroacoustics and aeroelasticity. As a result, there is considerable interest in modeling unsteady, nonuniform flows and in predicting their near-field pressure distribution and far-field acoustics.

Interest in unsteady aerodynamics began shortly after the first powered flight. Torsional tail flutter was widespread on military biplanes at the beginning of World War I. More aeroelastic instabilities appeared as airplane speed and maneuverability increased. The basic mechanisms associated with lift generation of bodies flying at uniform speed were then well understood. Lifting airfoils were modeled as a distribution of *bound* vortices. The early treatments of unsteady airfoil theory were hence mainly concerned with extending this model to account for unsteady effects, namely, the apparent mass and wake effects. This was accomplished by complementing the bound vortices with a distribution of *free vortices* extending in the wake. This aerodynamic theory has been developed by a number of authors, in particular, Wagner [1], Glauert [2], Theodorsen [3], Kussner [4], and von Karman and Sears [5].

The linear aerodynamic theory which evolved from this simple model accounts for the principal effects of a real fluid. It must then be understood as an asymptotic approximation to the Navier-Stokes equations in the limiting case of a high Reynolds number. Although the viscosity of the fluid is explicitly neglected, the theory accounts for its effects by the imposition of the Kutta condition at the trailing edge and the vortex shedding in the wake. By 1935, the linear theory was successful in explaining aeroelastic instabilities in potential flows [3].

In his 1938 doctoral dissertation, William R. Sears [6] studied the aerodynamics of thin airfoils in nonuniform motion. As an application of the general theory he had developed, he analyzed the fundamental problem of a flat plate in a gust. He noted that, since the upstream vortical disturbance is convected by a uniform mean flow, it preserves its upstream structure. The flow arising from the interaction between this disturbance and the flat plate is then the result of the blockage effect caused by the vanishing of the normal velocity at the surface of the body. Using the general formula for the lift that he had previously developed,

Sears then derived his celebrated expression for the response of an airfoil to a transverse sinusoidal gust [7].

The linear theory was later extended to compressible subsonic flows by Possio [8] and Reissner [9]. As for the incompressible case, these flows are described by constant coefficient linear equations. Moreover, the boundary-value problems for flows around oscillating bodies or flows around bodies with upstream disturbances are identical except for the boundary condition along the surface of the body. Hence, the analytical methods and numerical procedures developed for flutter related analyses are directly applicable to the gust response problem.

Other analyses of unsteady and nonuniform flows were concerned with the distortion and interaction of upstream disturbances with bluff bodies. This interaction invariably produces significant changes in the flow structure near the body and often completely alters the flow behavior by causing instability [10] and/or separation [11]. Often the intensity of the flow nonuniformities is small, thus in many applications, they can be considered as small disturbances to the mean flow. Such flows are called *weakly rotational flows.* The first treatment of this kind is due to Lighthill [12], who analyzed the interaction of an imposed steady upstream vorticity field with a cylinder. This analysis revealed that, as a result of the infinite stretching of the vorticity at the stagnation point, both the vorticity and the spanwise component of the velocity are singular along the surface of the cylinder.

The structure of the upstream flow nonuniformities is usually complex and characterized by multiple scales. The organized distortions, such as wakes, are inherently large scale disturbances. The random distortions are associated with turbulence. Numerous observations have shown that at high speed the greater part of the turbulent energy resides in a group of large eddies which are stable, persisting structures and which only slowly transfer energy to smaller eddies without significant change of structure [13]. As a result, the large eddies account for most of the undesirable effects of unsteady flows. This also implies that, for weakly rotational flows with organized or large–eddy structure, dissipative effects will, in general, be small and that the nonlinear interaction between different disturbances will be weak. It is therefore possible to neglect nonlinear effects and to use a linear approach to investigate large structure weakly rotational flows. In this approximation, the different upstream disturbances are transported and, in the process, distorted primarily by the variation of the mean flow. The equations of motion can thus be linearized about the mean flow[1]. This

[1]This linearization may not be valid for a transonic flow with a strong shock wave.

brings about a considerable simplification in the mathematical treatments of such flows. The effect of the mean flow on convected eddies was first attempted by Prandtl [14] and later by Taylor [15]. Ribner and Tucker [16], and Batchelor and Proudman [17] studied the distortion of turbulence by a sudden contraction in the stream. The latter introduced the term *rapid distortion* to characterize this linear approach to the study of turbulent motions. Hunt [18] extended this theory to analyze the turbulent flow around a two-dimensional circular cylinder.

Goldstein and Atassi [19] developed the first aerodynamic analysis for a two-dimensional periodic gust interacting with a lifting airfoil. The analysis shows that the distortion of the gust by the mean flow of the airfoil causes a significant variation in both the amplitude and phase of the unsteady vortical velocity. This effect is found to have a significant influence on the aerodynamic response of the airfoil [20].

The approach used by Goldstein and Atassi is conceptually related to the rapid distortion theory of turbulence. Both linearize the flow equations about the mean flow. However, the rapid distortion theory focuses on the modification of the incoming turbulence structure by the mean flow and uses statistical methods to calculate velocity correlations and energy spectra. While in unsteady aerodynamics, one is more concerned with upstream organized disturbances and the velocity and pressure fields which result from their distortion by the mean flow and interaction with the body.

For compressible flows, Goldstein [21] showed that the linearized Euler equations can be further simplified for potential mean flows. He decomposed the unsteady velocity into a vortical part whose expression is a known function of the imposed upstream disturbance and an irrotational part governed by a nonconstant coefficient wave equation. Atassi and Grzedzinski [22] introduced a modified form of Goldstein's decomposition making it suitable for numerical computations. Numerical solutions using this approach were obtained for a three-dimensional periodic gust in subsonic flow for a single airfoil by Scott and Atassi [23], and for a two-dimensional cascade geometry by Hall and Verdon [24], and Fang and Atassi [25]. The computation time required for these solutions is less than a minute on scientific workstations. The results of the numerical computations indicate that both the near field unsteady aerodynamic loading and the far field radiated sound are strongly affected by the upstream gust parameters and the nonuniformity of the mean flow around the body.

This linear approach accounts for some essential physical effects of weakly nonuniform

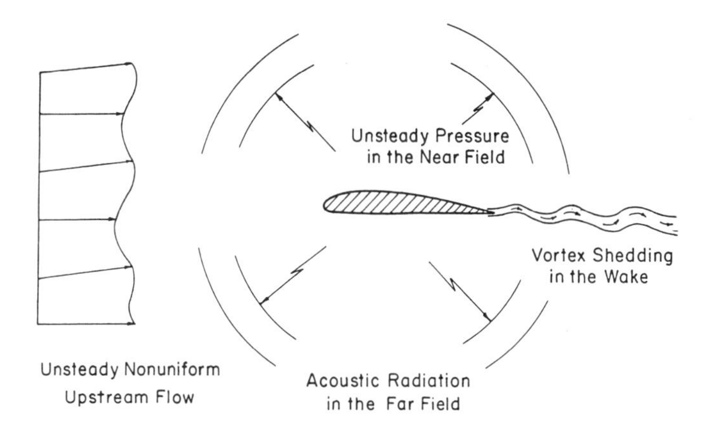

Figure 1. Airfoil in unsteady nonuniform flow.

flows around real lifting bodies (Figure 1). Its numerical implementation produces highly accurate, fast and robust codes particularly suitable for multidisciplinary applications such as forced vibrations, tone and broadband noise, and turbulence structure analysis. On the other hand, linear methods have certain limitations. They do not account for nonlinear effects such as inertial and viscous interaction between eddies, flow separation, or shock-boundary layer interaction.

Unsteady transonic flows exhibit nonlinear characteristics particularly when a strong shock wave induces a large separation region. As a result, extensive research has been conducted in recent years to develop time accurate Euler and Navier-Stokes procedures for nonlinear unsteady flows. These procedures are essentially based on schemes developed for steady flows in which a sequence of iterative solutions are marched forward in time until a convergent steady-state solution is reached. For unsteady vortical flows with acoustic radiation, stringent requirements must further be imposed to reduce dispersion and dissipation errors, thus leading to higher order numerical schemes [26]. The computational requirements associated with these numerical simulations continue to hinder their use in aeroelastic and aeroacoustic applications.

This lecture will review the early pioneering work in linear unsteady aerodynamic theory.

It will then present the recent developments associated with the mean flow distortion effects. Finally, the nonlinear characteristics of unsteady transonic flows are reviewed and recent results of linear and nonlinear computations are presented.

2 LINEAR THEORIES

Linear unsteady aerodynamics considers *weakly* non-uniform flows around non-lifting thin bodies placed parallel to the upstream mean flow. The fluid is further assumed to be inviscid and non-heat conducting. Taken together, these assumptions lead to the approximation wherein (a) the mean flow is uniform, and (b) the unsteady disturbances are small so that in the equations of motion their products can be neglected. These conditions imply that vortical and entropic disturbances are uniformly convected by the mean flow and thus preserve their upstream structure. The flow velocity, pressure, density and entropy can hence be linearized about their constant mean values

$$\vec{V}(\vec{x},t) = \vec{U}_\infty + \vec{u}(\vec{x},t) \tag{1}$$

$$p(\vec{x},t) = p_0 + p'(\vec{x},t) \tag{2}$$

$$\rho(\vec{x},t) = \rho_0 + \rho'(\vec{x},t) \tag{3}$$

$$s(\vec{x},t) = s'(\vec{x},t) \tag{4}$$

where $\vec{x} = \{x_1, x_2, x_3\}$ is the position vector with the x_1-axis in the direction of the upstream mean velocity $\vec{U}_\infty$, and t is time. Because the mean flow velocity $\vec{U}_\infty$ is constant, the unsteady flow quantities $\vec{u}(\vec{x},t)$, $p'(\vec{x},t)$, $\rho'(\vec{x},t)$ and $s'(\vec{x},t)$ are governed by constant coefficient linear equations.

The linear character of the governing equations of motion and the preservation of the upstream structure of the incoming disturbances suggest the following splitting of the total unsteady velocity

$$\vec{u}(\vec{x},t) = \vec{u}_\infty + \nabla\phi \tag{5}$$

where $\vec{u}_\infty$ is the imposed upstream rotational disturbance velocity and $\nabla\phi$ is an irrotational velocity which must satisfy the impermeability condition along the body surface

$$\nabla\phi \cdot \vec{n} = -\vec{u}_\infty \cdot \vec{n} \tag{6}$$

where $\vec{n}$ represents the outward unit normal to the body surface. Far upstream $\nabla\phi$ must vanish for incompressible flows or represent outgoing acoustic waves for compressible flows.

To complete the boundary-value problem for the potential ϕ, one further imposes the condition that the pressure is continuous across the wake streamline. Thus, in the linear approximation, the mathematical problem for *weakly* rotational flows is reduced to solving the boundary-value problem for the potential function ϕ. This leads to the important conclusion that, in linear unsteady aerodynamics, the analysis of *weakly* rotational flows is the same as that of an oscillating body in a uniform mean flow but with different boundary condition at the body surface.

The early developments dealt with incompressible flows. In this case, the only independent parameter is the ratio of the two lengths, ℓ and ℓ' characterizing the body dimension and the wave length of the incoming vortical disturbance, respectively. This results in a considerable simplification of the analytical treatments and leads to closed form analytical solutions for some basic unsteady aerodynamic problems.

For compressible flows, pressure variations are propagated in the fluid with a finite velocity. The unsteady flow will then also depend on the Mach number, and closed form analytical solutions are only available for supersonic motions and for asymptotic values of the parameters in subsonic flows.

2.1 Incompressible Flows: The Early Developments

For incompressible flows, the irrotational part of the unsteady velocity was usually assumed to be harmonic. The flow was then modeled using the concepts of circulation theory and the method of conformal mapping [27]. Thus, as for steady thin airfoil theory, the body is equivalent to a system of *bound* vortices distributed along its surface and a Kutta condition is imposed at the trailing edge. However, in unsteady flows, the circulation Γ around a cross section of the body changes with time in response to flow nonuniformities, and for every change of Γ a vortex is shed from the trailing edge. By Kelvin's circulation theorem, the strength of the shed vortex is equal but opposite in sign to the change in Γ. The vortices detached from the trailing edge are usually called the *free* vortices since they are convected downstream at the constant mean flow velocity. They induce a velocity along the body surface and thus have a marked influence upon the flow around the body. The magnitude of their induced velocity is calculated by conformally mapping the flat plate representing the airfoil into a circle. For a continuously changing nonuniform flow, a vortex sheet extends along the wake line. At every instant, the intensity of the vortex sheet is determined by

Kelvin's circulation theorem. Thus the vortex sheet contains a recorded history of the fluid motion and the fluid acts as if it had a *memory*. The intensity of the vortex sheet is directly proportional to the strength of the upstream flow nonuniformities, and thus may be much stronger than the vortex sheets produced by the viscous boundary layers on each side of the body, which are neglected in the linear analysis.

Note that this simple model accounts for the essential features associated with the unsteady motions of real fluids. The first remarkable prediction of linear theory was that the lift acting upon an airfoil experiencing a sudden change of angle of attack will only asymptotically reach its steady state result. This important result was derived by Wagner [1], who established that real fluids respond with delay to outside excitations. This development was followed by the work of Theodorsen [3] on the oscillating airfoil and that of Kussner [4] on the sharp edged gust.

2.2 The von Karman-Sears Legacy

By the late 1930s, the essential physical phenomena associated with unsteady airfoil theory were well understood. However, the expressions derived for the aerodynamic forces were rather too complex to directly convey their physical significance.

Von Karman and Sears [5] developed a unified theory for *weakly* nonuniform flows that eliminated the unnecessary mathematical complications. Their intention was "to make the airfoil theory of non-uniform motion more accessible to engineers by showing the physical significance of the various steps of the mathematical deductions, and to present the results of the theory in a form suitable for immediate application to certain flutter and gust problems." The theory still represented the airfoil and its wake by a distribution of bound and free vortices. However, the lift and moment acting upon a two-dimensional airfoil were calculated directly from simple physical considerations of change of momentum of a system of vortices. The theory recovered the results of Theodorsen [3] for oscillating airfoils and gave general formulas for the lift as the sum of three parts:

$$L' = L'_0 + L'_a + L'_w \tag{7}$$

and a similar expression for the moment. In Eq. (7) L'_0 is the quasi-steady lift which is proportional to the bound vorticity on the airfoil, L'_a represents the contribution of the apparent mass and its expression is given in terms of the bound vorticity distribution, and L'_w is the

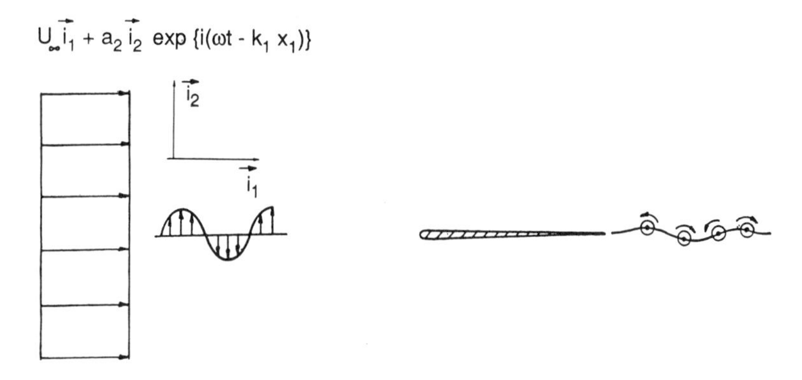

Figure 2. Thin airfoil in a transverse gust.

wake induced lift whose expression is given in terms of the free vorticity distribution in the wake.

The von Karman-Sears theory achieved a considerable simplification of the analysis and became the standard approach for incompressible nonuniform flows. It also paved the way for the work of Sears [7] on the response of a thin airfoil to a transverse sinusoidal gust. This problem is illustrated schematically in Figure 2. The upstream gust velocity was taken to be

$$\vec{u}_\infty = a_2 \vec{i}_2 \; exp[i\omega(t - x_1/U_\infty)] \tag{8}$$

where a_2 is the magnitude of the gust, $\vec{i}_2$ is the unit vector perpendicular to the airfoil, ω is the gust angular frequency, and $U_\infty = |\vec{U}_\infty|$. Sears [7] derived the following analytical expression for the lift per unit span

$$L' = \pi \rho_0 c U_\infty a_2 S(k_1) e^{i\omega t}. \tag{9}$$

In Eq. (9) c denotes the airfoil chord length, ρ_0 represents the fluid density and $S(k_1)$ is the well-known spiral Sears function which is given by

$$S(k_1) = \frac{2}{\pi k_1 [H_0^2(k_1) - i H_1^2(k_1)]} \tag{10}$$

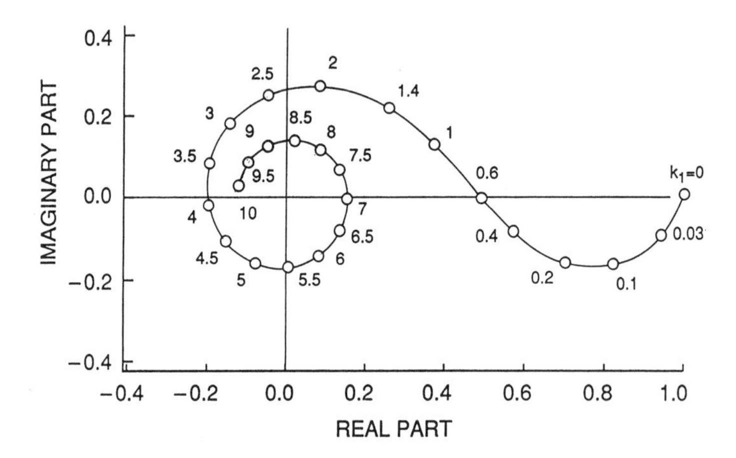

Figure 3. Vector diagram showing the real and imaginary parts of the Sears function versus the reduced frequency k_1.

$k_1 = (\omega c)/(2U_\infty)$ is the reduced frequency, and $H_0^2(k_1)$ and $H_1^2(k_1)$ are the Hankel functions of the second kind of orders 0 and 1, respectively. Figure 3 is a vector diagram showing the real and imaginary parts of $S(k_1)$ as a function of k_1. It is seen that as k_1 increases, the magnitude of the lift decreases continuously. At low reduced frequency, the phase of $S(k_1)$ first lags behind the gust phase and then leads it by a progressively greater amount. There is no critical gust frequency which produces abnormally large forces on the rigid airfoil. The moment of the aerodynamic forces with respect to the airfoil mid chord is

$$\mathcal{M}' = L' \cdot c/4. \tag{11}$$

Thus the total lift acts at all times and for all frequencies at the quarter chord point of the airfoil.

The Sears function can be approximated to within a few percent over most of its range by [28]

$$S(k_1) \approx \frac{exp\{ik_1[1 - \frac{\pi^2}{2(1+2\pi k_1)}]\}}{\sqrt{1 + 2\pi k_1}}. \tag{12}$$

There has been surprisingly little experimental verification of the Sears function. However, low-frequency ($k_1 \ll 1$) data for oscillating airfoils collected by Acum [29] show discrepancies of the order of 10 percent when compared with the Theodorsen function.

The von Karman–Sears approach was used by Graham [30] to obtain the solution for the oblique gust problem and by Whitehead [31] for a cascade of airfoils. Both formulated the problem in terms of a singular integral equation and obtained numerical solutions. Asymptotic analytical solutions have been obtained by Filotas [32] in the limits of high and low wave numbers.

2.3 Compressible Subsonic Flows

When the mean flow Mach number is not too small, the speed of sound c_0 is comparable in magnitude to the convective velocity U_∞. Pressure disturbances no longer propagate symmetrically but travel at different speeds in different directions. The compressible subsonic problem was first analyzed by Possio [8]. He formulated it in terms of a plane wave expansion and derived a singular integral equation for the pressure jump across the airfoil. This integral formulation has been widely used because it is particularly suitable for asymptotic analysis and for obtaining numerical solutions. However, certain essential results regarding the properties of such flows can be obtained by analyzing the partial differential equations governing the disturbance fields. In addition, this provides a basis for comparison with other theories where the governing equations have variable coefficients and for which it is no longer possible to formulate the problems in terms of integral equations.

The flow disturbances are governed by the linearized Euler equations

$$\frac{D_0\rho'}{Dt} = -\rho_0 \nabla \cdot \vec{u} \tag{13}$$

$$\rho_0 \frac{D_0\vec{u}}{Dt} = -\nabla p' \tag{14}$$

$$\frac{D_0 s'}{Dt} = 0 \tag{15}$$

where

$$\frac{D_0}{Dt} \equiv \frac{\partial}{\partial t} + U_\infty \frac{\partial}{\partial x_1}. \tag{16}$$

For an ideal gas, the entropy s' is related to the pressure p' and density ρ' by the equation of state

$$s' = c_v \frac{p'}{p_0} - c_p \frac{\rho'}{\rho_0} \tag{17}$$

where c_p and c_v are the specific heats at constant pressure and volume, respectively.

Kovasznay [33] has shown that because the governing equations, Eqs. (13) - (15), are constant coefficient linear equations, the flow disturbances can be split into distinct vortical, potential, and entropic modes obeying three independent differential equations. This result, known as the *splitting theorem*, generalizes the velocity decomposition in Eq. (5).

To derive Kovasznay's result, we begin by splitting the unsteady velocity into vortical and irrotational parts.

$$\vec{u} = \vec{u}^{(v)} + \nabla\phi \tag{18}$$

with $\vec{u}^{(v)}$ satisfying

$$\nabla \cdot \vec{u}^{(v)} = 0 \tag{19}$$

and

$$\frac{D_0 \vec{u}^{(v)}}{Dt} = 0 \tag{20}$$

Substituting Eq. (18) into (14) and integrating gives

$$p' = -\rho_0 \frac{D_0 \phi}{Dt} \tag{21}$$

It can then be readily shown that the potential ϕ satisfies the linear convective wave equation

$$\frac{1}{c_0^2}\frac{D_0^2 \phi}{Dt^2} - \nabla^2 \phi = 0 \tag{22}$$

where c_0 is the speed of sound in the uniform mean flow.

Thus the vortical velocity $\vec{u}^{(v)}$ is solenoidal, purely convected, and completely decoupled from pressure fluctuations. $\nabla\phi$ is usually referred to as the *potential velocity* or the *acoustic velocity*, and is directly related to the pressure. The two parts of the unsteady velocity are coupled only by the impermeability condition, Eq. (6), along the body surface. The entropy s' is purely convected and only affects the density ρ' through the equation of state, Eq. (17). Note that the splitting of the unsteady velocity is not unique but as was shown by Goldstein [34], the difference between any two splittings is a field which is both irrotational and solenoidal.

If there are no incident acoustic waves, Eqs. (15) and (20) can be immediately integrated, and we get

$$s' = s'_{\infty}(x_1 - U_{\infty}t, x_2, x_3) \tag{23}$$

$$\vec{u}^{(v)} = \vec{u}_{\infty}(x_1 - U_{\infty}t, x_2, x_3). \tag{24}$$

The entropy and the vortical velocity are determined in terms of their upstream imposed values. The mathematical problem is then reduced to finding the function ϕ which must satisfy the convective wave Eq. (22) with the surface condition

$$\frac{\partial \phi}{\partial x_2} = -u_{\infty 2}(x_1 - U_{\infty}t, 0, x_3), \quad -\frac{c}{2} \leq x_1 \leq \frac{c}{2}, \; x_2 = 0 \tag{25}$$

where $u_{\infty 2}$ is the component of $\vec{u}_{\infty}$ along the x_2-axis (Figure 4). In the far field,

$$\nabla\phi \rightarrow 0 \text{ as } x_1 \rightarrow -\infty. \tag{26}$$

In addition, as in the case of incompressible flow, the unsteady velocity must satisfy Kelvin's theorem of the constancy of circulation in a circuit moving with the fluid and the Kutta condition at the trailing edge. Thus a vortex sheet extends downstream from the trailing edge. Across this vortex sheet, the pressure and the normal velocity are continuous, but ϕ has a discontinuity $\Delta\phi$. Applying Eq. (21) on both sides of the vortex sheet and using the pressure continuity across the vortex sheet, we obtain

$$\frac{D_0}{Dt}\Delta\phi = 0. \tag{27}$$

Equations (22), and (25) – (27), with the condition that ϕ is continuous in the x_1-direction at the trailing edge, completely define the boundary-value problem for ϕ.

This boundary-value problem is almost identical to that encountered in oscillating airfoils except that Eq. (25) replaces the condition prescribing the oscillatory motion of the body surface. Therefore, in the linear approximation, all methods and procedures developed for small oscillatory motions equally apply to the gust response problem.

Without loss of generality, we can consider a single Fourier component to represent the incident disturbance

$$\vec{u}_{\infty} = \vec{a}\; exp\{i(\omega t - \vec{k} \cdot \vec{x})\}. \tag{28}$$

and obtain the general result by superposition. Equation (28) represents an incident plane vortical wave whose magnitude is $\vec{a} = \{a_1, a_2, a_3\}$ and which propagates in the direction $\vec{k} = \{k_1, k_2, k_3\}$ as shown in Figure 4. Note that since the upstream disturbance is convected by the mean velocity, $\omega = k_1 U_{\infty}$. Equations (19) and (24) imply that

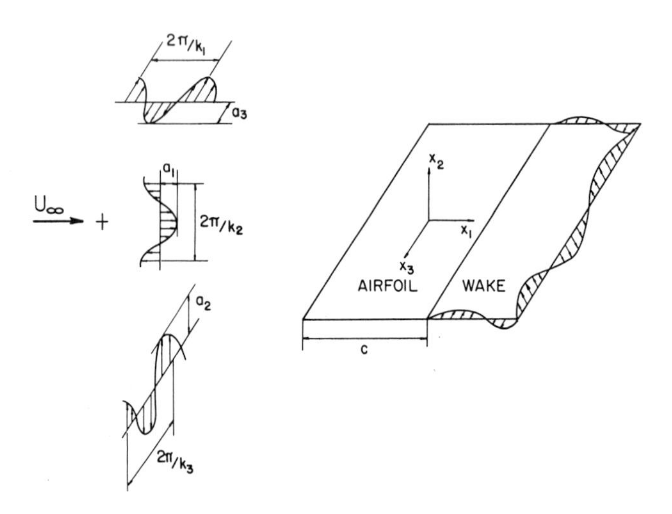

Figure 4. Flat plate in a three-dimensional gust.

$$\vec{a} \cdot \vec{k} = 0. \tag{29}$$

Equation (27) can now be integrated to obtain

$$\Delta\phi = C_0\, exp\{i[\omega t - k_1(x_1 - c/2) - k_3 x_3]\} \tag{30}$$

where $C_0 e^{i\omega t}$ is the jump of ϕ at the trailing edge. Reissner [9] simplified Eq. (22) by introducing the Prandtl-Glauert coordinates $\vec{\tilde{x}} = \{\tilde{x}_1, \tilde{x}_2, \tilde{x}_3\}$, where

$$\begin{aligned} \tilde{x}_1 &= x_1 \\ \tilde{x}_2 &= \beta_\infty x_2 \\ \tilde{x}_3 &= \beta_\infty x_3 \end{aligned} \tag{31}$$

and the transformation

$$\tilde{\phi} = \phi\, exp\{-i(\omega t + K_1 M_\infty x_1)\} \tag{32}$$

where M_∞ is the upstream mean flow Mach number, $\beta_\infty = \sqrt{1 - M_\infty^2}$ and $K_1 = \omega/(\beta_\infty^2 c_0)$. The potential function $\tilde{\phi}$ satisfies the Helmholtz equation

$$(\tilde{\nabla}^2 + K_1^2)\tilde{\phi} = 0 \tag{33}$$

with the boundary conditions

$$\frac{\partial\tilde{\phi}}{\partial\tilde{x}_2} = -\frac{a_2}{\beta_\infty} exp\{-i(\frac{k_1\tilde{x}_1}{\beta_\infty^2} + K_3\tilde{x}_3)\} \; for \; -\frac{c}{2} < x_1 < \frac{c}{2} \tag{34}$$

$$\Delta\tilde{\phi} = \Delta\tilde{\phi}_{t.e.} \, exp\{-i(\frac{k_1\tilde{x}_1}{\beta_\infty^2} + K_3\tilde{x}_3)\} \; for \; x_1 > \frac{c}{2} \tag{35}$$

where $\tilde{\nabla}$ represents the del operator in the Prandtl-Glauert coordinates, $K_3 = k_3/\beta_\infty$ and $\Delta\tilde{\phi}_{t.e.}$ is the jump of $\tilde{\phi}$ at the trailing edge.

For an airfoil of an infinite span and constant chord length c, it is possible to factor out the dependence on x_3 by introducing the new dependent variable

$$\varphi = \tilde{\phi} \, exp\{ik_3x_3\}. \tag{36}$$

We also non-dimensionalize all lengths and velocities with respect to $c/2$ and U_∞, respectively. The function φ satisfies the two-dimensional Helmholtz equation

$$\frac{\partial^2\varphi}{\partial\tilde{x}_1^2} + \frac{\partial^2\varphi}{\partial\tilde{x}_2^2} + (K_1^2 - K_3^2)\varphi = 0 \tag{37}$$

This equation has two types of solutions depending on whether the quantity $(K_1^2 - K_3^2)$ is positive or negative. For $K_1^2 > K_3^2$, the far field consists of outwardly propagating acoustical waves and the flow is called *supercritical.* On the other hand, for $K_1^2 < K_3^2$, the far field decays exponentially and the flow is called *subcritical.* Every solution depends on the set of three parameters (M, k_1, k_3). Graham [35] constructed similarity rules which show that there is a subset of solutions of each type differing by a constant multiple and depending only on two independent parameters.

(a) Subcritical Flows: $K_1 < K_3$

The governing equation has the form

$$\frac{\partial^2\varphi}{\partial\tilde{x}_1^2} + \frac{\partial^2\varphi}{\partial\tilde{x}_2^2} - \chi^2\varphi = 0 \tag{38}$$

Similar solutions of this type are such that χ =constant and k_1/β^2 = constant. Every member of this group can be obtained from one of them, say, $K_1 = 0$, corresponding to the incompressible $(M = 0)$ oblique gust case.

(b) Supercritical Flows: $K_1 > K_3$

The governing equation has the form

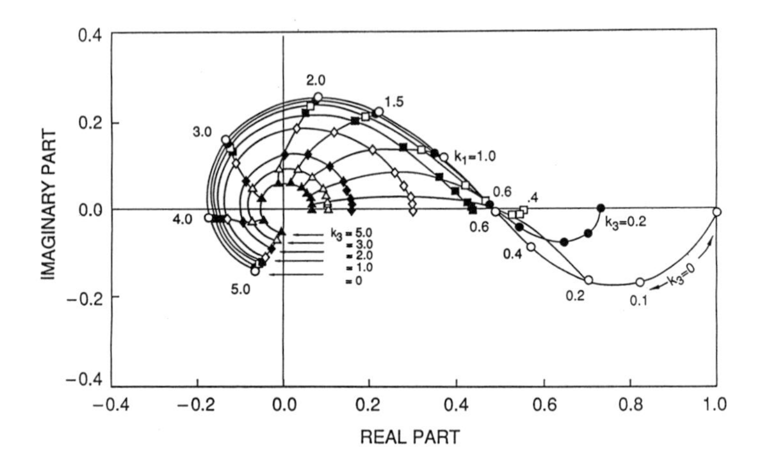

Figure 5. Vector diagram showing the real and imaginary parts of the response function $\mathcal{S}(k_1, k_3, 0)$ versus k_1 for an oblique gust at various k_3. Lines of constant k_1 are shown for $k_1 = 0.2, 0.6, 1.0, 1.5, 2.0, 3.0, 4.0, 5.0$.

$$\frac{\partial^2 \varphi}{\partial \tilde{x}_1^2} + \frac{\partial^2 \varphi}{\partial \tilde{x}_2^2} + \chi^2 \varphi = 0 \tag{39}$$

Similar solutions of this type are again such that χ =constant and k_1/β^2 = constant. Every member of this group can be obtained from one of them, say, $K_3 = 0$, corresponding to the compressible two-dimensional case.

For $K_1 = K_3$, both sets of solutions are similar to the incompressible two-dimensional case studied by Sears.

The expression for the lift per unit span can be cast as in Eq. (9)

$$L' = \pi \rho_0 c U_\infty a_2 \mathcal{S}(k_1, k_3, M) e^{i\omega t}. \tag{40}$$

The airfoil reponse function $\mathcal{S}(k_1, k_3, M)$ is a generalized Sears function which reduces to $S(k_1)$ for $k_3 = 0$, and $M = 0$. $\mathcal{S}(k_1, k_3, M)$ can be obtained from either $\mathcal{S}(k_1, k_3, 0)$, $S(k_1)$ or $\mathcal{S}(k_1, 0, M)$, depending on whether K_1 is smaller, equal or larger than K_3. Figure 5 shows the variation of the response function for an oblique gust in incompressible flow at various values of the spanwise wave number k_3. Lines of constant k_1 are also shown. Note that as k_3 increases, the magnitude of the unsteady lift is always reduced. This effect is more significant at low reduced frequencies where a change in the spanwise wave number from 0 to 1.0 reduces the lift by 40 to 70 percent depending on the value of k_1. The gust front makes an angle $\Lambda = tan^{-1}(k_3/k_1)$ with the airfoil span direction. Adamczyk [36] has shown that the problem of a transverse gust about an airfoil with a sweep angle Λ is reducible to that of an oblique gust convected at a speed equal to the mean flow velocity component normal to the airfoil span direction. Hence, Figure 5 also suggests that the magnitude of the gust response will be considerably reduced for swept airfoils. Sweep can then be used as a means to control and reduce unwanted fluctuating aerodynamic forces.

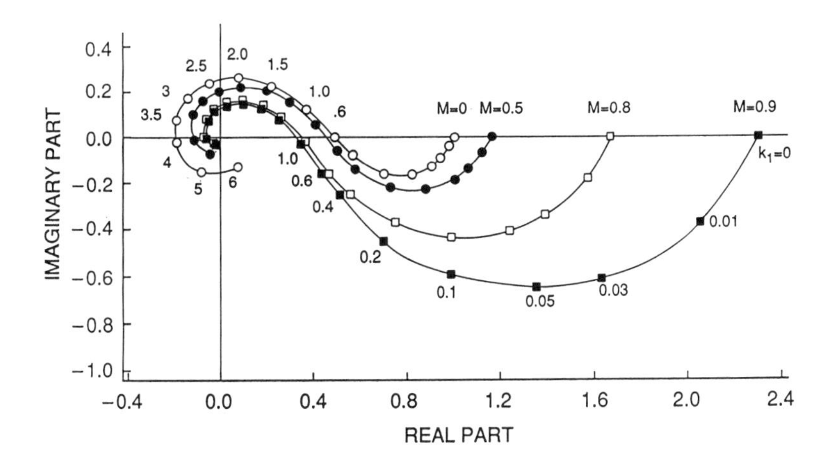

Figure 6. Vector diagram showing the real and imaginary parts of the response function $\mathcal{S}(k_1, 0, M)$ versus k_1 for a transverse gust at various M.

Figure 6 is a vector diagram showing the real and imaginary parts of the response function versus the reduced frequency k_1 for a two–dimensional gust ($k_3 = 0$) at various Mach numbers. The incompressible case ($M = 0$) corresponds to the Sears function. The limiting values of the response functions along the real axis corresponding to $k_1 = 0$ are equal to $1/\beta_\infty$ as expected from the Prandtl–Glauert theory. We note that at higher Mach number the magnitude of the airfoil response function increases at low reduced frequency and decreases at high values of k_1. An asymptotic expansion of $\mathcal{S}(k_1, k_3, M)$ as $k_1 \to \infty$, for finite M, yields $\mathcal{S}(k_1, k_3, M) = O(1/k_1)$, while for $M \to 0$, $S(k_1) = O(1/\sqrt{k_1})$. This discrepancy is a result of the fact that the incompressible Sears theory must be considered as an asymptotic

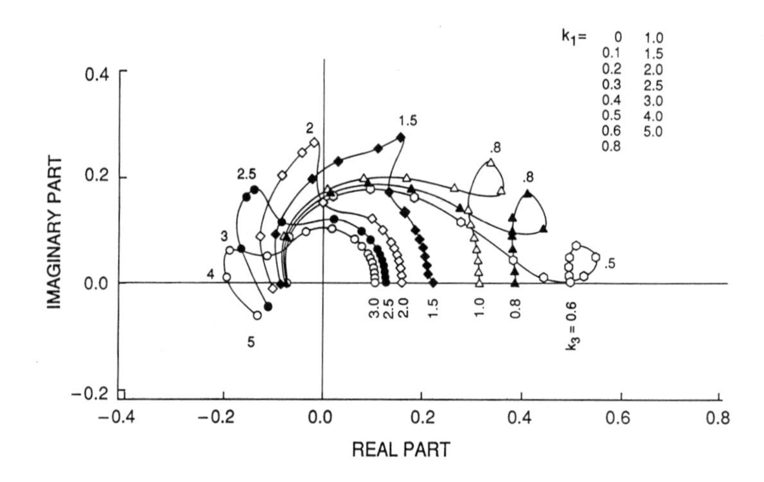

Figure 7. Vector diagram showing the real and imaginary parts of the response function $\mathcal{S}(k_1, k_3, 0.8)$ versus k_1 for an oblique gust at various k_3.

theory in the limit of $k_1 M \to 0$ and k_1 finite.

For a compressible oblique gust, dramatic changes take place in the response function. Figure 7 is a vector diagram showing the real and imaginary parts of the response function versus k_1 at $M = 0.8$ for various values of k_3. At lower spanwise wave numbers, k_3, compressibility effects dominate, while at larger k_3 the unsteady flow changes from supercritical to subcritical. As this occurs, the magnitude of the lift is very sensitive to changes in k_1 while its phase changes are rather small. This produces the loops and spikes observed in the response function diagram.

Because of the broad application of the theory, numerous asymptotic forms were developed to obtain closed-form analytical approximations. We briefly mention the work of Osborne [37], Amiet [38] , and Graham and Kullar [41] on compressibility effects; and Amiet [39], and Martinez and Widnall [40] on the high frequency approximation.

Extension to a linear cascade has been carried out by Whitehead [42] and Fleeter [43] , and to a three-dimensional body geometry by Namba[44].

3 MEAN FLOW DISTORTION THEORIES

The mean motion of a real flow is generally not uniform especially when it is around a lifting body. Unsteady aerodynamic theories which account for the mean flow distortion of the incident vortical disturbances but neglect their nonlinear inertial and viscous effects are herein referred to as *mean flow distortion* theories. The fluid is still assumed to be inviscid and non-heat conducting, and the intensity of the upstream rotational disturbances is small. These assumptions lead to the linearization of Euler's equations of motion about a nonuniform mean flow. The unsteady part of the flow is thus governed by *nonconstant* coefficient linear equations.

This approach is justified only if (a) the flow is *weakly* rotational, (b) the time taken by a fluid particle to traverse a region in which the mean velocity changes (convective time) is much less than the time needed for the vortical disturbances to change from their own inertial and viscous forces (Lagrangian time), and (c) the integral scale ℓ' associated with the disturbances is much larger than the boundary-layer thickness δ so that the interaction with boundary-layer eddies can be neglected. These conditions can be summarized as follows.

$$u'_\infty \ll U_\infty \tag{41}$$

$$\frac{\ell}{U_\infty} \ll \frac{\ell'}{u'_\infty} \tag{42}$$

$$\ell' \gg \delta \tag{43}$$

where ℓ is a representative length of the body, and u'_∞ is a characteristic velocity of the incident vortical disturbances. These conditions are the same as those used in the rapid distortion theory of turbulence. For more details, we refer the reader to a recent review article by Savill [45].

For oscillatory transonic flows, the linearization of the equations of motion is also subject to the restriction ([47], [48])

$$k_1 \gg |1 - M_\infty|. \tag{44}$$

The governing equations for an inviscid and non-heat conducting flow can be written as

$$\frac{D\rho}{Dt} + \rho\nabla\cdot\vec{V} = 0 \tag{45}$$

$$\rho\frac{D\vec{V}}{Dt} = -\nabla\cdot p \tag{46}$$

$$\frac{DS}{Dt} = 0 \tag{47}$$

where $\vec{V}$, p, ρ and S are the velocity, pressure, density and entropy, respectively, and $D/Dt \equiv \partial/\partial t + \vec{V} \cdot \nabla$ is the material derivative. To these equations, one should add the equation of state of the fluid, $\rho = \rho(p, S)$. These equations must be supplemented with continuity conditions for the pressure and the normal velocity across the wake line and, for transonic and supersonic flows, with the Rankine-Hugoniot jump conditions across shock waves. We also suppose that any shock wave taking place in the fluid will be weak. Hence, the rise of entropy across a shock wave which is of $O[(M-1)^3]$ [46] can be neglected. We further assume that the entropy of the upstream mean flow is constant. The entropy is therefore taken to be constant everywhere in the mean flow.

The flow velocity, pressure, density and entropy can then be linearized about the mean flow quantities:

$$\vec{V}(\vec{x},t) = \vec{U}(\vec{x}) + \vec{u}(\vec{x},t) \tag{48}$$

$$p(\vec{x},t) = p_0(\vec{x}) + p'(\vec{x},t) \tag{49}$$

$$\rho(\vec{x},t) = \rho_0(\vec{x}) + \rho'(\vec{x},t) \tag{50}$$

$$S(\vec{x},t) = S_0 + s'(\vec{x},t). \tag{51}$$

The mean flow quantities $\vec{U}(\vec{x}), p_0(\vec{x})$, and $\rho_0(\vec{x})$ are governed by the equations

$$\nabla \cdot (\rho_0 \vec{U}) = 0 \tag{52}$$

$$\rho_0(\vec{U} \cdot \nabla)\vec{U} = -\nabla p_0. \tag{53}$$

In what follows we suppose that analytical or numerical solutions to Eqs. (52) – (53) are available and that all mean flow quantities are known. The equations governing the unsteady flow quantities can then be obtained by linearizing Eqs. (45) – (47) and subtracting out the mean flow Eqs. (52) – (53) to obtain

$$\frac{D_0\rho'}{Dt} + \rho'\nabla \cdot \vec{U} + \nabla \cdot (\rho_0 \vec{u}) = 0 \tag{54}$$

$$\rho_0(\frac{D_0\vec{u}}{Dt} + \vec{u} \cdot \nabla\vec{U}) + \rho'\vec{U} \cdot \nabla\vec{U} = -\nabla \cdot p' \tag{55}$$

$$\frac{D_0 s'}{Dt} = 0 \tag{56}$$

where $D_0/Dt \equiv \partial/\partial t + \vec{U} \cdot \nabla$ is the material derivative associated with the mean flow.

Far upstream, the velocity $\vec{V}(\vec{x},t)$ and the entropy s' must be of the form

$$\vec{V}(\vec{x},t) = \vec{U}_\infty + \vec{u}_\infty(\vec{x} - \vec{i}_1 U_\infty t) \quad as \quad x_1 \to -\infty \tag{57}$$

$$s'(\vec{x},t) = s'_\infty(\vec{x} - \vec{i}_1 U_\infty t) \quad as \quad x_1 \to -\infty \tag{58}$$

where $\vec{i}_1$ is a unit vector in the direction of the upstream mean velocity $\vec{U}_\infty$. $\vec{u}_\infty$ and s'_∞ can be any functions of their arguments with the restriction that $\vec{u}_\infty$ be solenoidal,

$$\nabla \cdot \vec{u}_\infty = 0. \tag{59}$$

For the important case of flows which exist for all time, we can often represent the upstream disturbances $\vec{u}_\infty(\vec{x} - \vec{i}_1 U_\infty t)$ and $s'_\infty(\vec{x} - \vec{i}_1 U_\infty t)$ by the Fourier integrals

$$\vec{u}_\infty(\vec{x} - \vec{i}_1 U_\infty t) = \int \vec{a}(\vec{k})\ exp[-i\vec{k} \cdot (\vec{x} - \vec{i}_1 U_\infty t)] d\vec{k}$$

$$s'_\infty(\vec{x} - \vec{i}_1 U_\infty t) = \int b(\vec{k})\ exp[-i\vec{k} \cdot (\vec{x} - \vec{i}_1 U_\infty t)] d\vec{k}.$$

Then since Eqs. (54) – (56) are linear, we can find solutions for $\vec{u}$, p' and s' for any upstream disturbances $\vec{u}_\infty$ and s'_∞ simply by superposing solutions to the corresponding problem for the upstream harmonic disturbances

$$\vec{u}_\infty(\vec{x} - \vec{i}_1 U_\infty t) = \vec{a}\ exp[-i\vec{k} \cdot (\vec{x} - \vec{i}_1 U_\infty t)] \tag{60}$$

$$s'_\infty(\vec{x} - \vec{i}_1 U_\infty t) = b\ exp[-i\vec{k} \cdot (\vec{x} - \vec{i}_1 U_\infty t)]. \tag{61}$$

Therefore, for this class of flows, we need consider only harmonic disturbances. In this case, Eq. (59) implies that

$$\vec{a} \cdot \vec{k} = 0. \tag{62}$$

The unsteady velocity $\vec{u}$ must satisfy the impermeability condition along the body surface

$$\vec{u} \cdot \vec{n} = 0 \tag{63}$$

and the unsteady pressure and normal velocity must be continuous across the wake.

It is not possible, in general, to split the unsteady velocity into vortical and potential parts governed by separate equations as in the linear approximation. Mean flow distortion analyses of weakly rotational flows must therefore determine a rotational velocity field $\vec{u}$ satisfying a homogeneous boundary condition, Eq. (63), at the surface of the body and non-homogeneous conditions, Eqs. (60)–(61) at infinity. In contrast, for oscillating bodies the unsteady velocity field is potential and must satisfy specified conditions at the body surface and vanish at infinity.

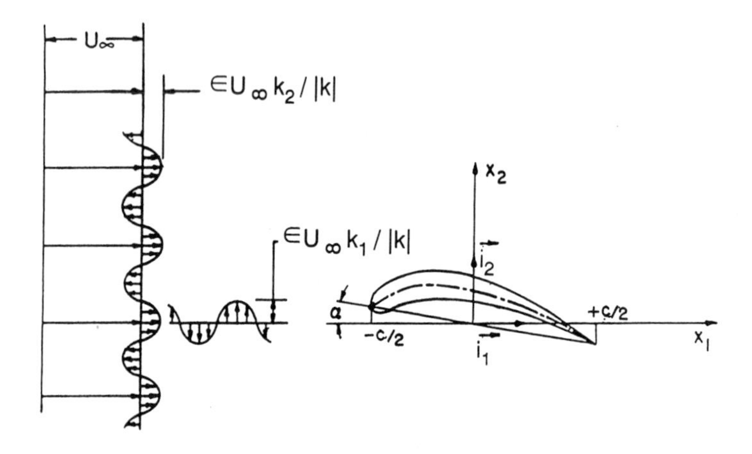

Figure 8. Lifting airfoil in a two-dimensional gust.

3.1 Lifting Airfoil in a Two-Dimensional Gust

The problem of a lifting airfoil passing through a gust pattern was first examined by Horlock [49]. Using a heuristic approach, Horlock partially accounted for the second order effects of small mean-flow incidence on the unsteady lift. His treatment, however, was incomplete in that he only accounted for the modified boundary condition at the airfoil surface while neglecting the coupling between the unsteady flow and the mean flow around the airfoil.

Goldstein and Atassi [19] were the first to introduce the concept of mean flow distortion effects in unsteady aerodynamics. They considered a lifting airfoil with chord length c placed at non-zero incidence to a uniform flow U_∞ in the x_1 direction (Figure 8). Far upstream, a two-dimensional harmonic gust ($a_3 = 0$) of amplitude $|\vec{a}| = \epsilon U_\infty$, where $\epsilon \ll 1$, is imposed on the flow. In order to obtain a relatively simple closed-form solution, the analysis was restricted to the case of an airfoil with small thickness, θ; angle of attack, α; and camber, m. A small parameter β is introduced to scale these steady flow disturbances caused by the airfoil, i.e., α, m and θ are of $O(\beta)$. Goldstein and Atassi[1] then constructed an asymptotic solution for the case where $\epsilon \ll \beta \ll 1$, using the theory of sectionally analytic functions and the method of matched asymptotic expansions. In this asymptotic theory, the solution of

[1]The authors would like to acknowledge the encouragements and insightful comments of Professor W. R. Sears during the course of this work (see appendix A).

$O(\epsilon)$ corresponds to the Sears solution. The effect of the mean flow distortion is accounted for by the solution of $O(\epsilon\beta)$. This solution fully accounts for the effect of distortion of the unsteady flow by the airfoil mean flow.

The distortion of the gust by the mean flow, which imparts a nonlinear character to the problem, would suggest that the effects of thickness, camber and angle of attack cannot simply be superposed. However, Atassi [20], revisiting the problem, has shown that these effects can, in fact, be linearly superposed. This important result is primarily due to the linearization of the mean flow, but it is also a consequence of the resulting *local* dependence of the outer solution on the linearized mean flow. Thus, for a thin airfoil with small camber and angle of attack moving in a periodic gust pattern, the unsteady lift $\mathcal{L}'$ can be constructed by linear superposition to the Sears lift L' (9) of three independent components: L'_α resulting from a non-zero angle of attack of the mean flow , L'_m resulting from the airfoil camber, and L'_θ produced by the airfoil thickness. The unsteady lift $\mathcal{L}'$ can therefore be written as

$$\mathcal{L}'(k_1, k_2, \alpha, m, \theta) = \frac{k_1}{k} L'(k_1) + \alpha L_\alpha(k_1, k_2) + m L_m(k_1, k_2) + \theta L_\theta(k_1, k_2) \qquad (64)$$

The expressions for L_α, L_m and L_θ are given in terms of Bessel and Hankel functions of the complex variable $k_1 + ik_2$. It is convenient to introduce a non-dimensional response function

$$\mathcal{R}(k_1, k_2, \alpha, m, \theta) = \frac{\mathcal{L}'(k_1, k_2, \alpha, m, \theta)}{\pi \rho_0 c U_\infty |\vec{a}| exp(i\omega t)} \qquad (65)$$

Figure 9 shows three vector diagrams of the response function $\mathcal{R}$ for three different thin airfoils having the same steady lift ($C_L \approx 2$). This is obtained by varying the mean flow angle of attack α and the airfoil camber m. The gust is propagating at 45° to the upstream mean flow. The angle of attack α is given in radians. It is remarkable to note that at low reduced frequencies, the magnitude of the unsteady lift is significantly reduced from about one to 0.25. This is mainly due to the effect of the chordwise component of the gust ($a_1 = -a_2$). On the other hand, at high frequencies, the unsteady lift is much larger than the corresponding values for the Sears function, particularly for airfoils with mean flow angle of attack. The significant differences between the unsteady lifts for the three airfoils with the same steady lift coefficient, indicate that the distortion effects of the mean flow cannot be lumped into one single parameter as the airfoil mean loading.

3.2 Vortical Disturbances of Potential Flows

The mean flow distortion approximation has been extensively used to study the response of turbulence to applied strains. However, since the focus of the present article is on unsteady

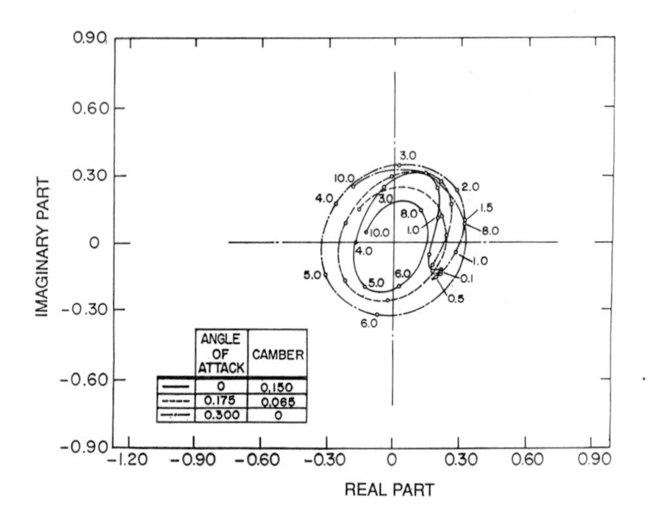

Figure 9. Vector diagram showing the real and imaginary parts of the response function $\mathcal{R}(k_1, k_2, \alpha, m, \theta)$ versus k_1 for three thin airfoils in a two-dimensional gust having the same steady lift. The gust is propagating at 45° to the mean flow.

aerodynamics, we only describe those methods which treat high speed compressible flows.

3.2.1 Goldstein's Splitting of the Disturbance Velocity

Goldstein [21] proposed a unified approach to small amplitude, unsteady, vortical and entropic disturbances imposed on steady, potential flows. The flow velocity, pressure, density, and entropy are linearized about the mean flow quantities as in Eqs. (48)–(58). In addition, the mean flow velocity $\vec{U} = \{U_1, U_2, U_3\}$ can be expressed in terms of a potential $\Phi(\vec{x})$ by

$$\vec{U} = \nabla\Phi. \tag{66}$$

Goldstein then shows that the unsteady perturbation velocity $\vec{u}$ can be written as

$$\vec{u} = \vec{u}^{(I)} + \nabla\phi \tag{67}$$

where $\vec{u}^{(I)}$ is a rotational disturbance whose expression is a known function of the imposed upstream disturbances, and $\nabla\phi$ is related to the perturbation pressure p' by

$$p' = -\rho_0 \frac{D_0\phi}{Dt}. \tag{68}$$

The rotational velocity $\vec{u}^{(I)}$ is given by

$$\vec{u}^{(I)} = \vec{u}^{(H)} + \frac{1}{2c_p} s'_\infty(\vec{X} - \vec{i}_1 U_\infty t)\nabla\Phi \tag{69}$$

where $\vec{u}^{(H)} = \{u_1^H, u_2^H, u_3^H\}$ is a solution to the first order homogeneous equation

$$\frac{D_0}{Dt}\vec{u}^{(H)} + \vec{u}^{(H)} \cdot \nabla\vec{U} = 0 \tag{70}$$

and must satisfy the upstream condition that $\vec{u}^{(I)} \to \vec{u}_\infty$ as $x_1 \to -\infty$. This completely determines the expression for $\vec{u}^{(H)}$ in terms of the upstream disturbance,

$$u_i^{(H)} = \vec{A}(\vec{X} - \vec{i}_1 U_\infty t) \cdot \frac{\partial \vec{X}}{\partial x_i} \tag{71}$$

where

$$\vec{A}(\vec{X} - \vec{i}_1 U_\infty t) = \vec{u}_\infty(\vec{X} - \vec{i}_1 U_\infty t) - \frac{U_\infty}{2c_p} s'_\infty(\vec{X} - \vec{i}_1 U_\infty t)\, \vec{i}_1. \tag{72}$$

The components of $(\vec{X} - \vec{i}_1 U_\infty t)$ are essentially the Lagrangian coordinates of the steady flow fluid particles. The components of $\vec{X} = \{X_1, X_2, X_3\}$ are defined as follows. $X_2(x_1, x_2, x_3)$ and $X_3(x_1, x_2, x_3)$ are independent integrals of the equations

$$\frac{dx_1}{U_1} = \frac{dx_2}{U_2} = \frac{dx_3}{U_3} = dt \tag{73}$$

such that

$$X_2 \to x_2, \quad X_3 \to x_3 \quad \text{as } x_1 \to -\infty. \tag{74}$$

The mean flow streamlines lie along the intersection of surfaces $X_2 =$ constant and $X_3 =$ constant. The equations for the mean flow streamlines in terms of x_1 can be written as

$$x_2 = y_s(x_1, X_2, X_3), \quad x_3 = z_s(x_1, X_2, X_3). \tag{75}$$

X_1/U_∞ is the Lighthill 'drift' function

$$\Delta(x_1, x_2, x_3) = \frac{x_1}{U_\infty} + \int_{-\infty}^{x_1} [\frac{1}{U_1[x'_1, y_s(x'_1, X_2, X_3), z_s(x'_1, X_2, X_3)]} - \frac{1}{U_\infty}]\, dx'_1. \tag{76}$$

The change in Δ between any two points of a streamline is equal to the time it takes a steady-flow fluid particle to traverse the distance between those two points.

Finally, the only unknown quantity is the perturbation potential ϕ which satisfies the linear inhomogeneous wave equation

$$\frac{D_0}{Dt}\left(\frac{1}{c_0^2}\frac{D_0\phi}{Dt}\right) - \frac{1}{\rho_0}\nabla\cdot(\rho_0\nabla\phi) = \frac{1}{\rho_0}\nabla\cdot(\rho_0\vec{u}^{(I)}) \tag{77}$$

where $c_0 = c_0(\vec{x})$ is the mean flow speed of sound. For a rigid body, the boundary condition at the body surface Σ is

$$\vec{n}\cdot\nabla\phi = -\vec{n}\cdot\vec{u}^{(I)} \quad \text{for} \quad \vec{x}\in\Sigma. \tag{78}$$

In addition[1]

$$\phi(\vec{x},t) \to 0 \quad \text{as} \quad x_1 \to -\infty. \tag{79}$$

Goldstein's decomposition of the velocity field reduces the methematical problem for this kind of *weakly* rotational motions to that of solving a single convected wave equation. This significantly simplifies the analytical treatments and the numerical procedures used to study such flows. For example, Goldstein's procedure leads to a single Poisson's equation for three-dimensional incompressible flows, while Hunt's approach [18] led to three such equations.

3.2.2 Modified Splitting for Flows Around Bodies

For streaming motions around bodies, there is usually a front stagnation point S on the body surface where the mean velocity vanishes. In this case, the 'drift' function Δ will develop a logarithmic singularity at S, that will remain along the entire body surface and its wake. Thus the Lagrangian coordinate $X_1 - U_\infty t$ will also be infinite. The quantities $\vec{A}(\vec{X} - \vec{i}_1 U_\infty t)$ and $s'_\infty(\vec{X} - \vec{i}_1 U_\infty t)$ will have singular arguments and will be indeterminate for periodic disturbances. Moreover, the rotational velocity $\vec{u}^{(I)}$, which is proportional to the gradient of Lagrangian coordinates, will have a singularity of order $1/n$ along the entire body surface, where n is the distance along the normal to the body.

Atassi and Grzedzinski [22] analyzed the flow near the stagnation point and noted that the vortex stretching at S, which is shown schematically in Figure 10, results in a singular behavior of order $1/n$ for the vorticity. Recognizing that the total unsteady velocity $\vec{u}$ cannot have such a strong singularity along the surface of the body, they concluded that the strong behavior of $\vec{u}^{(I)}$ must be cancelled by $\nabla\phi$. These features make it difficult to use Goldstein' approach when numerically calculating the unsteady distorted flow over bodies with a stagnation point.

---

[1] In a duct or a cascade, ϕ does not vanish if there are propagating acoustic waves.

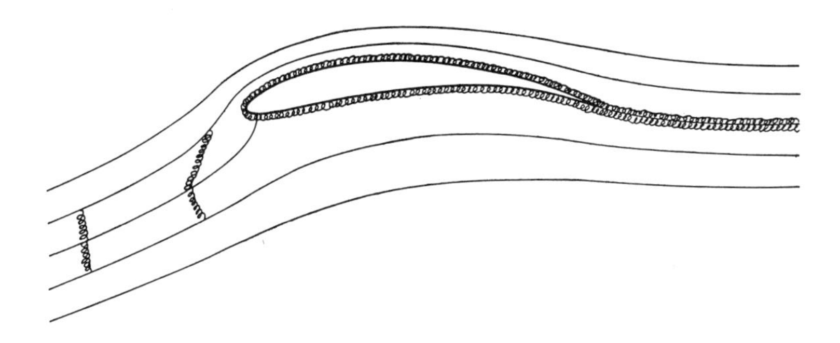

Figure 10. Vortex stretching at the stagnation point.

Atassi and Grzedzinski then showed that it is possible to find a potential velocity field that produces no pressure and that cancels the singular behavior of the streamwise and normal components of $\vec{u}^{(I)}$ along the body and its wake. The unsteady disturbance velocity can therefore be expressed as the sum of (i) a rotational part $\vec{u}^{(R)}$ that is a known function of the upstream disturbance and whose streamwise and normal components vanish along the entire body surface and its wake, (ii) an entropy dependent disturbance whose expression is a known function of the imposed upstream entropy disturbance, and (iii) an irrotational $\nabla\phi^*$ part whose potential ϕ^* must satisfy a linear nonconstant coefficient wave equation with a source term.

Thus we have the following partial splitting for the disturbance velocity field

$$\vec{u} = \frac{1}{2c_p} s'_\infty(\vec{X} - \vec{i}_1 U_\infty t)\nabla\Phi + \vec{u}^{(R)} + \nabla\phi^* \tag{80}$$

where the rotational velocity $\vec{u}^{(R)}$ satisfies the equation

$$\frac{D_0}{Dt}\vec{u}^{(R)} + \vec{u}^{(R)} \cdot \nabla\vec{U} = 0 \tag{81}$$

and the boundary conditions

$$\vec{u}^{(R)} \cdot \vec{\tau} = 0 \tag{82}$$

$$\vec{u}^{(R)} \cdot \vec{n} = 0 \tag{83}$$

along the surface of the body Σ and its wake $\mathcal{W}$. The potential function ϕ^* satisfies the equation

$$\frac{D_0}{Dt}\left(\frac{1}{c_0^2}\frac{D_0\phi^*}{Dt}\right) - \frac{1}{\rho_0}\nabla \cdot (\rho_0 \nabla \phi^*) = \frac{1}{\rho_0}\nabla \cdot (\rho_0 \vec{u}^{(R)}) - \frac{1}{2c_p}\frac{\partial s_\infty}{\partial t} \tag{84}$$

and, in view of Eqs. (82) and (83) and the fact that $\nabla\Phi \cdot \vec{n} = 0$, on Σ, the boundary conditions

$$\nabla\phi^* \cdot \vec{n} = 0 \quad on\ \Sigma \tag{85}$$

$$\tilde{\Delta}[\nabla\phi^* \cdot \vec{n}] = 0 \quad on\ \mathcal{W} \tag{86}$$

$$\nabla\phi^* \rightarrow \vec{u}_\infty - \vec{u}^{(R)} - \frac{1}{2c_p}s'_\infty \vec{U}_\infty \quad \text{as } x_1 \rightarrow -\infty \tag{87}$$

where $\tilde{\Delta}$ denotes the jump across the wake.

The homogeneous conditions along the surface of the body for the normal component of $\nabla\phi^*$, Eq. (85), and for the streamwise and normal components of $\vec{u}^{(R)}$, Eqs. (82) and (83), make this splitting of the unsteady velocity field particularly suitable for numerical calculations of subsonic and transonic (with weak shocks) flows. In such cases, the mean flow is determined from a nonlinear numerical code and an accurate resolution of the flow near the stagnation point is usually difficult to achieve. A numerical procedure based on the present method will not be very sensitive to the details of the mean flow near the stagnation point.

3.2.3 Application to Airfoils and Cascades

Numerical solutions to the boundary-value problem, Eqs. (84) – (87), in the frequency domain were obtained for a three-dimensional periodic gust, Eq. (28), in subsonic flow for a single airfoil by Scott and Atassi [23], and for a two-dimensional cascade geometry by Hall and Verdon [24], and Fang and Atassi [25]. The results of the numerical computations indicate that both the near field unsteady aerodynamic loading and the far field radiated sound are strongly affected by the upstream gust parameters and the nonuniformity of the mean flow around the body. The computation time required for these solutions is less than a minute per frequency on scientific workstations.

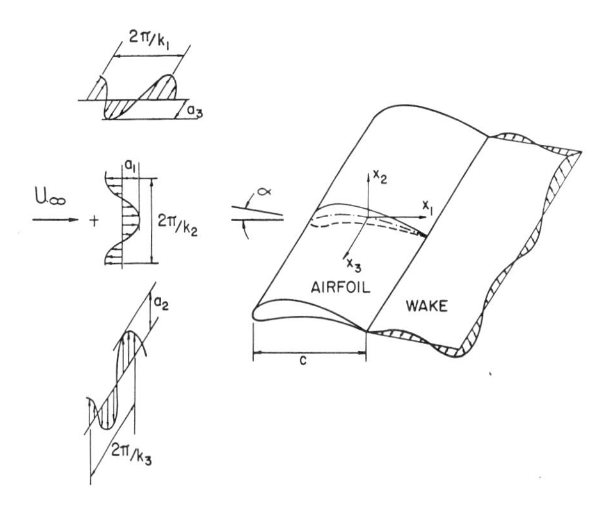

Figure 11. Lifting airoil in a three-dimensional gust.

To illustrate the effects of mean-flow loading and gust parameters, we consider the case of a lifting airfoil in a three-dimensional periodic gust (Figure 11). We examine the unsteady response function $\mathcal{R}$, Eq. (65), for two Joukowski airfoils, one with light mean loading and the other with high mean loading. Figures 12 and 13 show vector diagrams of the real and imaginary parts of $\mathcal{R}$, versus k_1 for the two airfoils in an oblique gust at various k_3 and $k_2 = k_1$. The significant difference between the results of the two figures underlines the considerable effect on the unsteady lift of the distortion of the incident gust by the airfoil mean flow. Figure 14 shows the unsteady lift for the same airfoil as in Figure 13 but for different gust upstream conditions. Comparison between Figures 13 and 14 points out the strong influence of the gust upstream parameters. These results indicate that the unsteady lift can be several times larger or smaller than that calculated for a reference flat plate depending on the gust upstream parameters, the airfoil geometry and mean flow angle of attack.

The radiated sound was *directly* calculated using the numerical solution of [23] by Patrick [50] to study the influence of airfoil shape and mean flow incidence on gust interaction noise. These calculations account for both the dipole effects resulting from the unsteady loading on the surface of the airfoil, and the quadrupole effects representing refraction and diffraction effects as well as the sound radiated from the surrounding unsteady flow. However, the numerical code [23] although accurate in the near field region, is not sufficiently accurate in

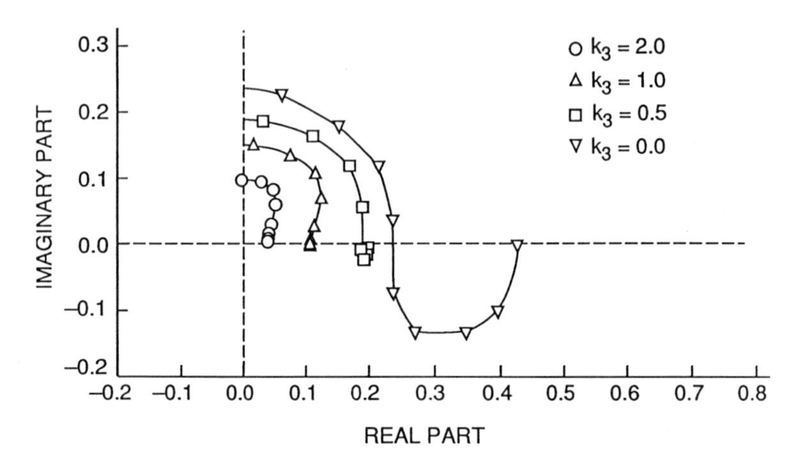

Figure 12. Vector diagram showing the real and imaginary parts of the response function $\mathcal{R}$ versus k_1 for an airfoil in an oblique gust ($k_2 = k_1$) at various k_3. The airfoil is at 5° incidence and has 5% camber and 5% thickness ratio. $M = 0.1$; $k_1 = 0, .05, .1, .25, .50, 1.0, 1.50, 2.0, 2.50$.

the far field. A Kirchhoff method is thus used to accurately obtain the sound pressure directivity. Figure 15(a) shows the pressure directivity, defined as the limit of $\{|p'|\sqrt{r}/(\rho_0 a_2 U_\infty)\}$ as $r \to \infty$, versus the polar angle for a symmetric airfoil, while Figure 15(b) represents only the dipole effects (r is the distance of the observation point to the airfoil). The difference between these two figures shows that quadrupole effects are significant and must be accounted for at values of K_1 larger than unity. The pattern of the total directivity changes significantly in the upstream direction for a two-dimensional gust ($k_2 \neq 0$) from that of a transverse gust. Figure 16 shows the effects of airfoil mean loading on the far-field acoustic directivity patterns. By comparing these directivity patterns with those resulting only from dipole distributions, one may conclude that the lobes seen in the figures are due to quadrupole effects and *not* to non-compact source effects.

An asymptotic analysis of Eq. (77) was also developed by Kerschen et al. [51] to study the far field of a thin airfoil in a gust in the high frequency limit. Their results indicate that in this limit, as for the case of a flat plate [40], the actual sound generation is mainly concentrated near the leading edge. The analysis also predicts changes of directivity similar to those observed in Figures 15(a) and 16(a).

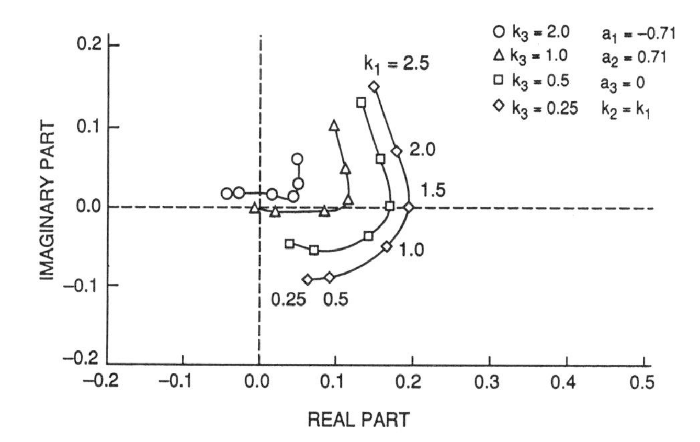

Figure 13. Vector diagram showing the real and imaginary parts of the response function $\mathcal{R}$ versus k_1 for a highly loaded airfoil in an oblique gust ($k_2 = k_1$) at various k_3. The airfoil is at 10° incidence and has 10% camber and 10% thickness ratio.

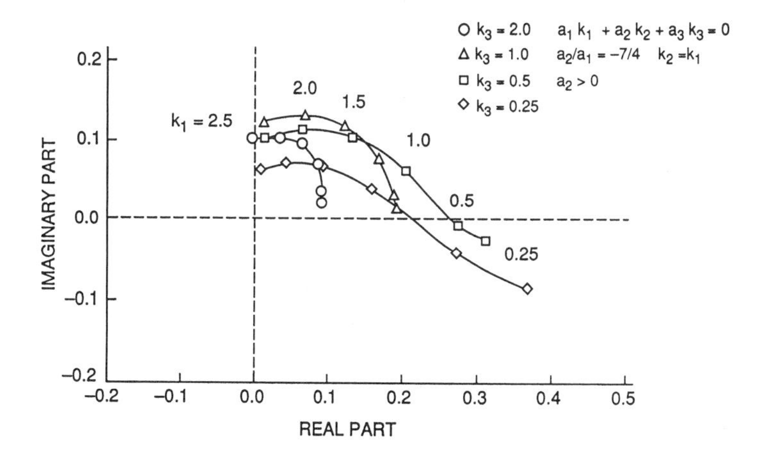

Figure 14. Vector diagram showing the real and imaginary parts of the response function $\mathcal{R}$ versus k_1 for the same airfoil as in Figure 13 but for different gust upstream conditions.

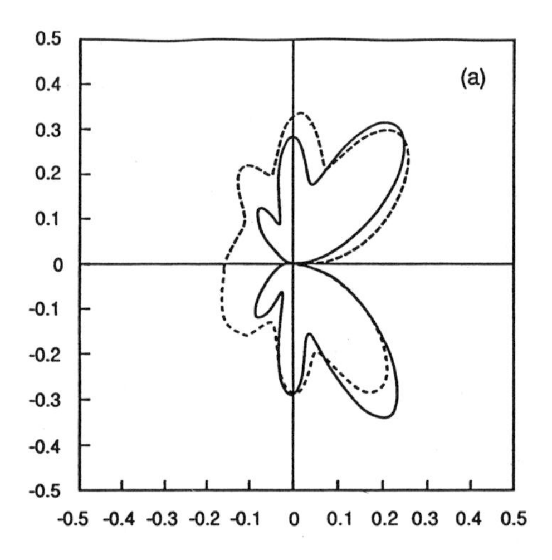

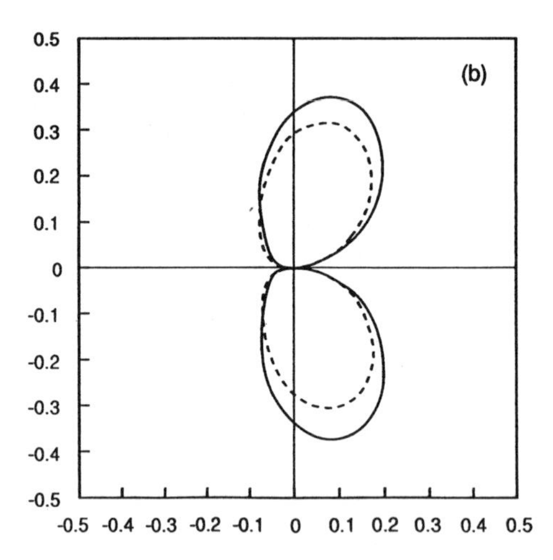

Figure 15. (a) Acoustic pressure directivity for a symmetric airfoil in a two-dimensional gust with .06 thickness ratio. $M = 0.7$, $k_1 = 3.0$. ——, $k_2 = 0$; - - -, $k_2 = 3.0$. (b) Dipole effects only.

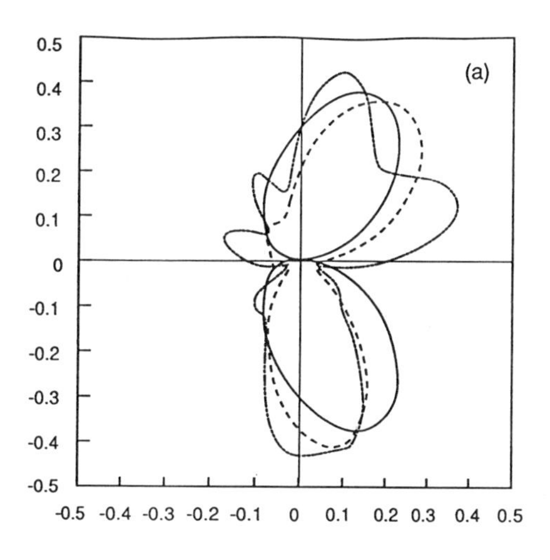

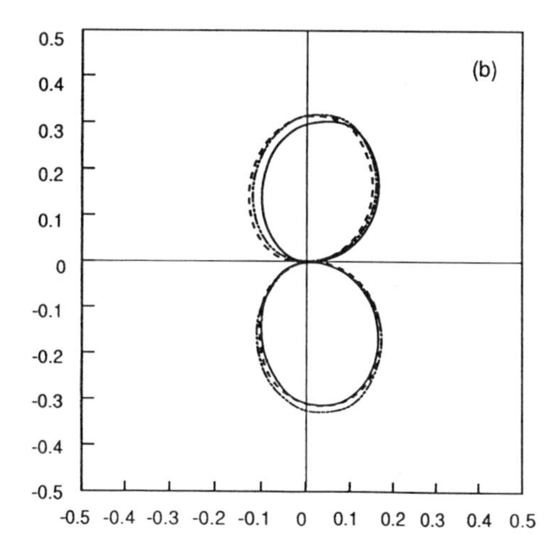

Figure 16. (a) Acoustic pressure directivity for a lifting airfoil in a transverse gust ($k_2 = 0$) with .12 thickness ratio. $M = 0.5$, $k_1 = 2.5$. ——, $m = 0$; - - -, $m = .02$; — — · · ——, $m = .04$. (b) Dipole effects only.

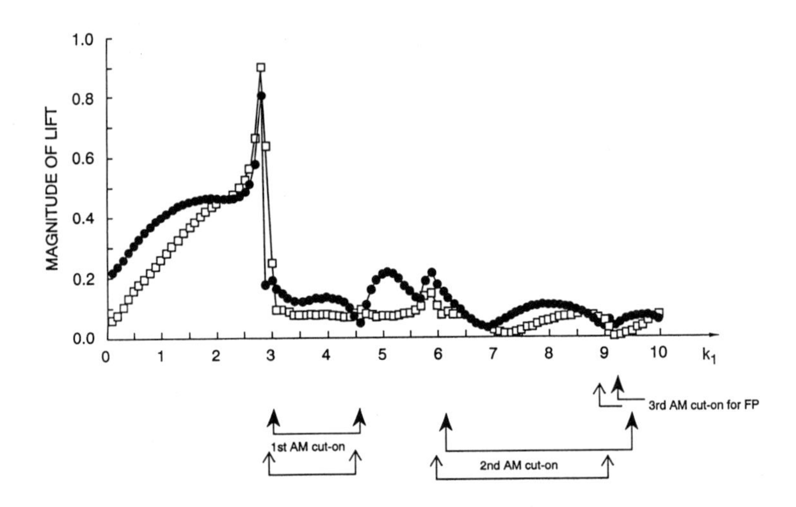

Figure 17. Magnitude of the response function $\mathcal{R}$ versus k_1 for an EGV cascade (squares) compared with that of a flat-plate cascade (circles). $M = 0.3$, $k_2 = k_1$.

Figure 17 shows the magnitude of the normalized unsteady lift, R, acting on a loaded cascade simulating an exit guide vane (EGV). This cascade consists of thick (12%), highly cambered (13%) modified NACA airfoils. It has a stagger angle of 15°, a blade spacing of 0.6 and operates at a prescribed inlet Mach number and inlet flow angle of 0.3 and 40°, respectively [24]. The unsteady lift for a flat-plate cascade with the same stagger and spacing is also shown for comparison. Note the large increase in the value of the lift as the first acoustic mode cuts on. The lift for the EGV cascade has smoother variations as acoustic modes cut-on and cut-off. Figure 18 shows the variation of the magnitude of the downstream propagating acoustic modes (normalized with repect to $\rho_0 a_2 U_\infty$) versus the frequency. The corresponding results for the flat-plate cascade are also plotted (solid) for comparison. Cascade loading appears to attenuate the variations of the magnitude of the acoustic modes near cut-on and cut-off conditions.

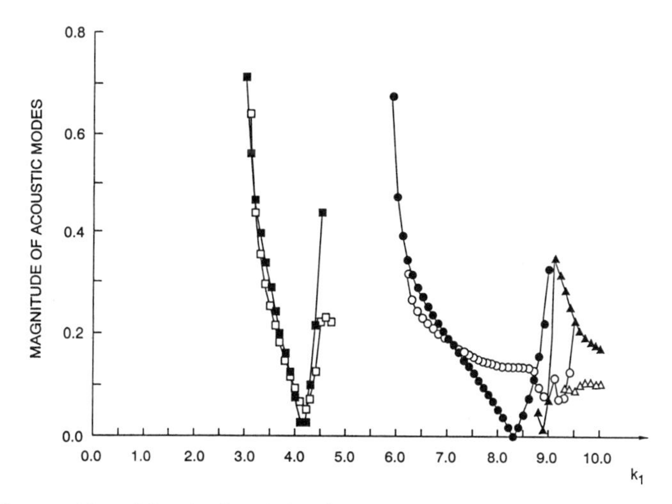

Figure 18. Magnitude of the downstream acoustic modes for an EGV cascade compared with those of a flat-plate cascade (solid). $M = 0.3$, $k_2 = k_1$.

4 UNSTEADY TRANSONIC FLOWS

We have so far only dealt with linearized treatments of unsteady flows around bodies with incident vortical disturbances. The underlying assumption is that they model, reasonably well, a class of real flows in which nonlinear effects are negligible. This is, for example, the case for attached subsonic flows. Transonic flows, on the other hand, are known for their mixed subsonic-supersonic flow patterns and may involve nonlinear processes such as shock wave-boundary layer interaction and strong coupling between steady and unsteady motions.

Extensive research has been conducted in recent years to develop computational methods for the simulation of unsteady nonlinear transonic flows. This effort has been primarily motivated by the need to accurately predict aeroelastic instabilities and flutter speeds of supercritical aircraft wings in transonic flight. As a result, these methods have been mainly developed for oscillating airfoils and other flutter related analyses. The numerical procedures are essentially based on schemes developed for steady flows in which a sequence of iterative solutions are marched forward in time until a convergent steady state solution is reached. Such methods provide a comprehensive modeling of the flow, but also require substantial

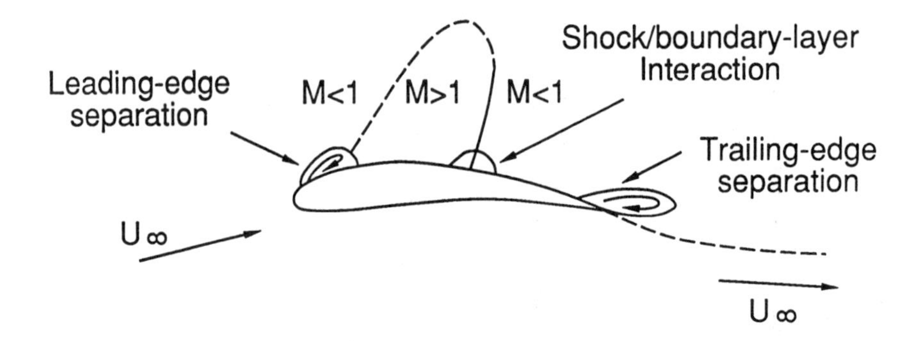

Figure 19. Airfoil in transonic flow.

computational resources. For a review of computational unsteady aerodynamics methods, we refer the reader to a recent review article by Verdon [57].

In aeroelastic and aeroacoustic applications, typical design analyses may require calculations of the unsteady aerodynamic near or far field pressure for more than 100 combinations of upstream Mach number, frequencies, gust parameters, or vibratory modes. These cannot be performed unless we develop highly efficient computational schemes. In what follows we examine the nonlinear character of transonic flows and the validity as well as the restriction of extending the mean flow distortion approach to this technologically important class of flows.

4.1 Nonlinear Character of Unsteady Transonic Flows

We first note that the only available data are for oscillating airfoils. We thus examine the essential features associated with these flows. Our review is essentially based on the experiments of Tijdeman and his group [52] in The Netherlands and the measurements of Davis and Malcolm [53] and [54] in the United States. More details are given in the review article by Tijdeman and Seebass [55].

Figure 19 shows a schematic of an airfoil in a transonic flow. The upstream flow is subsonic, but on the suction side of the airfoil there is a supersonic region which starts at the sonic line (dashed line) and is terminated by a shock wave. In addition to this mixed, subsonic-supersonic flow, there are, generally, three regions of viscous-inviscid interaction: a leading edge separation bubble, a trailing edge separation zone, and a shock–boundary layer interaction zone. This configuration suggests a strong interaction between the steady and unsteady flows. The nonlinear character of the flow mainly depends on the strength of the shock wave. It is, however, also reflected by the development of the supersonic zone which expands as the upstream Mach number or the angle of attack is increased. Experiments indicate that the shock-induced separation starts when the local Mach number is about 1.3. A strong shock may eventually lead to an extensive separated zone. Observations show that small variations in incidence may lead to considerable changes in the pressure distribution, shock position and shock strength, thus indicating a strong nonlinear behavior. Another consequence of this strong sensitivity of the flow to small changes is that the effects of viscosity can no longer be ignored; even for attached flows, the presence of the boundary-layer changes the effective contour of the airfoil.

We may then conclude that for a *strong* shock wave with an extensive separation region at its foot, viscous effects are significant and the unsteady flow is essentially *nonlinear*. For a *weak* shock wave, the flow remains essentially attached to the airfoil surface and the viscous-inviscid interaction affects mainly the shock region.

An important indication of the strength of nonlinear effects associated with a flow is to examine the time histories of the local pressure and the overall lift and moment for an airfoil performing a sinusoidal oscillation. For attached subsonic and supersonic flows, both the pressure and the overall loading show almost sinusoidal variation around their mean values, indicating a linear relationship between the displacement of the airfoil and the unsteady pressure. This implies that for such flows the problem of an oscillating airfoil can be decomposed into a steady problem and an unsteady problem. The equations of motion and the boundary conditions can thus be linearized about the mean flow of the airfoil.

For transonic flows, the unsteady pressure no longer exhibits this sinusoidal behavior, particularly near the shock region. However, Schippers [56] gave a Fourier analysis of his unsteady pressure measurements. This analysis shows the distribution along the airfoil surface of the amplitude of the higher harmonics of the unsteady pressure signals. The higher harmonics have small amplitude and do not noticeably contribute to the overall lift and only slightly to the moment. This is in spite of a strong nonlinear pressure signal at the foot of the shock.

The motion of the shock makes a significant contribution to the unsteady flow patterns and the overall unsteady loads. Optical flow studies by Tijdeman [52] have shown that three different types of shock motions can be distinguished depending on the upstream Mach number and the airfoil geometry: (a) a sinusoidal shock motion for a well developed shock, (b) an interrupted shock motion and (c) an upstream propagated shock. However, for a well developed shock, the shock motion takes place nearly sinusoidally with an amplitude proportional to that of the oscillatory motion of the airfoil.

We now examine the effects of frequency and amplitude of the oscillations. The linear relation between the unsteady loads and the amplitude of the airfoil motion strongly depends on the frequency. Observations indicate that the amplitudes of the shock motions are largest at low frequency. Thus we can expect that the nonlinear character of unsteady transonic flows will manifest itself mainly at low and moderate frequencies. Note that for aircraft wing applications, the reduced frequencies of interest are in the range of $k_1 < 0.5$, while for turbomachinery applications, they are mainly for $k_1 > 0.5$.

4.2 Comparison Between Linear and Nonlinear Euler Computations

The previous observations suggest that for unsteady transonic flows with *weak* shocks nonlinear viscous effects are not strong and that the equations of motion can be linearized about the mean flow of the airfoil. In order to assess further the validity of linearized calculations, we present a comparison between recent linear and nonlinear Euler calculations for a cascade geometry. These results were provided by Verdon, Huff and Ayer [58] for the *Tenth Standard Configuration* [59], initially proposed by Verdon. The linear calculations were obtained using LINFLO, a linearized unsteady aerodynamic code developed by Verdon and Hall [60]. The mean flow in LINFLO is assumed to be potential. The nonlinear computations were obtained from NPHASE, a fully nonlinear unsteady flow code developed by Huff et al. [61] . Thus, this comparison will also assess the effects of shock generated vorticity in the inviscid flow which are accounted for in NPHASE but neglected in LINFLO. The cascade blades are modified NACA 0006 airfoils with a circular-arc camber line. The height of the circular-arc camber line at the airfoil midchord is 0.05. The blade spacing is unity, and the cascade stagger angle is 45°. The reduced frequency is 0.5. The blades are undergoing an in-phase torsional vibration of 2° at midchord.

We first examine the steady flow. Figure 20(a) shows the Mach number distribution along the blade for an upstream Mach number, $M_\infty = 0.7$ and a steady flow angle of $\Omega_\infty = 55^o$ calculated using LINFLO and NPHASE. The flow is subsonic along the entire blade surface. Figure 20(b) shows the same blade at transonic speed with $M_\infty = 0.8$ and $\Omega_\infty = 58^o$. On both figures, the solid line corresponds to the potential flow calculations and the dashed line to the Euler solver results. For subsonic flow, the two codes give identical results. For transonic flows, the Euler solver gives a slightly smaller Mach number near the leading edge and the Euler shock is also slightly aft of the potential solver shock. These discrepancies may be attributed to the fact that the potential code neglects the vorticity behind the shock but also to the fact that in the leading edge area where the steady flow has strong gradients, numerical viscosity in the Euler code causes dissipation and produces vorticity; both of which tend to smooth the high velocity gradient in this area.

Figure 21 shows a comparison for the magnitude and phase of the first harmonic unsteady pressure difference distribution for the subsonic NACA 0006 cascade undergoing an in-phase torsional oscillation of amplitude $\alpha = 2^o$ at $k_1 = 0.5$ about the midchord. The upstream Mach number $M_\infty = 0.7$ and upstream flow angle $\Omega_\infty = 55^o$. The solid line represents the calculation from LINFLO , and the dashed line those of NPHASE. The two codes are in excellent agreement with each other.

Figure 22 shows a similar comparison to that of Figure 21 but for the unsteady pressure distribution in a transonic flow, with $M_\infty = 0.8$ and $\Omega_\infty = 58^o$. We see discrepancies mainly in the regions of strong interaction near the shock and the leading edge. However, since LINFLO uses a potential steady code, the difficulties near the leading edge may result from this code. Figure 23 shows the same comparison but for the unsteady pressure difference distribution. The large discrepancy near the leading edge is strongly diminished and the only major difference is near the shock location. The linear code shows an unsteady pressure discontinuity at the shock location because this location is fixed at the mean flow shock location in the linear analysis. The motion of the shock is accounted for by transferring the shock conditions to the mean location using the concept of analytical continuation and Reimann surfaces. In the nonlinear calculations, the shock is free to move and thus the results in Figures 22 and 23 show some time-average of the unsteady pressure first harmonic. We presume that the results of the linearized and nonlinear calculations could be much closer if the unsteady pressure were averaged in the linear calculations to take into account the

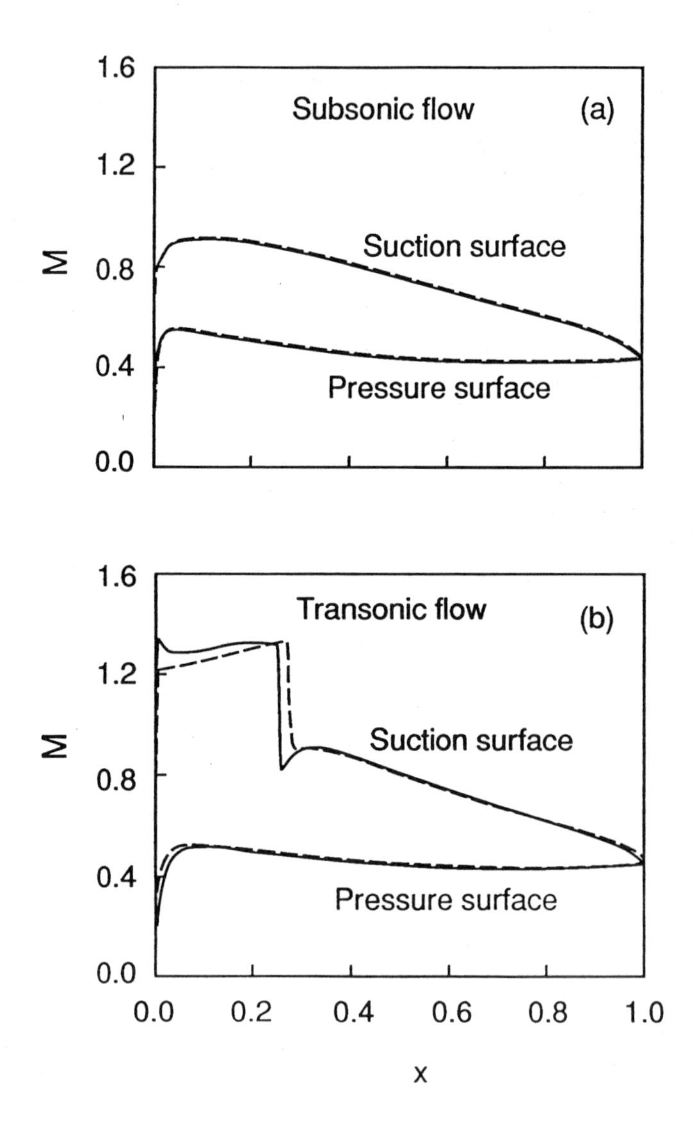

Figure 20. Surface Mach number for subsonic flow (a), and transonic flow (b). ——, potential code; - - - -, Euler code.

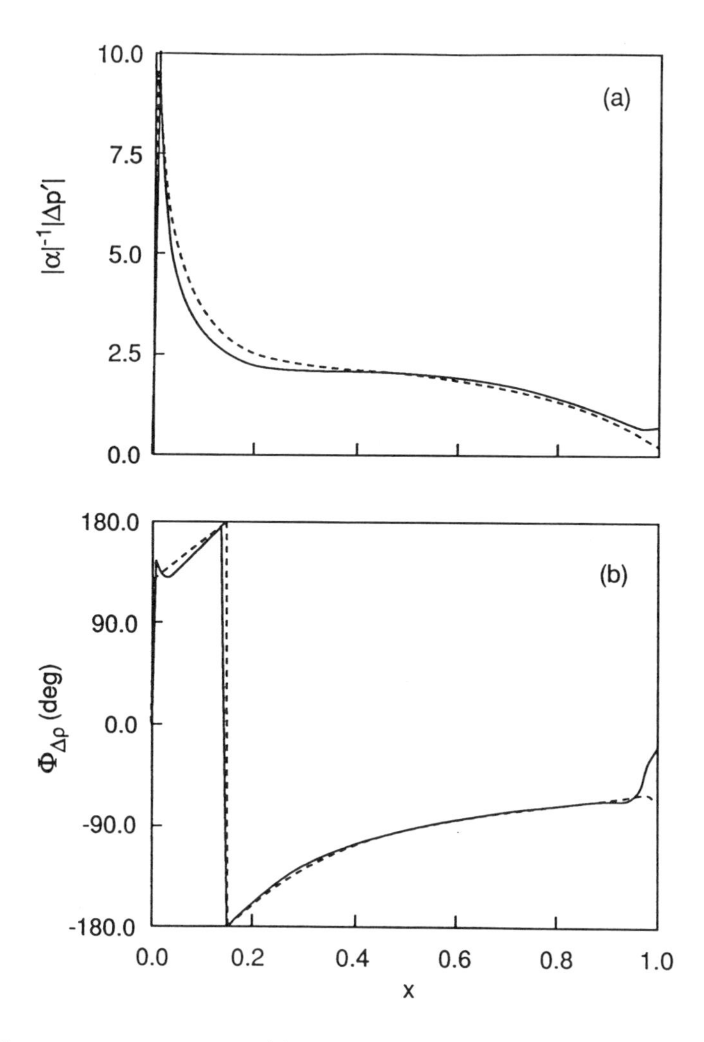

Figure 21. Magnitude (a) and phase (b) of the first harmonic unsteady pressure difference distribution for the subsonic NACA 0006 cascade undergoing an in-phase torsional oscillation of amplitude $\alpha = 2^o$ at $k_1 = 0.5$ about midchord; $M_\infty = 0.7$, $\Omega_\infty = 55^o$. ——, linearized analysis (LINFLO); - - - -, nonlinear analysis (NPHASE).

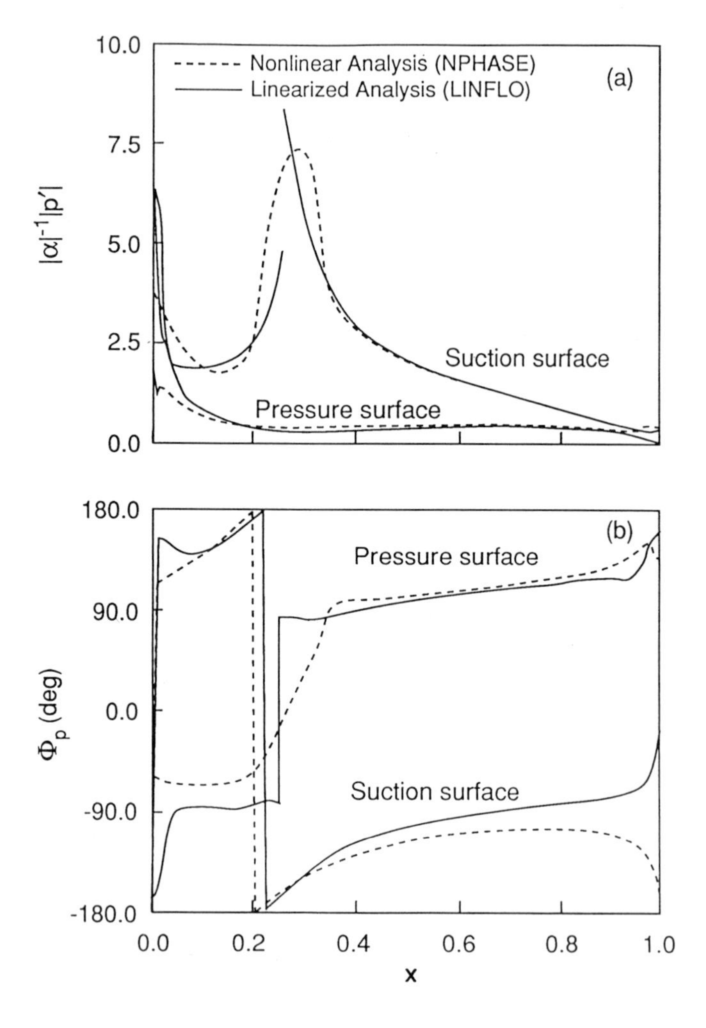

Figure 22. Magnitude (a) and phase (b) of the first harmonic unsteady pressure distribution for the transonic NACA 0006 cascade undergoing an in-phase torsional oscillation of amplitude $\alpha = 2^o$ at $k_1 = 0.5$ about midchord; $M_\infty = 0.8$, $\Omega_\infty = 58^o$.

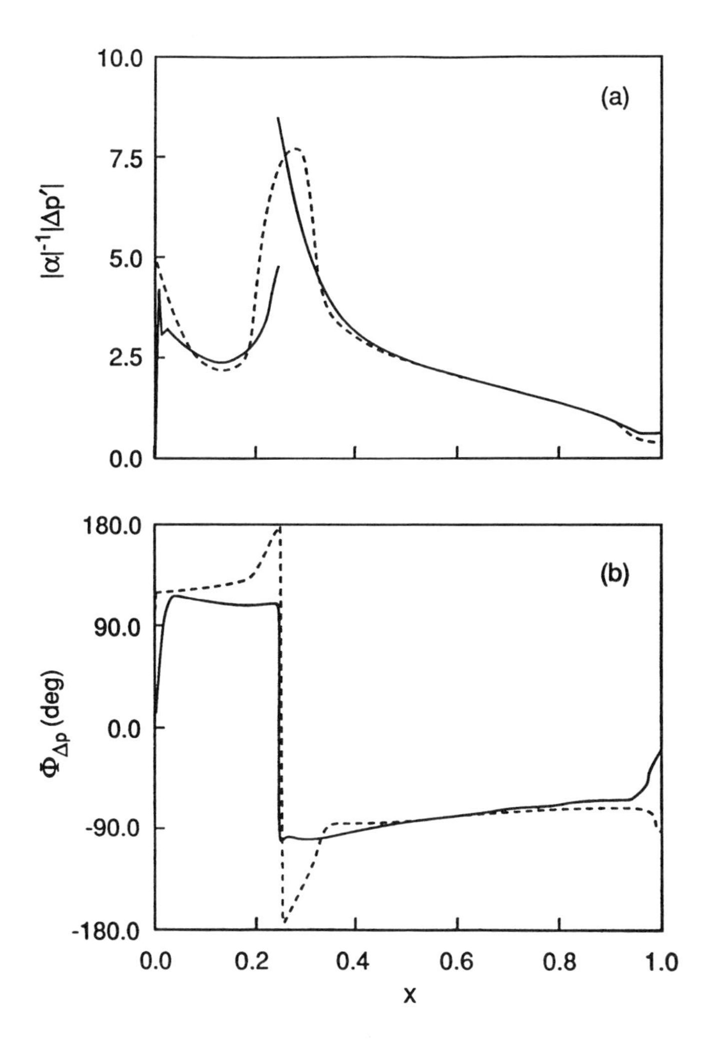

Figure 23. Magnitude (a) and phase (b) of the first harmonic unsteady pressure difference distribution for the transonic NACA 0006 cascade undergoing an in-phase torsional oscillation of amplitude $\alpha = 2^o$ at $k_1 = 0.5$ about midchord; $M_\infty = 0.8$, $\Omega_\infty = 58^o$. ——, linearized analysis (LINFLO); - - - - , nonlinear analysis (NPHASE).

excursion of the shock.

5 CONCLUSIONS

The linear unsteady aerodynamic theory accounts for the essential effects associated with the unsteady motion of a real fluid around a thin streamlined body, namely, the apparent mass and wake effects. The physical insight of von Karman and Sears brought to this theory a simple and elegant formulation which has since influenced the thinking of many generations of aerodynamicists. Today, as aerodynamicists continue to investigate the complex issues associated with unsteady flows, the legacy of von Karman and Sears remains as important as ever.

For lifting bodies, the mean flow distortion theory accounts for the distortion of the incident rotational disturbances by the nonuniform mean flow. This effect is found to have a significant influence on the unsteady aerodynamic force along the body surface and the sound radiated in the far field. Experiments and nonlinear computations suggest that this linear theory accurately models attached subsonic and supersonic flows, and that it also represents a good model for transonic flows with weak shock waves. Its numerical implementation for a two-dimensional geometry produces accurate and efficient codes with a computation time of about one minute on scientific workstations. As a result, this approach can be used for aeroacoustic and aeroelastic optimization analyses. It can also be extended to model, with moderate computation time, three-dimensional geometries for engineering applications.

Finally, transonic flows with strong shock–boundary layer interaction exhibit strong nonlinear effects and must be modeled using the Navier-Stokes equations. However, such modeling requires substantial computational resources. In addition, the accuracy of the nonlinear methods largely depends on the imposition of accurate inflow-outflow boundary conditions and the validity of the turbulence modeling. Both of these issues are not yet adequately resolved. Such methods should be used mainly as research tools and whenever nonlinear effects strongly impact the unsteady aerodynamics.

ACKNOWLEDGMENTS

The author would like to thank Mr. Dennis L. Huff and Dr. Joseph M. Verdon for providing the results of the comparison between linear and nonlinear computations. The author is also grateful to Dr. Marvin E. Goldstein for reading the manuscript and for his insightful comments. The work of the author cited in this article was supported by NASA Lewis Research Center, the Office of Naval Research and the Air Force Office of Scientific Research.

APPENDIX A

COLLEGE OF ENGINEERING
CORNELL UNIVERSITY
ITHACA, N. Y. 14850

SIBLEY SCHOOL OF MECHANICAL
AND AEROSPACE ENGINEERING
UPSON AND GRUMMAN HALLS

August 30, 1973

Dr. Hafiz Atassi
National Aeronautics and Space Administration
Lewis Research Center
Cleveland, Ohio 44135

Attention: 5453

Dear Hafiz:

Thank you for your interesting letter of 1 August.

For many years I have harbored the ambition to carry out the calculation that you and Dr. Goldstein are doing, namely to linearize about a steady flow around an airfoil instead of linearizing about a parallel flow. I am sure that it's correct in principle, and I think your use of the ϕ, ψ plane is a good way to carry it out. In principle you should be able to carry out all the cases that we did for the plane airfoil, such as some sort of sinusoidal gusts.

For comparison, you may be able to use the beautiful results of Giesing, who has done unsteady airfoil cases numerically, without any linearization at all.

The Horlock work always bothers me because I'm afraid we've already neglected terms of the same order ($\alpha\epsilon$, where α is incidence and ϵ is small perturbation) in setting up the whole theory. I guess, if it is useful, it should be considered a kind of empirical theory. As you imply, if $\alpha\epsilon$ is needed, he should really linearize about the flow around an airfoil at incidence α , as you do - but it's a lot harder to do!

With very best regards and love to Maria and Olivier,

Sincerely,

W. R. Sears

WRS:AA

REFERENCES

1 Wagner, H.: Uber Die Entstehung des Dynamischen Auftriebes von Tragflugeln, *Br. ARC., R.& M.* **1242**, 1925.

2 Glauert, H.: The force and Moment on an Oscillating Airfoil, *ZAMM* **5**(1), 17-35, 1929.

3 Theodorsen, Th.: General Theory of Aerodynamic Instability and the Mechanism of Flutter, *NACA Tech. Rep.* 496, 1935.

4 Kussner, H. G.: Zusammenfassender Bericht uber den Instationaren Auftrieb von Flugel, *Luftfahrtforschung* **13**, 410-424, 1936.

5 von Karman, Th. and Sears, W. R.: Airfoil Theory for Nonuniform Motion, *J. Aero. Sci.* **5** (10), 379-380, 1938.

6 Sears, W. R.: *A Systematic Presentation of the Theory of Thin Airfoils in Non-Uniform Motion*, Ph.D. Thesis, California Institute of Technology, 1938.

7 Sears, W. R.: Some Aspect of Non-stationary Airfoil Theory and its Practical Application, *J. Aero. Sci.* **8** (3), 104-108, 1941.

8 Possio, C.: L'Azione Aerodinamica sul Profilo Oscillante in un Fluido Compressible a Velocita Ipsonora, *L'Aerotecnica* t. XVIII, fasc. 4, 1938.

9 Reissner, E.: On the Application of Mathieu Functions in the Theory of Subsonic Compressible Flow Past Oscillating Airfoils, *N.A.C.A. TN* no. 2363, 1951.

10 Goldstein, M. E.: The Evolution of Tollmien-Schlichting Waves Near a Leading Edge, *J. Fluid Mech.* **127**, 59-81, 1983.

11 Goldstein, M. E., Leib, S. J. and Cowley, S .J.: Distortion of a Flat-Plate Boundary Layer by Free-Stream Vorticity Normal to the Plate, *J. Fluid Mech.* **237**, 231-260, 1992.

12 Lighthill, J. M.: Drift, *J. Fluid Mech.* **1**, 31-53, 1956.

13 Townsend, A. A.: *The Structure of Turbulent Shear Flow*, Cambridge University Press, 1976.

14 Prandtl, L.: Attaining a Steady Air Stream in Wind Tunnels, *NASA TM* 726, 1933.

15 Taylor, G. I.: Turbulence in a Contracting Stream, *ZAMM* **15**, 91-96, 1935.

16 Ribner, H. S. and Tucker, M.: Spectrum of Turbulence in a Contracting Stream, *NACA Rep.* 1113, 1953.

17 Batchelor, G. K. and Proudman, I.: The Effect of Rapid Distortion of a Fluid in a Turbulent Motion, *Quart. J. Mech. Appl. Math.* **1**, 83-103, 1954.

18 Hunt, J. C. R.: A theory of Turbulent Flow Round Two-dimensional Bluff Bodies, *J. Fluid Mech.* **61**, 625-706, 1973.

19 Goldstein, M. E. and Atassi, H. M.: A Complete Second-order Theory for the Unsteady Flow About an Airfoil Due to a Periodic Gust, *J. Fluid Mech.* **74** (4), 741-765, 1976.

20 Atassi, H. M.: The Sears Problem for a Lifting Airfoil Revisited – New Results, *J. Fluid Mech.* **141**, 109-122, 1984.

21 Goldstein, M. E.: Unsteady Vortical and Entropic Distortions of Potential Flows Round Arbitrary Obstacles, *J. Fluid Mech.* **89** (3), 433-468, 1978.

22 Atassi, H. M. and Grzedzinski, J.: Unsteady Vortical and Entropic Disturbances of Streaming Motions Around Bodies, *J. Fluid Mech.* **209**, 385-403, 1989.

23 Scott, J. R. and Atassi, H. M.: A finite Difference Frequency-domain Numerical Scheme for the Solution of the Linearized Euler Equations, *Computational Fluid Dynamic Symposium on Aeropropulsion*, *NASA CP* 3078, 55-104, 1991.

24 Hall, K. C. and Verdon, J. M.: Gust Response Analysis for a Cascade Operating in Nonuniform Mean Flow, *AIAA J.* **29** (9), 1463-1471, 1991.

25 Fang, J. and Atassi, H. M.: Compressible Flows with Vortical Disturbances Around a Cascade of Loaded Airfoils, *Unsteady Aerodynamics, Aeroacoustics and Aeroelasticity of Turbomachines and Propellers*, Editor, H. M. Atassi, Springer–Verlag, 149-176, 1993.

26 Hardin, J. C. and Hussaini, M. Y.: *Computational Aeroacoustics*, Springer–Verlag, 1993.

27 Durand, W. F.: *Aerodynamic Theory*, Vol. 2, Julius Springer, Berlin, 1935.

28 Giesing, J. P., Rodden, W. P. and Stahl, B.: Sears Function and Lifting Surface Theory for Harmonic Gusts, *J. Aircraft* **7**, 252-255, 1970.

29 Acum, W. E. A.: The Comparison of Theory and Experiment for Oscillating Wings, *AGARD Manual on Aeroelasticity*, Part II, Chap. 10, 1968.

30 Graham, J.M.R.: Lifting Surface Theory for the Problem of an Arbitrary Yawed Sinusoidal Gust Incident on a Thin Aerofoil in Incompressible Flow, *Aeron. Quart.* **21**, 182-198, 1970.

31 Whitehead, D. S.: Force and Moment Coefficients for Vibrating Airfoils in Cascade, *Br. ARC, RM* no. 3254, 1960.

32 Filotas, L. T.: Response of an Infinite Wing to an Oblique Sinusoidal Gust, Basic Aerodynamic Noise Research, NASA SP no. 207, 231-246, 1969.

33 Kovasznay, L.S.G.: Turbulence in Supersonic Flow, *J. Aero. Sci.* **20** (10), 657-674, 1953.

34 Goldstein, M. E.: *Aeroacoustics*, McGraw-Hill, 1976.

35 Graham, J.M.R.: Similarity Rules for Thin Aerofoils in Non–stationary Subsonic Flows, *J. Fluid Mech.* **43** (4), 753-766, 1970.

36 Adamczyk, J. J.: Passage of a Swept Airfoil Through an Oblique Gust, *J. Aircraft* **11** (5), 281-287, 1974.

37 Osborne, C.: Unsteady Thin-airfoil Theory for Subsonic Flow, *AIAA J.* **11** (2), 205-209, 1973.

38 Amiet, R. K.: Compressibility Effects in Unsteady Thin-Airfoil Theory, *AIAA J.* **12** (2), 253-255, 1974.

39 Amiet, R. K.: High Frequency Thin-airfoil Theory for Subsonic Flow, *AIAA J.* **14** (8), 1076-1082, 1976.

40 Martinez, R. and Widnall, S. E.: Unified Aerodynamic Theory for a Thin Rectangular Wing Encountering a Gust, *AIAA J.* **18** (6), 636-645, 1980.

41 Graham, J. M. R. and Kullar, I.: Small Perturbation Expansions in Unsteady Aerofoil Theory, *J. Fluid Mech.* **83** (2), 209-224, 1977.

42 Whitehead, D. S.: Vibration and Sound Generation in a Cascade of Flat Plates in Subsonic Flow, *Br. ARC, RM* no. 3685, 1972.

43 Fleeter, S.: Fluctuating Lift and Moment Coefficients for Cascaded Airfoils in Nonuniform Compressible Flow, *J. Aircraft* **10** (2), 93-98, 1973.

44 Namba, M.: Three Dimensional Flows, *Unsteady Turbomachinery Aerodynamics*, Editors, M. F. Platzer and F. O. Carta, *AGARD AG* no. 298, 1987.

45 Savill, A. M.: Recent Developments in Rapid Distortion Theory. *Ann. Rev. Fluid Mech.* **19**, 531-575, 1987.

46 Courant, R. and Friedrichs, K. O.: *Supersonic Flow and Shock Waves*, Interscience, 1948.

47 Miles, J. W.: *Potential Theory of Unsteady Supersonic Flow*, Cambridge University Press, 1959.

48 Landahl, M. T.: *Unsteady Transonic Flow*, Pergamon, 1961.

49 Horlock, J. H.: Fluctuating Lift Forces on Airfoils Moving Through Transverse and Chordwise Gusts, *J. Basic Engng, Trans. ASME* **90**, 494-500, 1968.

50 Patrick, S. M.: *The Acoustic Directivity from Airfoils in Nonuniform Subsonic Flows*, M.S. Thesis, University of Notre Dame, 1993.

51 Kerschen, E. J., Tsai, C. T. and Myers, M. R.: Influence of Airfoil Shape and Incidence Angle on High-frequency Gust Interaction Noise, *Unsteady Aerodynamics, Aeroacoustics and Aeroelasticity of Turbomachines and Propellers*, Editor, H. M. Atassi, Springer–Verlag, 765-782. 1993 .

52 Tijdeman, H.: Investigations of the Transonic Flow Around Oscillating Airfoils, *NLR* TR 77090 U, NLR, Amesterdam, The Netherlands, 1977.

53 Davis, S. S. and Malcolm, G. N.: Experiments in Unsteady Transonic Flow, *AIAA Paper* 79-769, *Proc. AIAA/ASME/ASCE/AHS 20th Structure, Structural Dynamics, and Materials Conference*, St. Louis, MO., 192-208, 1979.

54 Davis, S. S. and Malcolm, G. N.: Unsteady Aerodynamics of Conventional and Supercritical Airfoils, *AIAA Paper* 80-734, *Proc. AIAA/ASME/ASCE/AHS 21th Structure, Structural Dynamics, and Materials Conference*, St. Louis, MO., 417-433, 1980.

55 Tijdeman, H. and Seebass, R.: Transonic Flow Past Oscillating Airfoils, *Ann. Rev. Fluid Mech.* **12**, 181-222, 1980.

56 Schippers, P.: Results of Unsteady Pressure Measurements on the NLR 7301 Airfoil with Oscillating Control Surface, *NLR* TR 78124C, 1978.

57 Verdon, J. M.: Unsteady Aerodynamic Methods, *Unsteady Aerodynamics, Aeroacoustics and Aeroelasticity of Turbomachines and Propellers*, Editor, H. M. Atassi, Springer–Verlag, 3-42, 1993.

58 Verdon, J. M., Huff, D. L. and Ayer, T. C.: Personal Communication, 1993.

59 Atassi, H. M.: *Unsteady Aerodynamics, Aeroacoustics and Aeroelasticity of Turbomachines and Propellers*, Springer–Verlag, 1993.

60 Verdon, J. M. and Hall, K. C.: Development of a Linearized Unsteady Aerodynamic Analysis for Cascade Gust Response Predictions, *NASA Contractor Rep.* 4308, 1990.

61 Huff, D. L., Swafford, T. W. and Reddy, T. S. R.: Euler Flow Predictions for an Oscillating Cascade Using a High Resolution Wave-Split Scheme, *ASME Paper* GT-198, 1991.

Vortex Drift: A Historical Survey

Nicholas Rott

1 INTRODUCTION

I have the great privilege and pleasure to present a lecture at this Symposium honoring Bill Sears on his eightieth birthday. Actually I am doubly privileged as I already had an occasion to give a talk honoring Bill Sears, six years ago (1987) in Ithaca. The title of my talk at that occasion was simply "Vortices". This terse and pithy title was intended as a joke and it was, I am sure, also understood as such. However, it gave me a chance to touch upon a wide variety of topics. For instance, I could reminisce about times at the Graduate School of Aeronautical Engineering at Cornell University in the mid-fifties, when, under the guidance of Bill Sears, discussions about vortices was topic A. It would be tempting for me to repeat myself and to reminisce again recalling the fun and the excitement we have all shared at that time. It has led us, like many contemporaries, to learn more about free vortices - including, let us say, tethered vortices - and their applications in aeronautics. But instead of repeating myself I would like to continue with more on this subject, and consider an interesting property of free vortex regions: they can drift. I feel that it is justified to reconsider this problem going back to its historical roots.

Many excuses can be made for such an approach, but in the present case I would like to mention only two. First, I think that to talk on developments which span many epochs is well suited for a birthday symposium. Second, I hope to show that our understanding of vortex drift is greatly facilitated by following the historical developments.

The history of vortex theory begins in 1858, with the seminal treatise of Helmholtz [1]. The fundamental "three theorems" of Helmholtz, together with the Kelvin theorem of 1869, are explained in every textbook and taught in every fluid mechanics course so that I feel I do not have to say anything about them. But Helmholtz proceeded in his 1858 paper to the discussion of two examples, and here is where my survey begins.

Helmholtz established in his examples the foundations of vortex dynamics in two - and three dimensions. First he considered the flow generated by a line vortex of infinite length and uniform strength, and his second example was a circular axisymmetric line vortex. The difficulties associated with the two problems are vastly different. Helmholtz's treatment of the simple planar flow case is very brief. He established the basic law of motion for a free vortex which was later (in 1877) formalized by Kirchhoff [2] for **n** concentrated vortices moving in an unbounded plane of an ideal fluid. Kirchhoff gave a systematic treatment of the invariants of the motion including a streamfunction for the vortex paths, which is often called Kirchhoff's Hamiltonian (although Hamiltonians are not very often called streamfunctions). What was already noted by Helmholtz in 1858 was that the position of the centroid of the vortex system is fixed (in the system in which the fluid is at rest at infinity), unless, naturally, the total sum of the vortex strengths is zero. This is an important remark, as will be seen later.

The basic fact about the dynamics is that every particular vortex **k** of the **n** concentrated vortices moves with the velocity induced at its position by all other vortices, while the influence of the velocity field of the vortex **k** on itself is nil. It seems that most authors, beginning with Helmholtz, considered this statement to be self-evident - which is difficult to disprove - while others support it by arguments which I thought are lacking of rigor. This brought me almost into fights with some very esteemed colleagues. So now I have changed my tactics and I maintain that I know a proof which may or may not be better than any other proof but even if it is not needed, it illuminates the issues.

I heard this proof as a student, in a lecture given by Jakob Ackeret, when he was showing us the curious paths of the point vortices in plane ideal flow calculated by Gröbli [3] in 1877. Ackeret remarked that the direct explanation of these paths does *not* follow from a simple law of a mutual force acting between the vortices. The law of their motion says that they move in such a way that the force on every one of them is zero. They move so that they avoid the occurrence of a Joukowski force. They could sustain such a force only if there would be a cylindrical body in their center, with an arbitrary cross-section like a circle or an airfoil. Such a bound vortex can have any position and speed if we are able to sustain a force on a central body which may well be shrunk to a point. Thus the velocity of a free vortex follows from a force calculation using, for instance, the Blasius formula, and by setting the force equal to zero.

Ackeret's reference to Joukowski reminds us that the first consideration of bound vortices happened in the era of the first flight of the Wright brothers. It led to Prandtl's conceptual model of the airplane as the superposition of bound and free vortex elements. However, in airplane aerodynamics the in-depth consideration of free vortex motion became a secondary problem in comparison to the bound-vortex part. In plane flow, thin airfoil theory is an outstanding example. Fortunately for airplane aerodynamics it is much simpler than the theory of free vortex sheets, where in the delicately balanced fluid motion the forces vanish at any time and everywhere.

Back now to Helmholtz. He considers only the motion of free vortices, and his theorem that the centroid of such a system (*if* it exists for non-vanishing net vorticity) is at rest relative to the fluid at infinity can be viewed as a corollary to the Kutta-Joukowski theorem. He considers the absence of self-induction for concentrated vortices as self-evident. As we will see, Helmholtz often derives his results using a set of concentrated vortex elements as a model, and makes later the transition to a continuous distribution of vorticity. Here he asks the question whether for a fieldpoint, which is embedded in a continuous vortical region, the effect of the immediate neighbors remains sufficiently small for the existence of the integral over all elements. He reassures us that this is all right and makes a reference to a paper by Gauss from the year 1839. Today we accept the viewpoint of Helmholtz without asking any questions.

2 AXISYMMETRIC FLOW: KINEMATICS

Helmholtz proceeded to the second example for his theory of incompressible vortical flow, namely, the circular axisymmetric vortex line, with the resolve to use its streamfunction ψ as a basic solution in the same sense as he used the function log r in plane flow. He distributed vortex elements uniformly along a circle with the center at x' and a radius r' and determined the streamfunction ψ that gives the velocities u and v along a circle with the center at x and the radius r (see Fig. 1a). Let the vortex strength, defined as the surface integral of the vorticity over the infinitesimal cross- section of the vortex line at x', r' be called m (= half the circulation). Then Helmholtz's streamfunction is (using, however, essentially the definitions, notations and sign conventions of Lamb [4])

$$\psi = -\frac{m}{\pi}\sqrt{r\,r'}\;G(k) \tag{1}$$

where

$$k^2 = \frac{4\,r\,r'}{(r + r')^2 + (x - x')^2} \tag{2}$$

and G is an elliptic function with the modulus k, to be expressed by Legendre's F and E. The velocity components are, expressed by G and its logarithmic derivative $H = kdG/dk$:

$$vr = \frac{\partial \psi}{\partial x} = \frac{m}{\pi} \sqrt{r\,r'}\; H \frac{x - x'}{(r + r')^2 + (x - x')^2} \tag{3}$$

$$-\,ur = \frac{\partial \psi}{\partial r} = \frac{m}{2\pi} \sqrt{\frac{r'}{r}} \left[G + H \frac{r^2 - r'^2 + (x - x')^2}{(r + r')^2 + (x - x')^2} \right] \tag{4}$$

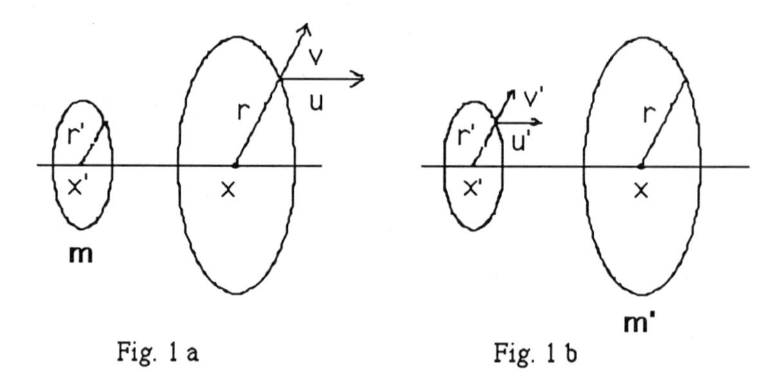

Figure 1. Notation used for the Helmholtz reciprocity relations.

Inspecting the symmetries of these expressions, we understand that Helmholtz is led to the consideration of the following combination:

$$ur^2 - vxr = \frac{\psi}{2} - \frac{m}{2\pi} \sqrt{r\,r'}\; H \frac{r^2 - r'^2 + x^2 - x'^2}{(r + r')^2 + (x - x')^2} \tag{5}$$

wherein the second term changes its sign when the unprimed and the primed coordinates are exchanged, just as in the expression (3) for vr. Unfortunately Helmholtz makes an algebraic mistake and states erroneously that (5) is a formula expressing $ur^2 - vxr/2$. In the following analysis, we ignore this mistake and reconsider it only to assess its effects on the final results.

Now the stage is set for the reciprocity relations of Helmholtz, which he obtained by exchanging the role of the primed and the unprimed coordinates. Let now a vortex of strength m' be situated along the circle x, r and produce velocities u', v' on the circle x', r' (Fig. 1b). Then the following relation follows from (3):

$$m' v r + m v' r' = 0 \tag{6}$$

We note that the perfect antisymmetry of the two terms that is expressed by (6) is obtained only after (3) has been multiplied by m'. (Analogous operations are also needed in the derivation of Kirchhoff's Hamiltonian, as well as in many other applications.) The same idea applied to (5) gives

$$m' (u r^2 - v x r) + m (u' r'^2 - v' x' r') = \frac{1}{2} (m' \psi + m\psi') \tag{7}$$

Now Helmholtz replaces m by use of the circumferential component σ of the vorticity vector ω (σ is chosen here to emphasize its non-cartesian character):

$$m' = \sigma \, dx \, dr, \qquad m = \sigma' \, dx' \, dr' \tag{8}$$

and integrates the expressions (6) and (7) over a meridional x-r plane. Then the distinction between primed and unprimed quantities disappears completely and the result that follows from (6) is

$$\iint v r \sigma \, dx \, dr = 0 \tag{9}$$

while (7) leads to

$$2 \iint (u r^2 - v x r) \sigma \, dx \, dr = \iint \psi \sigma \, dx \, dr \tag{10}$$

To assess the significance of these results, we consider the analogous formulas that were obtained later by Lamb [4] for an arbitrary velocity field **u** of an incompressible fluid (div **u** = 0) and the attendant vorticity vector ω = rot **u**, in a cartesian system x, y, z. He derives the formula

$$\int [\mathbf{u} \times \omega] \, dV = 0 \tag{11}$$

where dV = dx dy dz. This equation reduces to (9) when applied to axisymmetric flow. Lamb's more general result puts the purely kinematic character of this "basic" statement in full evidence. We can consider Lamb's method of derivation - essentially a series of partial integrations - as well known and will only quote his results.

Lamb also derives the general three-dimensional counterpart of the Helmholtz equation (10) and calls the set of relation that he obtains "useful". Rarely does Lamb use such a word from everyday language in his book [5], and I think we are justified to accept this word and use it when referring to the following results:

$$\frac{1}{2}\int \mathbf{u}^2\, dV = \int (\mathbf{u}\, [\mathbf{r} \times \omega])\, dV \tag{12}$$

The left-hand side here is the kinetic energy for which there also exists (as div $\mathbf{u} = 0$) a relation with the vector streamfunction **A**, defined by $\mathbf{u} = \text{rot}\, \mathbf{A}$:

$$\frac{1}{2}\int \mathbf{u}^2\, dV = \frac{1}{2}\int (\mathbf{A} \cdot \omega)\, dV \tag{13}$$

(This relation was already given by Helmholtz [1].) The resultant useful relation is obtained by eliminating the kinetic energy between (12) and (13):

$$\int (\mathbf{u}\, [\mathbf{r} \times \omega])\, dV = \frac{1}{2}\int (\mathbf{A} \cdot \omega)\, dV \tag{14}$$

When (14) is applied to axisymmetric flow by use of the scalar Stokes streamfunction, it gives the useful relation (10) of Helmholtz.

It is noted that the appearance of the kinetic energy in these formulas does not indicate that energy considerations are necessary or even helpful for our analysis. The formulas (12) and (13) express instantaneous kinematic relationships valid for any incompressible medium. Helmholtz shows that the kinetic energy is conserved in ideal fluids, but frequently his use of the relation (13) is only an expression of his preferences in notation. However, the appearance of the kinetic energy indicates that the whole analysis is valid only if it is carried out in a system in which the fluid at infinity is at rest .

3 THE DRIFT

We have seen that Lamb's approach was successful in obtaining results of greater generality so that we can ask ourselves whether it is worthwhile to follow the argumentations of Helmholtz any further. I hope to show that the answer is yes. To save the systematic appearance of our analysis we can permit ourselves a little anachronism and consider the derivation of the Helmholtz equations (9) and (10) as an ingenious shortcut to Lamb's results (11) and (14) for the special case of axisymmetric flow.

The important tool that is used by Helmholtz for the further development of his theory has already been introduced. He considers his integrals either as a sum to be extended over a series of thin individual vortex rings of the strength m, or as a continuous integral in space where m is replaced by σ dx dr. Thus his basic result (9) can also be written in the discrete form

$$\Sigma\, m_i\, v_i\, r_i \;=\; 0 \tag{15}$$

for a region that is broken up into many rings of sufficiently small masses m_i. Now Helmholtz applies to every ring his law, namely, that they are formed for any time by the same fluid particles, so that

$$v_i = \frac{dr_i}{dt} \tag{16}$$

and (15) can be integrated with respect to time to give

$$\Sigma\, m_i\, r_i^2 \;=\; \text{const} \tag{17}$$

Going back to the continuous representation we have

$$\begin{aligned} \iint r^2\, \sigma\, dx\, dr \;&=\; \text{const.} \\ &=\; 2\,I \end{aligned} \tag{18}$$

The interpretation of I as the flow impulse was given later by Lamb [4].

We are now ready for the consideration of the drift, which is defined as the velocity of a proper centroid. This means that we have to define both "proper" and "centroid". We all know that the centroid is a point that can be defined for the integrand of any definite integral, but only if the definite integral *does not vanish*. We are inclined to forget the importance of this restriction because masses in mechanics are always positive. This is *not* true in vortex dynamics, and a system with the total "mass" zero behaves in certain important aspects very differently from a system which does have a net "mass". Helmholtz gave here an early warning, which I have already quoted. He has shown that if there is a non-vanishing total vorticity, then there exists a space-fixed centroid. It does not move and thus there is no drift. If the total vorticity vanishes, then the simplest non-zero integral is the impulse. Then the proper - that is the simplest - centroid which can be defined is based on the impulse elements which form the integrand in (18). Such a point moves, and its velocity is the drift. We note that if the impulse also vanishes, then the drift becomes meaningless again.

Nevertheless, Helmholtz proceeded to define a flow constant that is apparently even simpler than (18) as it uses only "masses":

$$\Sigma\, m_i = \iint \sigma\, dx\, dr = \text{const.} = \mathcal{M} \tag{19}$$

and in our hierarchy of simplicity (19) seems to get ahead of the impulse integral (18). This is, however, a misconception that arises when we forget that both the total vorticity and the impulse integrals are vectors. The vanishing of the volume integral of the vorticity vector ω is expressed by

$$\int \omega\, dV = 0 \tag{20}$$

but this is automatically fulfilled for axisymmetric flow. The statement (20) is more general; it is a prerequisite for the validity of the general impulse formula given by Lamb [4]. A perfect analogy to this situation is encountered in the plane case if we restrict our attention to flows with a symmetry line. Then the vanishing of the total net circulation is automatically satisfied, but the centroid of the masses in a halfplane can be defined. This would be the analogue of a centroid based on $\mathcal{M}$ given by (19).

We can interpret the use of $\mathcal{M}$ by Helmholtz as a support for the "propriety" of the choice of the centroid. He defines an equivalent radius R (say) of a ring where all the mass can be concentrated, so that $\mathcal{M}\, R^2 = \Sigma\, m_i\, r_i^2$. Lamb actually follows him in this tradition, and a historical review has to deal with both flow invariants, (18) and (19). However, as R also remains a constant, only the x-component of the centroid motion matters. It can (and will) be calculated without ever using $\mathcal{M}$. - The reason for this lengthy discussion of an apparently minor point should become clear later.

To calculate the position of the centroid of the expression (18) for 2 I on the x-axis, we consider the moment S of its integrand perpendicular to the x axis, with respect to the origin of x:

$$S = \iint x\, r^2\, \sigma\, dx\, dr \tag{21}$$

Then the x-coordinate of the centroid, x_c, is

$$x_c = \frac{S}{2I} \tag{22}$$

and the drift is the velocity U of the point x_c:

$$2IU = \frac{dS}{dt}. \tag{23}$$

For fastest results, we follow Helmholtz and replace (21) by

$$S = \Sigma x_i r_i^2 m_i \tag{24}$$

then we differentiate the Lagrangian coordinates x_i, r_i with respect to time and revert the sums into their continuous form. The result of this operation, which can be called the Lagrangian differentiation of S, is the basic formula for the drift:

$$2IU = \iint u r^2 \sigma \, dx \, dr + 2 \iint v x r \sigma \, dx \, dr \tag{25}$$

a result to be confirmed presently by a different derivation. Actually every author, beginning with Helmholtz, glosses over this formula as apparently useless: it can not be applied unless the velocity u is given in the system in which the fluid at infinity is at rest. Replacing u by $u + u_o$ changes U to $U + u_o$, and this makes no sense. The conclusion is that we have to have $u(\infty) = 0$ in the system in which (25) is going to be applied; this is part of the result (25). In many applications, however, this system is not known a priori .

Actually Helmholtz and every author after him is ready to cope with this difficulty, with the help of what we have called the useful relation. In the formulation (10) for axisymmetric flow, we see that the useful relation connects a combination of the two integrals in (25) with an integral over the streamfunction. Thus we can eliminate the integral over u, which causes the difficulties. The result is

$$4IU = \iint \psi \sigma \, dx \, dr + 6 \iint v x r \sigma \, dx \, dr \tag{26}$$

This second form of the final result is naturally subject to the same restriction on the system in which it can be applied as (25). The condition is now that the streamfunction ψ has to give zero velocity at infinity. However, if we determine the vector streamfunction by integrals over the vorticity vector like Helmholtz did, then we are assured that the velocity at infinity is zero [as the condition (20) is fulfilled]. Besides, whenever the streamfunction is determined first, (26) has very significant practical advantages. - We note that had we not corrected the algebraic error of Helmholtz, we would have obtained a 5 instead of a 6 as the coefficient of the last integral in (26).

In the application of (26) to the thin vortex ring, Helmholtz uses the following approximation to the streamfunction (1):

$$\psi = \frac{m}{\pi} r \log (8r/s) \tag{27}$$

where

$$s^2 = (r - r')^2 + (x - x')^2 \tag{28}$$

At this degree of approximation, the last term in (26) is not needed. Helmholtz uses the approximation $I = (1/2)\mathcal{M} R^2$, where R is the mean radius of the ring [see (18) and (19)], and estimates the order of magnitude of the r.h.s. integral in (26) by replacing ψ (r, s) by the constant value ψ (R, a), where a is the radius of the ring cross-section. This would lead to the following formula (not written out by Helmholtz):

$$U = \frac{m}{2\pi R} [\log (8R/a) + c] \tag{29}$$

where c is a constant of order 1. To calculate c, an O(1) correction is needed to the term log (8r/s) in the streamfunction (27), and the algebraic error in (26) has to be corrected. The subsequent analysis is far from simple. The first formula for the drift that contains the correct value of c, for a thin ring with uniform vorticity σ, was given by Lord Kelvin, in a letter appended to the translation of Helmholtz's paper (by P. G. Tait [6], published in 1867). He found $c = -1/4$. Unfortunately for the historians, Kelvin stated his result without proof. The first correct answer with published proof was given by Hicks [7] in 1885, who used for his theory a system of toroidal coordinates. It is our intention to consider in this historical survey only work that is relevant to the theory of the drift in its full generality; applications to the vortex ring are only secondary for our purpose. Thus we do not discuss the work of Hicks any further and will omit (with apologies) any references that do not fulfill our criterion. The approach to the drift problem that was initiated in the paper of Helmholtz was continued and significantly extended by Lamb [4]. His proof of Kelvin's formula is the first that makes full use of the classical basic relations that we have collected at the beginning of this survey. Following the same ideas, Saffman [8] determined in 1970 the constant c for an arbitrary vorticity distribution $\sigma(s)$ within a thin ring. (The restriction to very thin rings can be relaxed; such work was initiated by Fraenkel [9].) These are apparently the final conclusion that can be drawn from the classical approach. In his 1970 paper, however, Saffman also started a new era of modern developments.

4 MODERN DEVELOPMENTS.

The new era was signaled by Saffman's discovery [8] that drift theory can be generalized in such a way that it becomes valid for viscous flow. I have presented here the derivation of the classical results avoiding intentionally the use of statements that are restricted in their validity to inviscid flow. In viscous flow, the kinetic energy is not constant. Neither does the total vortex "mass" contained in all vortex ring elements remain constant. There is a dissipation of vorticity that can be expressed as an integral along the axis of symmetry, just as in plane symmetrical flow the dissipation in a halfplane can be expressed as the diffusion across the symmetry line. In both cases, the total vector sum of the vorticity still remains constant, namely, equal to zero. The only non-zero flow constant considered thus far that retains its significance in viscous flow unchanged is the impulse, and the definition of the drift has been based on its centroid.

What becomes amply clear in hindsight is that the basic equations of the theory give only an instantaneous value of the drift velocity. This is fundamentally true also for inviscid flow, and a connection between impulse and drift that holds for more than an instant can be found only if the configuration is steady. This is the case for an inviscid vortex ring. A simple example where no universal connection between impulse and drift exists in inviscid flow is the fully integrable case treated by Gröbli [3] of two point vortices and their mirror images in a plane with a symmetry line. The impulse is constant but the drift velocity varies with the configuration. (Helmholtz [1] already gave a discussion of the leapfrogging of two coaxial vortex rings.)

Actually the notion of steadiness can be extended to include self-similar flows, which are, according to Cantwell, Coles and Dimotakis [10], steady for an observer equipped with zoom-optics. The basic example of the asymptotic drift in viscous slow flow falls into this category; it was treated by Cantwell and Rott [11] in a paper that appeared in 1988.

Saffman proved his discovery by verification in the equations of motion, and we are going to do essentially the same. Up to the results (25) and (26) the only arguments restricted in their validity to inviscid flow are the Helmholtz rules of Lagrangian differentiation and integration. However, we can reconsider the proofs using Eulerian differentiation, and (as there are no Lagrangian Navier-Stokes equations) at this occasion we can introduce the viscous term into the Helmholtz vorticity equation, a step never envisioned by Helmholtz in 1858. The only equations that need a new proof are numbered (18) and (25), while the strictly kinematic equations (9) and (10) remain valid. To prove (18), i.e., the constancy of the impulse, we now have to show that

$$\iint (\partial\sigma/\partial t)\, r^2\, dx\, dr = 0 \tag{30}$$

and (25) is proven if we can show that dS/dt obtained by the Eulerian differentiation of (21) is the same as the value obtained by Helmholtz:

$$\iint (\partial\sigma/\partial t)\, x\, r^2\, dx\, dr = \iint u\, r^2\, \sigma\, dx\, dr + 2\iint v\, x\, r\, \sigma\, dx\, dr \tag{31}$$

To make a long story short, we consider the equations (30) and (31) to be proven for inviscid flow and consider only the effect of the viscous terms in the equation for $\partial\sigma/\partial t$:

$$(\partial\sigma/\partial t)_{visc} - (\partial\sigma/\partial t)_{invisc} = \nu \left[\frac{\partial^2\sigma}{\partial x^2} + \frac{\partial^2\sigma}{\partial r^2} + \frac{\partial(\sigma/r)}{\partial r}\right] \tag{32}$$

Partial integrations show that both the operations $\iint ... r^2\, dx\, dr$ and $\iint ... x\, r^2\, dx\, dr$, when applied to the r.h.s. of (32), give zero. This proves the validity of all results obtained thus far in viscous flow. In particular, both (25) and (26) give the instantaneous drift velocity U by integrations over the velocity- and vorticity fields, at any given time t.

As the first application of his new philosophy of the drift, Saffman reconsidered the thin vortex ring and found that formally he only had to replace the ring radius a with the "viscous radius" $\sqrt{(4\nu t)}$ that is reached after the creation of the concentrated vortex at the time t = 0. (This result was already anticipated in a 1967 paper by Tung and Ting [12].)

Saffman also calculated a new value of c, which retained its old order of magnitude. The formal change is small, but the unsteady viscous spreading of the concentrated vorticity completely obliterates the classical effects in inviscid flow, where a steady solutions exists with a separating streamline between the vortical and the potential regions.

Next, the viscous problem was attacked in a numerical study carried out as thesis work at the Department of Aeronautics and Astronautics of Stanford University, in cooperation with the NASA Ames Research Center; it appeared as a NASA Technical Memorandum by Stanaway, Cantwell and Spalart [13] in 1988. The instantaneous drift was calculated at every time step by use of the formula (25); it was derived as the speed of the centroid, by Eulerian differentiation. The start of the scheme in a fluid at rest at the time $t = 0$ and the adherence to this system ensured that the relation (25) could always be used. It is clear in hindsight that whenever the "useful relation" is *not* needed, then the derivation of drift theory requires only very few lines.

As already noted, a paper on asymptotic drift by Cantwell and Rott [10] also appeared in 1988, and a reassuring check of their theory was found in the perfect agreement with the numerical work of Stanaway et al. In the Cantwell-Rott paper certain new dynamic interpretations of the drift were first attempted; the most important of them turned out to be that the drift is the velocity of the farfield dipole potential which represents the flow far from a finite region of vorticity if surrounded by a fluid at rest, both for inviscid and for viscous flow. The paper on the expansion of this idea is now being processed.

In closing I would like to note that drift theory as presented thus far has no relevance to flow problems where there *is* a force acting on the flow, so that vortical flows in wakes (as exemplified by the Kármán vortex street) are subject to completely different laws. Furthermore, the derivation of the results presented make it amply clear how important the assumption is that the fluid is at rest at infinity, and that potential flow prevails at sufficiently large distances from the vortical flow region. Also, we should not forget that many predictions of drift theory are poorly confirmed by experiments, as was found by Maxworthy [14]. I think the lesson is: let us look carefully at infinity, it might be closer than we think.

REFERENCES

1 H. Helmholtz: Ueber Integrale der hydrodynamischen Gleichungen, welche den Wirbelbewegungen entsprechen, *J. reine angew. Math.* (Crelle's) **55**, 25, 1858.

2 G. Kirchhoff: *Vorlesungen über Mathematische Physik.*, Teubner, Leipzig, 1877.

3 W. Gröbli: Specelle Probleme über die Bewegung geradliniger paralleler Wirbelfäden, *Vierteljahrsschrift der Naturforsch. Ges.* Zürich **22**, 37 & 129, 1877. See also H. Aref, N. Rott and H. Thomann: Gröbli's Solution of the Three-Vortex Problem, *Annual Rev. Fluid Mech.* **24,** 1, 1992.

4 H. Lamb: *Hydrodynamics.*, 6th ed., Dover, New York, 1932.

5 See Ref 4, p. 218. Lamb had this result (and made the same comments) already in the first edition of his book, then entitled *A Treatise on the Mathematical Theory of the Motion of Fluids*, Cambridge Univ. Press, 1879.

6 H. Helmholtz: On Integrals of the Hydrodynamical Equations which Express Vortex Motion, translated by P.G. Tait; letter by Sir W. Thomson appended, *Phil Mag.* (4) **33**, 485, 1867.

7 W. M. Hicks: On the Steady Motion and the Small Vibrations of a Hollow Vortex, *Phil. Trans. A* **175**, 183, 1885.

8 P. G. Saffman: The Velocity of Viscous Vortex Rings, *Studies in Appl. Mech.* **49**, 371, 1970.

9 L. E. Fraenkel: Examples of Steady Vortex Rings of Small Cross-Section in an Ideal Fluid, *J. Fluid Mech.* **51**, 119, 1972.

10 B. Cantwell, D. Cole & P. Dimotakis: Structure and Entrainment in the Plane of Symmetry of a Turbulent Spot, *J. Fluid Mech.* **87**, 641, 1978.

11 B. Cantwell and N. Rott: The Decay of a Viscous Vortex Pair, *Phys. Fluids* **31**, 3213, 1988.

12 C. Tung and L. Ting: Motion and Decay of a Vortex Ring, *Phys. Fluids* **10,** 901, 1976.

13 S. K. Stanaway, B. J. Cantwell & P. R. Spalart: A Numerical Study of Viscous Vortex Rings using a Spectral Method, NASA TM 101041, 1988.

14 T. Maxworthy: Some Experimental Studies of Vortex Rings, *J. Fluid Mech.* **81**, 465, 1977.

Determining Unsteady 2D and 3D Boundary Layer Separation

Leon L. Van Dommelen
Szu-Chuan Wang

ABSTRACT

This paper explains the differences between steady and unsteady separation processes. It proposes several techniques to diagnose unsteady separation, from the quasi-steady case to fully unsteady. Procedures range from simple applications of the MRS conditions to complete Lagrangian solutions of the 3D Lagrangian boundary layer equations. They should be of interest for experimental, computational, and theoretical fluid dynamicists.

1 INTRODUCTION

In recent years, much has been learned about how boundary layers that are initially attached to a solid surface detach from it. It demonstrated dramatic differences between the steady and unsteady processes with which the boundary layer vorticity separates from the wall. For the steady, two-dimensional, laminar, fixed wall case, Prandtl [1] already pointed out that separation is indicated by a point of zero wall shear. At that point the wall streamline bifurcates, creating a free streamline that moves significantly away from the wall and encloses a downstream region of recirculating flow.

Consequently, an abundance of simple symptoms establishes that steady two-dimensional separation occurs: bifurcation of the streamlines, zero wall shear, reversed flow, and recirculation. For many it is hard to accept that in the unsteady case there would be NO such simple, unambiguous, signs of separation. However, the combined numerical and analytical evidence is quite overwhelming that there is indeed none. Unsteady separation must be diagnosed by much more careful considerations.

The next sections address some methods that work or do not work in determining whether unsteady separation occurs. First the criteria for steady separation are discussed in more detail, since it is often tempting to describe unsteady separation as quasi-steady. Caution is needed when doing so. Subsequent sections examine what can be done if the separation is truly unsteady.

2 QUASI-STEADY SEPARATION

Even for steady two-dimensional laminar separation, the seemingly intuitive bifurcation of the wall streamline is only a small part of the total picture. As first shown by Goldstein [2], it is quite possible for the wall streamline to depart from the surface without a dramatic increase in the distance of the boundary layer vorticity from the surface. The difficulty is that the angle between the bifurcated streamline and the surface would be very small; the distance between the detached streamline and the wall would not be able to dominate the typical thickness of the boundary layer itself. When Goldstein's theory applies, it can imply a dramatic motion of the boundary layer away from the wall, but only if the streamwise scales become rapidly shorter.

The theoretical description of steady, laminar, two-dimensional separation, the Sychev-Smith [3,4] theory, illustrates even better the limitations of the point of zero wall shear as an indication of separation. According to the theory, the boundary layer fluid is deflected away from the wall by a small region, called 'lower deck'. In the lower deck, the wall shear assumes all values, from values indicative of typical attached boundary layers in the first part to reversed values of similar magnitude downstream. While there is one point of zero shear, where the wall streamline bifurcates, there is no larger increase in boundary layer penetration at this point than at the other points in the region [5].

So while it is not strictly incorrect to say that steady laminar separation occurs at zero shear/bifurcating wall streamline/flow reversal/recirculation, it is much like saying that France is where the Eiffel tower is. The actual separation occurs in a small region, not at a point, and the typical values of the shear are neither zero nor small.

A difficulty that the Sychev-Smith model faces is that it is difficult to verify: experimentally these flows turn unsteady above a certain Reynolds number, and numerical solutions to the steady Navier-Stokes equations also have limitations with respect to the Reynolds number. Yet no true progress is likely for turbulent flows, for which the governing equations are uncertain, if the simpler laminar flow solution is not understood first. There at least the equations to be satisfied are known. The difficulty may be less pronounced in the unsteady transient case, in which small disturbances have only a limited time available to grow.

Besides this point, there are no known objections to any of the usual steady separation criteria for, say, steady laminar separation from a circular cylinder. This flow should have symmetrically positioned points of zero wall shear at which the wall streamline bifurcates, and near which the boundary layer vorticity clearly detaches from the wall.

Yet the case of quasi-steady flow is not that clear-cut. For example, unsteadiness m
be introduced by giving the cylinder an extremely small rotational motion. Even simplier, w
might give a constant small rotational motion, to create a steady separation, but with a slight motion between the location of separation and the wall. In fact, steady separating flows with moving walls have often been used as simple examples of unsteady separation processes: when viewed in a coordinate system moving with the local wall velocity, it becomes an unsteady separation from a fixed wall. This simple idea goes back to at least the work of Moore [6].

Clearly, if the solution to the steady Navier-Stokes equations is properly posed, a sufficiently small rotational speed should not measurably change the flow. Yet, as Van Dommelen [7] points out, the position of the streamline bifurcation points is not well behaved: one bifurcation point jumps completely out of the separation region, the other to a different position in the separation zone. Since the physical flow does not change for small rotational velocity, we must conclude that the streamline bifurcation points are no longer directly related to separation.

In a system in which the wall is fixed, separation would of course be indicated by bifurcation of the wall streamline. Yet that is hardly a satisfactory criterion, since the wall might not be rigid, in which case each location would have its own coordinate system. The criterion of zero shear is the least ambiguous, if the deviations from steady separation are small. Notice that it is the criterion that makes least sense; zero wall shear is but one point in a separation zone with varying shear. A more sensible criterion would be to take the location of most rapid variation of the wall shear: at least this criterion describes how fast particles near the wall move away from it. Zero wall shear does not.

Some authors use shape parameters to study separation. However, it should be pointed that the unsteady separation structure makes predictions about such parameters. Theoretically the boundary layer thickness at which a fraction f of the external flow velocity is achieved is:

$$\delta_f \sim \mathrm{Re}^{-1/2} / \left(t_s - t\right)^{1/4}$$

The displacement and momentum thicknesses would be, using steady definitions,

$$\delta^* \approx -\left(1+\frac{|u_s|}{U}\right)\delta_f$$

$$\theta \approx -\left(1+\frac{|u_s|}{U}\right)\frac{|u_s|}{U}\delta_f$$

in which $|u_s|$ is the upstream propagation velocity of the separation structure. Note that the predicted momentum thickness is negative. Such expressions might offer some guidance; however, it should not be expected that they have any real accuracy at practical Reynolds numbers. The difficulty is that they ignore a diffusive thickness of the boundary layer that is likely to be considerable. Even for steady flow, the generality of shape parameters is uncertain: the Sychev-Smith theory does not restrict the velocity profile at separation, although it predicts a rapid growth in the ratio of displacement thickness to momentum thickness downstream of it.

3 UNSTEADY SEPARATION REDUCIBLE TO STEADY FLOW

As mentioned in the previous section, steady separation from a moving wall provides a simple non-trivial example of unsteady separation. Any steady separation viewed in a moving system becomes an unsteady separation, but steady separation from a moving wall is needed to allow the position of separation to propagate along the wall [6].

For the downstream moving wall case, an equivalent to the Sychev-Smith theory was developed by Sychev [8], Van Dommelen and Shen [9] and Elliott, Cowley and Smith [10]. The quasi-steady criteria of the previous section do not apply: the shear at separation is reversed and not zero, and there is no bifurcation of the wall streamline. Streamlines in the system in which the separation is steady are believed to bifurcate near separation, but slightly downstream of the region where the actual separation occurs. The actual separation process does not take the form of a streamline bifurcation, but of a rapid increase in streamline spacing.

For the alternate case that the wall moves upstream, the various theories [6,10,11] show considerable disagreement. For example, Van Dommelen [7] suggests that in this case, there is no bifurcation of the streamlines near separation at all.

4 UNSTEADY SEPARATION

From the previous sections, it should be evident that none of the steady criteria is likely to have any relevance for truly unsteady separation. Clearly, the streamlines have very little meaning in any case. The boundary layer vorticity propagates along the particle paths, not along the streamlines. For unsteady separation, zero wall shear has no relevance, nor do seemingly interesting streamline bifurcations, nor does reverse flow, nor does recirculation.

All this was pointed out most clearly by Moore [6], Rott [12], and Sears [13] back in the 1950s. The reason streamlines are still widely presented in unsteady separating flows is that existing numerical and experimental methods produce them, not because they are physically meaningful. For many authors, it is still hard to accept that these interesting, intuitive streamline topologies are not in any way related to the important physical processes occurring in unsteady separation.

A sketch of the streamlines during the development of unsteady separation is given in Figure 1.

Figure 1. Schematic of the asymptotic structure of unsteady separation at very high Reynolds numbers.

In this sketch, the Reynolds number has been assumed very high so that the features stand out clearly. That unsteady separation at high Reynolds numbers takes this form was established numerically for several flows, but especially for the example of a circular cylinder that is impulsively set into motion in a direction normal to its axis. These results were obtained by a multitude of authors using the boundary layer equations and a variety of numerical methods [14-20]; these different methods agree to high accuracy. It also can be shown that the boundary layer equations do remain valid sufficiently long to describe the initial separation process as sketched. Because of the overwhelming numerical evidence, and because the numerically observed flow development can fully be explained using the method of matched asymptotic expansions [7], we must reasonably conclude that the solution of the Navier-Stokes equations at sufficiently high Reynolds numbers is indeed as sketched.

Some authors do report conflicting results for the same problem. Wang [21] obtained results suggesting new flow developments at about the time and location that separation occurs in the accepted computations [14-20]. However, there are noticeable differences in quantitative values as well as major differences in resulting interpretation: in Wang's results the separation is caused by envelope formation of the wall streamlines, which requires a singular wall shear. Neither Wang's results, nor those of [22] or those of [23], has received independent verification.

The consequences of the accepted numerical solution are profound. For example, consider the criteria for steady separation. Zero wall shear forms when the cylinder has moved 1/6th diameter. The streamlines show a recirculating `wake' beyond that time. Yet there is no separation: the boundary layer remains at the surface regardless how much the Reynolds number is increased, and eventually becomes negligibly thin. Mathematically, when the Reynolds number is increased, the boundary layer vorticity remains within a distance proportional to the inverse square root of the Reynolds number to the wall. Since this is the same as the boundary layer thickness, there is never a discernable spacing between boundary layer vorticity and the wall.

Such a thin reversed-flow boundary layer at the wall is qualitatively evident in experimental results [24], as well as in numerical results such as in Figure 2. Although difficult to discern in Figure 2, reversed vorticity (or shear) near the wall implies that there is a long very thin region of recirculation inside the boundary layer over much of the rear half of the boundary layer. Some authors refer to such recirculating streamlines as a 'vortex', but the vorticity distribution shows only thin layers of vorticity. Since zero wall shear is the position where the wall streamline bifurcates, and the bifurcated wall streamline implies reversed and recirculating flow, all these criteria are satisfied, but they are not relevant to unsteady separation.

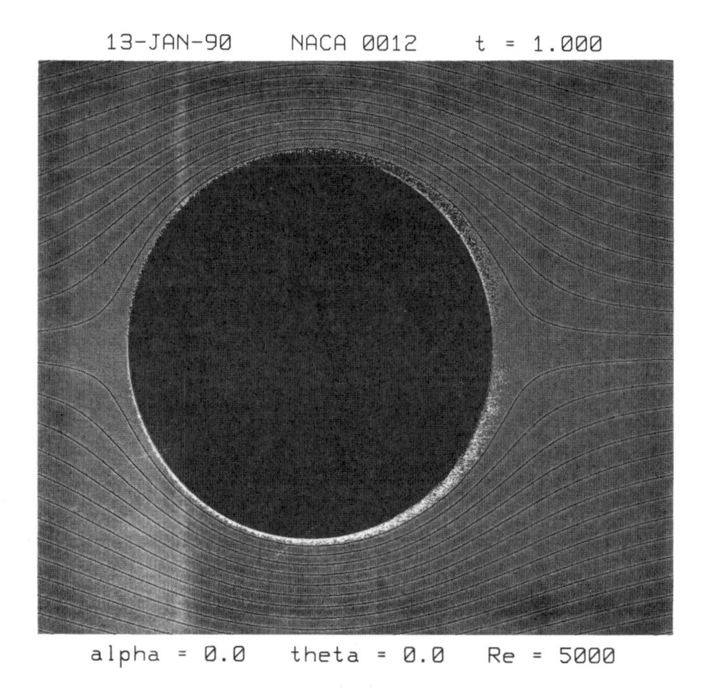

Figure 2. Vorticity field for a circular cylinder impulsively set into motion at a Reynolds number 10,000, after one radius of motion, illustrating zero wall shear, a 'free streamline,' a 'recirculating wake' and 'eddies' or 'vortices,' all without separation.

Much later, after 3/4 diameter motion rather than 1/6, the boundary layer vorticity is ejected away from the surface. Still the process is not related to zero wall shear; it is localized in a relatively small ejection zone downstream of the point of zero wall shear, and upstream of the rear stagnation point. Neither of these two streamline bifurcation points is in the region where the ejection occurs. While a point of closed streamlines eventually occurs within the ejection zone, this point, too, is not in the region that causes the actual ejection. The region causing the ejection, (called MRS region, for Moore, Rott, and Sears), is located lower in the boundary layer. More precisely, the MRS region is vertically centered around the position of maximum reversed flow velocity in the separation velocity profile. Since the maximum reversed flow velocity is non-zero, the local streamlines do not intersect; they are roughly parallel instead.

The differences between steady and unsteady separation had been predicted earlier on philosophical grounds by Sears and Telionis [25], who expanded and generalized earlier work by Moore [6]. In fact, their theories motivated the numerical studies of the problem of the circular cylinder that eventually uncovered the separation process. Sears and Telionis introduced a new requirement for analysis of unsteady separation: examination of the limiting behaviour of the flow when the Reynolds increases without bound.

The need for such a more general setting arises since there is no 'magic' indicator akin to streamline bifurcation or zero wall shear in unsteady separation. From an individual unsteady experiment or computation we cannot easily decide whether separation occurs. To make an unambiguous determination, we must examine what happens to the flow when the Reynolds number increases: increasing the Reynolds number has the effect of sharpening the flow features. Thus the boundary layer becomes thinner, which makes its position and shape much more distinct.

We might consider a thought-experiment in which we would raise the Reynolds number arbitrarily for the circular cylinder. If we could do so, the appearance of the flow would be as first derived by Van Dommelen and Shen [26] using Lagrangian analysis, and independently verified in Eulerian description by Van Dommelen [7] and Elliot, Cowley, and Smith [10]. According to those results, at the instant that the cylinder has moved over, say, one radius, the boundary layer would show a 'closed recirculating wake' over much of the rear half of the cylinder, with points of zero wall shear at the upstream ends. Yet while for increasing Reynolds number this boundary layer would become vanishingly thin, we would not observe a separation between its position and the wall. Figure 2 gives Navier-Stokes results for this time at a Reynolds number 10,000.

On the other hand, if we would examine the flow after one half radius of additional motion, at about the separation time, raising the Reynolds number would show with increasing clarity that there is a small region at which the upper part of the boundary layer is ejected upwards away from the wall (cf. Figure 1). While the thickness of that ejected layer of vorticity is still comparable to the attached boundary layer on the remainder of the cylinder, its distance from the wall would be significantly larger. Thus we would see a local boundary layer that is much thicker than elsewhere on the cylinder, and that resembles a free vortex layer separated from the wall by a thick region of little vorticity. The thick region of little vorticity is the MRS region that actually causes the ejection. Below the MRS region, we also would see some reverse vorticity that remains at the wall.

The further we would raise the Reynolds number, the sharper the picture would become. If we could raise the Reynolds number to infinity, the sharp features would become a singularity when presented using boundary layer scalings. This, then, is the singularity hypothesis of Sears and Telionis, often misrepresented by authors unfamiliar with the limiting processes involved. As Sears and Telionis did point out, there is probably no singularity at finite Reynolds numbers; the singularity is a feature of infinite Reynolds number theory. Yet, though not physical, the existence of the singularity at infinite Reynolds number is the sole unambiguous sign of separation. It exaggerates the actual strong, but finite breakup of the attached boundary layer.

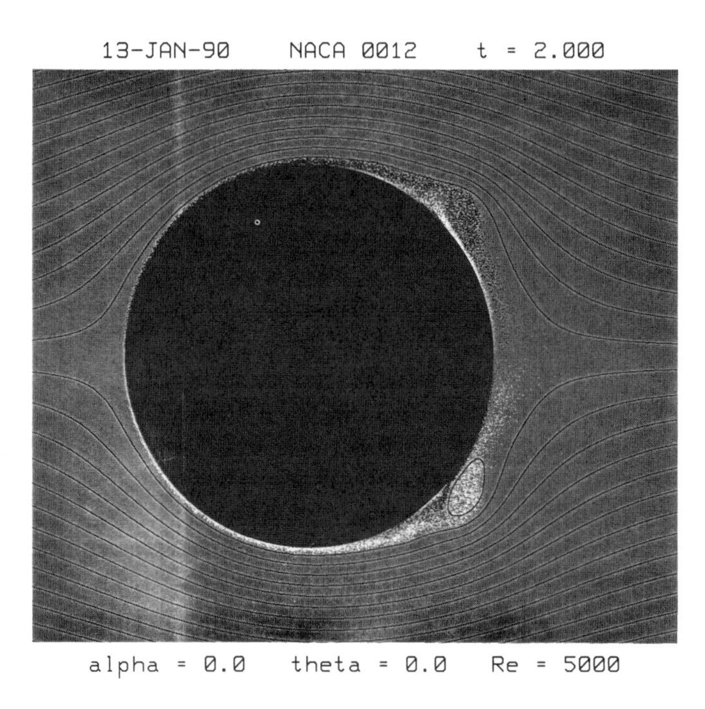

Figure 3. Vorticity field for the same circular cylinder as Figure 2, but after one diameter of motion illustrating the flow field after separation has occurred.

One might wonder how much of this evolution at very large Reynolds numbers is relevant at practical Reynolds numbers. From the numerical values, it is readily estimated that for relatively low Reynolds numbers the boundary layer approximation breaks down long before the separation process would become as clear as sketched in Figure 1. However, such an argument is probably needlessly pessimistic. While the initial separation process may not be very dramatic, its consequences certainly are, and one would not expect that such major flow changes would totally disappear for smaller Reynolds numbers unless diffusion is very large indeed. To illustrate this, Figure 3 shows again the cylinder at Re=10,000, but now after one diameter motion, somewhat beyond the theoretical time of separation at infinite Reynolds number. There is clear evidence of an initial local break-up of the boundary layer that is even more distinct after 1.25 diameter motion in Figure 4.

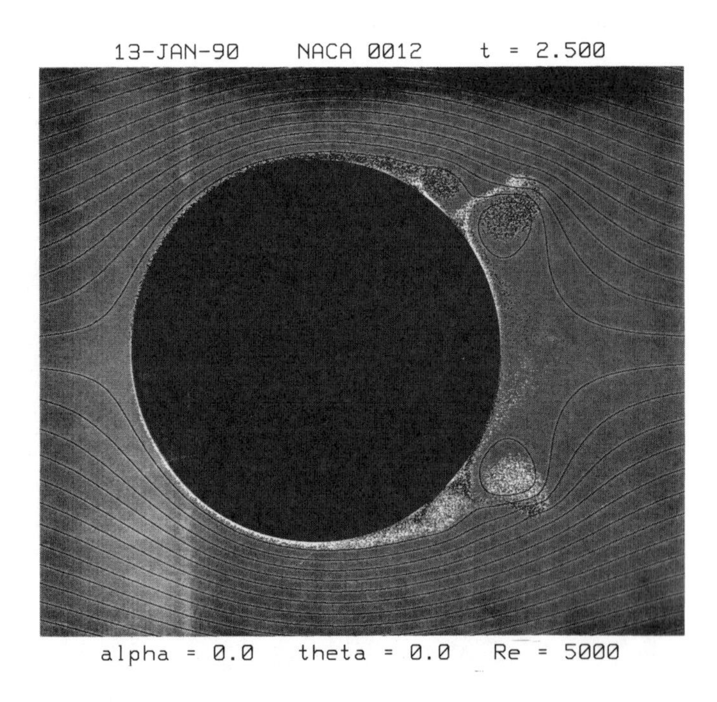

Figure 4. Vorticity field for the same circular cylinder as Figure 2, but after 1.25 diameter of motion.

5 INDICATIONS OF UNSTEADY SEPARATION

The question how to diagnose unsteady separation from limited data at relatively low Reynolds is of considerable interest. The most direct approach is of course to increase the Reynolds number while keeping all other parameters constant. However, that may not be realistic. Instead there is a number of distinct features that are easily checked. While the MRS region causing the separation is not at any special position with respect to the streamline topology or wall shear, its location is certainly not arbitrary.

First, Sears and Telionis postulated that the MRS region would satisfy two so-called MRS conditions. One of these conditions fixes the vertical position within the boundary layer: it states that the local vorticity vanishes:

$$\mathrm{Re}^{-1/2}\frac{\partial u}{\partial y}\approx 0$$

Practically speaking, it implies that the MRS region is at the position of the maximum reversed velocity in the boundary layer velocity profile. This has subsequently been verified by the numerical solutions.

The horizontal position of separation along the wall is constrained by a second MRS condition; it states that the velocity with which separation structure propagates along the wall equals the particle velocity in the MRS region

$$u\approx\frac{dx_s}{dt}$$

Yet this condition is only useful for special flows, such as for steady separation from a moving wall, or for separation in semi-similar unsteady flows, for which the speed of propagation of the separation structure is known. In general, the second MRS condition is of little help. It does not help to look for bifurcation of streamlines in moving systems; in a system moving with a particle at zero vorticity, the streamlines will bifurcate regardless whether separation occurs.

To determine the streamwise location of separation in the fully unsteady case, the detailed structure derived by Van Dommelen and Shen [7,26] may be used. It predicts that the forming MRS region is located at the position of most rapid streamwise increase of the maximum reversed flow velocity. Consequently, along the maximum reversed flow line the actual position of the MRS region is indicated by a maximum value to

$$-\frac{\partial u}{\partial x}=G$$

The maximization should be restricted to the maximum reversed flow line: much larger values of G may occur in higher parts of the boundary layer. The condition also implies that the determinant of the second order derivatives of u vanishes on the zero vorticity line, which may be used as an alternate indicator.

The time and location where separation actually occurs may be estimated from

$$t_s = t + \frac{1}{G}$$

$$x_s = x + u_s(t_s - t)$$

However, these estimates can only be used in a limited range of times sufficiently close to separation that the convective effects have started to dominate the local flow, but not so close that the boundary layer approximation has lost all meaning. Note that the first estimate predicts a singularity in G at the separation time.

A completely different approach is to track particle deformation. Unsteady separation is characterized by strong particle deformations that can be used to decide whether separation occurs. This method was used by Van Dommelen and Shen [14] to show numerically that separation does indeed exist.

It should be noted that it is probably impossible to predict a forming unsteady separation from the wall shear distribution, limiting streamlines, streamlines close to the surface, etc. The reason is the vorticity layer that remains behind at the wall, Figure 1, that effectively isolates the flow near the wall from what is going on in the MRS region above it. Signs of separation at the wall are only possible when the separation has sufficiently advanced that interaction effects come into play. Such effects are the subject of current research [18,19,27,28,29], and little is known as yet. However, when interaction does appear, it may be too late to do much about it. This was dramatically illustrated for a spinning sphere by Van Dommelen [30].

6 INDICATIONS OF UNSTEADY 3D SEPARATION

Diagnosing the three-dimensional case presents some additional difficulties beyond those discussed in the previous section. An initial three dimensional unsteady separation occurs along some describing line on the surface, and the direction of this describing line is in principle arbitrary. Contour lines of constant boundary layer thickness may have a typical appearance as sketched in Figure 6. The initial separation zone is restricted to a small segment of the describing line and has an even smaller typical dimension in the direction normal to the line. As a result, in cross sections normal to this line the separation structure appears quasi-two-dimensional, as in Figure 5.

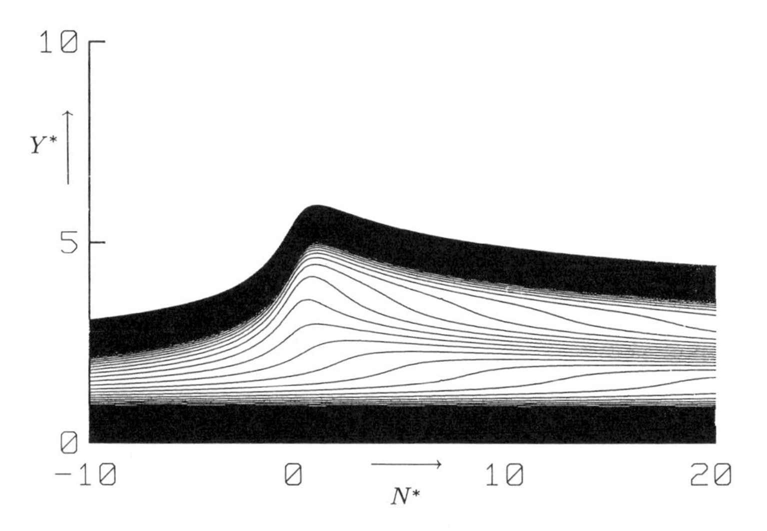

Figure 5. Asymptotic lines of constant vorticity near unsteady separation in scaled local coordinates.

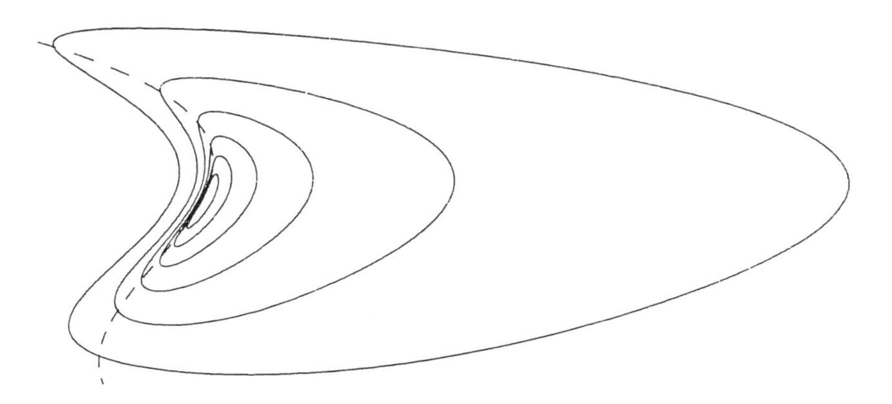

Figure 6. Typical lines of constant boundary layer thickness for an initial three-dimensional unsteady separation.

The MRS conditions derived by Van Dommelen and Cowley [31] are similar to the two-dimensional case. Both velocity profiles eventually will have a local extremum at the MRS position. Thus any velocity profile may theoretically be used to fix the vertical position of the MRS region. However, determination of the position based on the velocity component in the direction normal to the describing line will be significantly more accurate. This may require some iteration, since the direction of the describing line is of course not known a priori. The direction of the describing line may be estimated from the boundary layer displacement effects, or from the gradients of the velocity components in the MRS region: the gradients in the direction normal to the describing line are much larger than those parallel to it.

The position and time of separation may conveniently be determined using the invariance property of the projected velocity divergence:

$$-\left(\frac{\partial u}{\partial x}+\frac{\partial w}{\partial z}\right)=G$$

When symmetries exist at the location of separation, the appearance of the separation structure changes from what is shown in Figure 5. For various important cases, the modified structures are shown by Van Dommelen and Cowley [31].

7 DETERMINATION USING BOUNDARY LAYER COMPUTATIONS

Sears and Telionis [25] pointed out that an unambiguous determination of unsteady separation can be made by solving the boundary layer equations and examining the solution for singular behavior. Yet the case of the circular cylinder impulsively set into motion showed that this is not as simple as it may appear. Computations proved that even before separation, the boundary layer is sufficiently thick that a large number of mesh points are needed to describe it. In addition, the streamwise and timewise length scales to resolve are quite small, and if not enough resolution is provided, the structure can be lost completely. As a result, initially there was considerable confusion about the solution and whether it was singular or not.

This question was first resolved unambiguously by the Lagrangian computation of Van Dommelen and Shen [14]. A Lagrangian computation deals much better with the strong convection processes that cause both the thick boundary layer and the singularity itself. In effect, the Lagrangian mesh is solution adaptive for these effects: local resolution in the separation zone increases by an order of magnitude due to particle motion. Further, it works out that the Lagrangian solution itself is non-singular, so that there is no loss of order of accuracy near separation.

A Lagrangian solution has another unexpected benefit: a new and accurate criterion for separation. It proves that unsteady separation is indicated by vanishing derivatives in the mapping Jacobian, making the smooth Lagrangian solution singular in physical coordinates. The zero in the derivatives is relatively easy to verify numerically.

Since the solution of Van Dommelen and Shen, a considerable number of Eulerian computations have also verified the separation structure [15-20]. Yet the Lagrangian solution may still be the best one for several reasons. First, it has infinite resolution at the separation point. Second, there is no decrease in the order of accuracy at separation unlike in a conventional computation. Third, it is much less ambiguous, since it needs only verify that certain derivatives are zero. A conventional computation has to try to establish that certain derivatives are infinite.

Thus there is considerable interest in improved Lagrangian computation procedures. In the next section, we discuss some numerical issues involved.

8 LAGRANGIAN BOUNDARY LAYER COMPUTATIONS

While Lagrangian coordinates have decisive advantages in computing unsteady separation, they have their difficulties. As the Lagrangian mesh distorts, it gets increasingly difficult to discretise the equations without altering the logic of the mesh point ordering. One possibility is to do away with all structure of the computational points, for example by using random walk, as in Figures 2-4, or more recently, by using redistribution techniques [32].

Still, structured Lagrangian computations do have advantages for diagnosing separation, since smoothness and values of the derivatives with respect to the Lagrangian coordinates are important. One simple way to deal with mesh distortion on a regular mesh topology is periodic global redefinition. Essentially, the solution is interpolated back onto a numerically more acceptable point distribution. Periodic redefinition is powerful, but if the redefinition is performed too frequently, the advantages of the Lagrangian computation over a Eulerian might be lost unless the redefinition is performed with great care.

A second approach is to use computational techniques that can deal with the strong distortions. Such techniques are the topic of this section. The optimum procedure is presumably a combination of periodic redefinition with techniques that can handle strong mesh distortions, but little work in this area has been done yet.

On a regular mesh topology, strong mesh distortions take the form of greatly varying coefficients in the numerical solution. A simple example is the case of the circular cylinder. The coefficient of the leading order viscous derivative at the front stagnation point grows by a factor 400 between the time the motion starts and the time of separation. In the same time at the rear stagnation point, the coefficient decreases by a factor 400. This means that at the separation time, the leading coefficient varies by 5 orders of magnitude across the numerical domain.

Moreover, while in a conventional computation there is only one viscous derivative, in a Lagrangian computation all second order derivatives appear. As a result, the numerical methods more closely resemble those for typical elliptic problems rather than the more simple boundary layer schemes. Typically, a Lagrangian computation has to deal with sparse band matrices instead of block tridiagonal systems.

There are also problems associated with the first order terms. For while the first order convective terms have been eliminated by the Lagrangian transform, the discretization of the viscous terms introduces other first order derivatives. Their coefficients show large variations similar to the viscous terms. What this means numerically is that first order terms need to be discretized at very large equivalent Courant numbers. Moreover, the second order terms do not necessarily dominate the first order terms. For example, at separation the coefficients of the second order terms vanish more rapidly than those of the first order terms.

Despite these problems, it proves that accurate Lagrangian results can be obtained with standard discretizations such as Crank-Nicholson. The difficulty is in solving the implicit equations generated by such a discretization.

One possibility is iterative solution. A procedure formulated by Van Dommelen and Shen [14] converges despite sizable mixed derivatives, large Courant numbers and coupling with mesh motion. It was based on a zig-zag, diagonally dominant discretization of the first order terms and reversal of the sweep directing after each sweep. This procedure was effective, however, convergence becomes slow near separation and for fine meshes. Multigrid greatly improves convergence (unpublished results), but the improvements are not as dramatic as for other elliptic problems. The most likely reason is the difficulty of interpolating accurately between highly distorted meshes. Of course, multigrid also increases code complexity.

While Van Dommelen and Shen's iterations may be vectorized along diagonals, this is awkward and the vector length remains limited. Moreover, the generalization toward the Navier-Stokes equations is not obvious. Recently, ADI procedures have received attention. In principle, these can eliminate the need for iteration. A 'Beam & Warming' factorization was used by Van Dommelen [33] with only limited success; it proved to have only modest stability in two dimensions. It also can be shown to be unstable in 3D unless the Courant number is vanishingly small, an unacceptable condition in Lagrangian coordinates. The ADI iteration used by Peridier, Smith, and Walker [28] required extremely small time steps when the mesh distorted.

Recent work has focussed on procedures that work in three dimensions. It can be shown that the iterative procedure used by Van Dommelen and Shen [14] converges provided that the two momentum equations are iterated separately. It still allows multigrid.

Most ADI procedures can not handle either the large Courant numbers, or the large mixed derivatives in a Lagrangian computation. One exception is the time-split backward-time discretization following schemes proposed by Yanenko and D'Yakonov [34]. It takes the form

$$(1 + X)(1 + Y)(1 + Z)u^{n+1} = (1 - M)u^{n} + f$$

where n indicates the time plane, Xu, Yu, and Zu indicate the terms involving derivatives in the three mesh directions, and Mu stands for the terms involving mixed derivatives. However, this scheme is only first order accurate in time, and numerical experiments indicate that it suffers from severe factorization errors unless the time step is kept very small. Passive Richardson extrapolation on the time step brings some improvement. A significant improvement results from cycling the order of the factors, yet convergence remains somewhat erratic.

An alternate factorization has recently been formulated as

$$(1 + X + Y_1 + Z_1)(1 + Y_2)(1 + Z_2)(u^{n+1} - u^{n}) = Lu^{n}$$

where Lu^{n} indicates the explicit evaluation of the spatial terms and Y_1 and Z_1 are the first order derivative terms in the second and third mesh direction. Since these first derivatives are upwinded, the first factor is effectively explicit in the second and third direction. Not only is this scheme theoretically and experimentally stable in three-dimensions, even in two dimensions it is much more efficient than any previous Lagrangian scheme we have used. Work is currently in progress to numerically discover the first three-dimensional unsteady separation structure using this scheme.

9 CONCLUSION

Diagnosing unsteady separation is clearly more difficult than for steady flow. Yet, as seen, many indicators can be used, several relatively easily. It is worrisome that work keeps appearing that claims to study unsteady separation, but that does not attempt to relate the results to what is known unambiguously. It is easy to waste time and effort by attempting to verify ad-hoc criteria without a solid theoretical meaningful comparisons instead.

ACKNOWLEDGEMENT

The authors gratefully acknowledge support for this research by the AFOSR, F49629-89-C-0014.

REFERENCES

1 Prandtl, L.: Uber Fleussigkeitsbewegung bei sehr kleiner Reibung, (in Ludwig Prandtl gesammelte Abhandleungen, pp. 575-584, Springer-Verlag 1961), 1904.

2 Goldstein, S.: On Laminar Boundary-Layer Flow near a Position of Separation, *Q. J. Mech. Appl. Math.*, **1**, pp. 43-69, 1948.

3 Sychev, V.V.: Laminar Separation, *Izv. Nauk SSSR, Mekh. Zhidk.Gaza*, **3**, pp. 47-59, 1972.

4 Smith, F.T.: Laminar Flow of an Incompressible Fluid past a Bluff Body: the Separation, Reattachment, Eddy Properties and Drag, *J. Fluid Mech.*, **92**, pp. 171-205, 1979.

5 Van Dommelen, L.L. and Shen, S.F.: Separation from a Fixed Wall, Second Symposium On Numerical and Physical Aspects of Aerodynamic Flows, ed. T. Cebeci, pp. 393-402, Springer-Verlag, 1984.

6 Moore, F.K.: On the Separation of the Unsteady Laminar Boundary Layer, In *Boundary-Layer Research,* ed. H.G. Gortler, Springer, 1958.

7 Van Dommelen, L.L.: *Unsteady Boundary Separation*, Ph.D. thesis, Cornell University, 1981.

8 Sychev, V.V.: Asymptotic Theory of Non-Stationary Separation, *Izv. Akad. Nauk. SSSR, Mekh. Zhid. i Gaza,* **6**, pp. 21-32, 1979.

9 Van Dommelen, L.L. and Shen, S.F.: An Unsteady Interactive Separation Process. *AIAA J.,* **21**, pp. 358-362, 1983.

10 Elliott, J.W., Cowley, S.J. & Smith, F.T.: Breakdown of Boundary Layers: i. On moving surfaces, ii. In semi-similar unsteady flow, iii. In fully unsteady flow, *Geophys. Astrophys. Fluid Dynamics,* **25**, pp. 77-138, 1983.

11 Van Dommelen, L.L. & Shen, S.F.: Boundary-Layer Separation Singularities for an Upstream Moving Wall, *Acta Mech.,* **49**, pp. 241-254, 1983.

12 Rott, N.: Unsteady Viscous Flows in the Vicinity of a Separation Point, *Q. Appl. Math.,* **13**, pp. 444-451, 1956.

13 Sears, W.R.: Some recent developments in airfoil theory, *J. Aeronaut. Sci.,* **23**, pp. 490-499, 1956.

14 Van Dommelen, L.L. & Shen, S.F.: The Spontaneous Generation of the Singularity in a Separating Laminar Boundary Layer, *J. Comp. Phys.,* **38**, pp. 125-140, 1980.

15 Cowley, S.J.: Computer Extension and Analytic Continuation of Blasius' Expansion for Impulsive Flow past a Circular Cylinder, *J. Fluid Mech.,* **135**, pp. 389-405, 1983.

16 Ingham, D.B.: Unsteady Separation, *J. Comp. Phys.,* **53**, pp. 90-99, 1984.

17 Ece, M.C., Walker, J.D.A. & Doligalski, T.L.: The Boundary Layer on an Impulsively Started Rotating and Translating Cylinder, *Phys. Fluids,* **27**, pp. 1077-1089, 1984.

18 Henkes, R.A.W.M. & Veldman, A.E.P.: On the Breakdown of the Steady and Unsteady Interacting Boundary-Layer Description, *J. Fluid Mech.,* **179**, pp. 513-530, 1987.

19 Vasantha, R. & Riley, N.: On the Initiation of Jets in Oscillatory Viscous Flows, Proc. R. Soc. Lond. Series A, **419**, pp. 363-378, 1988.

20 Puppo, Gabriella: Ph. D. Thesis, Courant Institute, 1990.

21 Wang, K.C.: Unsteady Boundary-Layer Separation, Martin Marietta Lab., Baltimore, Maryland, USA, Tech. Rept. MML TR 79-16c, 1979.

22 Telionis, D.P. & Tsahalis, D. Th.: Unsteady Laminar Separation over Impulsively Moved Cylinders, *Acta Astronautica,* **1**, pp. 1487, 1974.

23 Cebeci, T.: Unsteady Boundary Layers with an Intelligent Numerical Scheme, *J. Fluid Mech.,* **163**, pp. 129-140, 1986.

24 Bourd, R. and Coutanceau, M.: The Early Stage of Development of the Wake behind an Impulsively Started Cylinder for 40<Re<10000, *J. Fluid Mech.,* **101**, pp. 583-607, 1980.

25 Sears, W.R. & Telionis, D.P.: Boundary-Layer Separation in Unsteady Flow, SIAM *J. Appl. Math.,* **23**, pp. 215, 1975.

26 Van Dommelen, L.L. and Shen, S.F.: The Genesis of Separation, Symposium on Numerical and Physical Aspects of Aerodynamic Flows, ed. T. Cebeci, pp. 293-311, Springer-Verlag, 1982.

27 Cowley, S.J., Van Dommelen, L.L., and Lam, S.T.: On the Use of Lagrangian Variables in Unsteady Boundary--Layer Separation, *Phil. Trans. R. Soc. Lond. Series A,* **333**, pp. 343-378, 1990.

28 Peridier, V.J., Smith, F.T. and Walker, J.D.A.: Vortex-Induced Boundary-Layer Separation. Part 1. The unsteady limit problem, *J. Fluid Mech.,* **232**, pp. 99-131, 1991.

29 Peridier, V.J., Smith, F.T. and Walker, J.D.A.: Vortex-Induced Boundary-Layer Separation, Part 2, Unsteady interacting boundary-layer theory, *J. Fluid Mech.,* **232**, pp. 99-131, 1991.

30 Van Dommelen, L.L.: On the Lagrangian Description of Unsteady Boundary Layer Separation, Part 2, The spinning sphere, *Journal of Fluid Mechanics,* **210**, pp. 627-645, 1990.

31 Van Dommelen, L.L. and Cowley, S.J.: On the Lagrangian Description of Unsteady Boundary Layer Separation, Part 1, General theory, *Journal of Fluid Mechanics,* **210**, pp. 593-626, 1990.

32 Van Dommelen, L.L.: A Vortex Redistribution Technique, FAMU/FSU College of Engineering Fluid Mechanics Research Laboratory report FMRL-TR 3, Florida, 1989.

33 Van Dommelen, L.L.: Lagrangian Techniques For Unsteady Flow Separation, *Forum on Unsteady Flow Separation*, ASME FED, **52**, pp. 81-84, 1987.

34 Marchuk, G. I.: *Handbook of Numerical Analysis,* ed. P. G. Ciarlet and J. L. Lions, North-Holland, 1990.

Tailless Aircraft Design — Recent Experiences

Ilan Kroo
Stanford University

ABSTRACT

Several methods for achieving longitudinal trim of tailless aircraft with a desired stability level have been employed over the century of experimentation with such designs. This paper illustrates these approaches in recent applications, focussing on the advantages and problems associated with three very different "flying wing" concepts. Analyses and flight tests have demonstrated that it is often possible to avoid the performance penalties once thought to be inherent in the tailless design.

Example projects include the development of an unswept tailless aircraft for high altitude long endurance missions, a moderately-swept, foot-launched sailplane, and an unstable, oblique all-wing design.

1 INTRODUCTION

It is an honor and a delight to contribute to this symposium in celebration of Bill Sear's eightieth birthday. Although I have met Professor Sears only a couple of times, I have studied and been inspired by his work in unsteady aerodynamics, supersonic aerodynamics, and tailless airplane design. So I do feel in a very real way one of Bill's students. I have chosen to talk about some experiences with tailless aircraft design, but feel a bit presumptuous giving such a talk to an audience that includes Bill Sears and Irv Ashkenas, some of the true pioneers in this field. So with their indulgence I'll try to describe some of the things that my colleagues and I have done in the last decade in tailless aircraft design.

This lecture will begin with a very short history and overview of some of the issues involved in tailless aircraft design, concentrating, not on the question of whether tailless aircraft are a good idea, but rather on how one designs a good tailless airplane. I will focus on three approaches to one of the fundamental problems of tailless design, that of longitudinal control, and will mention a few additional considerations in their design.

One may be puzzled by the fact that we see so few tailless airplanes. Although the tail of a commercial transport aircraft constitutes 25-35% of the wing area and pushes down with as much as 5% of the aircraft weight (~100 passengers with baggage), the horizontal tail has remained a prominent feature of modern aircraft and despite over thirty years of technological progress, the 707, rolled out in about 1954 and the A340 first flown in 1991, look very similar. This is not simply a reflection of aircraft manufacturers' conservatism, but an indication of the fact that horizontal tails are an efficient means of satisfying the requirement for longitudinal trim and control. These are not insignificant constraints. If one optimizes an airplane with respect to about 20 parameters (wing and tail geometry variables, engine size, and operational parameters) to minimize direct operating cost subject to constraints on range, engine-out climb, and field lengths, a design similar to that shown on the left part of figure 1 is produced [1]. With the same analyses, but letting the tail lift coefficient be very large and removing the stability constraint, the tail disappears, and the aircraft gross weight, fuel consumption, and D.O.C. go down by about 10%. Unfortunately, this aircraft cannot rotate on take-off (so the fuel savings is even larger), but these large changes suggest that the approach to aircraft trim and pitch control is a very important element of the design, with the conventional horizontal tail just one possible solution. The *best* solution depends strongly on additional considerations, but in many applications designs without horizontal tails have met with some success.

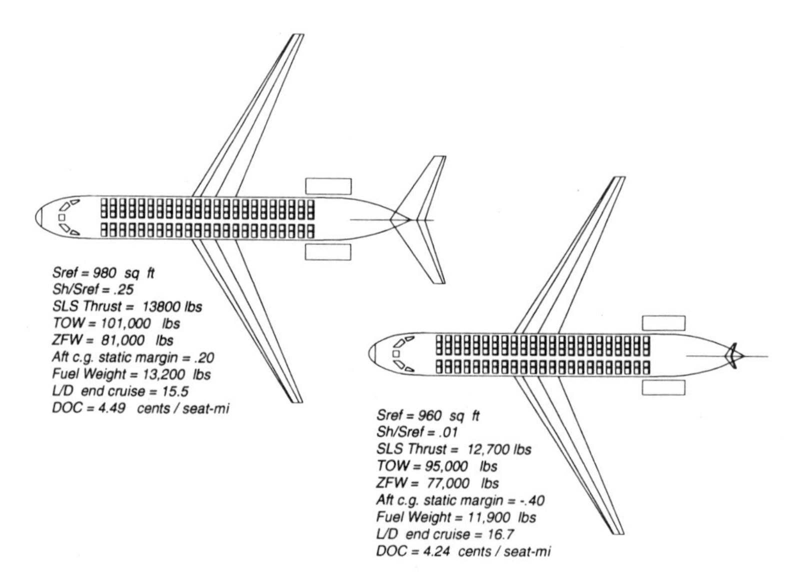

Figure 1. Two Aircraft Designed for Minimum D.O.C.
Right: No Stability or Control Constraints.

History

Tailless aircraft have been inspired by a variety of flying seeds and animals. From Zanonia seeds (Fig. 2) to Pterosaurs and modern birds that fly well with little or no tails, Nature seems not to have converged on the 707-like configuration and many airplanes have been based on these models [2].

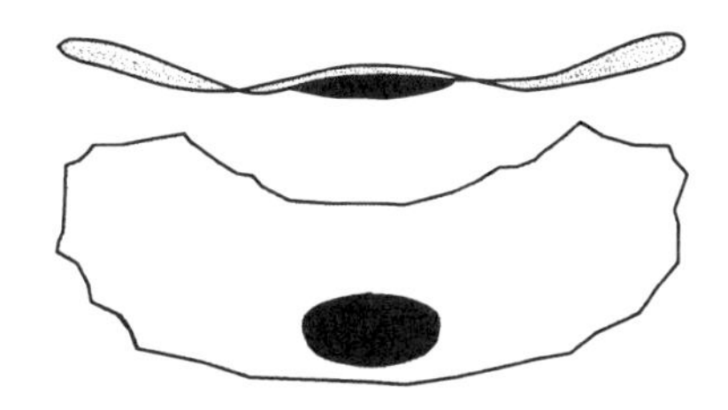

Figure 2. Seed of Zanonia macrocarpa, Span ≈ 6". An Inspiration to Early Aircraft Designers.

Figure 3 illustrates an array of tailless aircraft developed from the early 1900's to the 1950's. The figure suggests a number of approaches to the design of tailless aircraft. In the upper left corner of the figure is one of the early tailless gliders modeled on the Zanonia seed, built by Igo Etrich, and his father, Ignaz. The Etrichs had experimented with one of Otto Lilienthal's gliders which they purchased for 5£ after Lilienthal's death. Aided by Prof. F. Ahlborn, who had investigated the Zanonia seed and written a paper entitled "On the Stability of Aeroplanes" in 1897, the

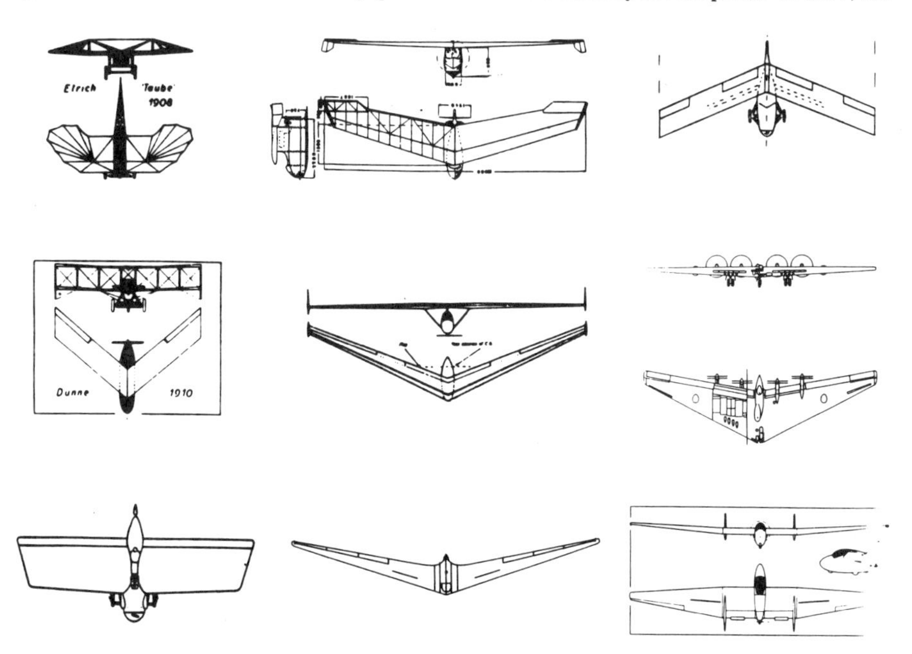

Figure 3. Historical Tailless Aircaft Illustrating Several Design Approaches.

Etrichs constructed what was probably the first stable airplane in about 1906. The Taube (Dove) followed several tailless designs by the Etrichs and several hundred were built by a number of companies in Germany and Austria [2,3].

Below the Etrich airplane is a picture of one of the Dunne tailless biplanes, circa 1910. J.W. Dunne designed several tailless biplanes and monoplanes using a combination of sweep and wash-out for inherent stability. The stability and controllability of these airplanes led to limited commercial success although the rather high parasite drag and inferior maneuverability discouraged their adoption for military applications, Dunne's original intention.

On the lower left portion of the figure, is one of Rene Arnoux's "flying plank" designs. Arnoux experimented with monoplanes, biplanes, pushers, tractors, low and mid-wing tailless designs all characterized by their use of unswept, nearly rectangular planform wings. From 1909 to 1923 Arnoux demonstrated many successful flights with a variety of airplanes based on this concept. The airplane shown here was an ambitious racer with a 320 h.p. engine that crashed during flight testing.

In the upper center of the figure is the Lippisch "Experiment 64" glider of 1925 incorporating Dunne's concept of sweep and twist for stability, but with more respectable performance. The concept evolved over the next ten years in the Stork series of gliders and many experiments by Alexander Lippisch on a variety of tailless designs. During this same time Professor G.T.R. Hill was developing the Pterodactyl series of tailless airplanes, originally based on the supposed planform shape of a pterosaur, and culminating in a 1934 military fighter, the Pterodactyl Mk V, with a 700 h.p. engine and tractor propeller. Forward sweep was also used for tailless aircraft as evidenced by the 1922 Landwerlin-Berreur racing monoplane of 1922, also powered by a 700 h.p. engine (upper right corner).

Inboard flaps, suggested by Lippisch, are seen in a drawing of the G.M. Buxton high-performance tailless sailplane of 1938 (center). The planform of this glider is similar to that of the Horten brothers, whose enthusiastic development of flying wing sailplanes and powered aircraft began in 1934. Shown below the Buxton glider is the Horten IV sailplane of 1942 with its aspect ratio of 21. The Horten sailplanes were considered to be competitive with conventional sailplanes during 1938-1939 when they were entered in several competitions, flown 180 miles and once climbed 21,000 ft above launch in a thermal. The Hortens continued their work during World War II including an aspect ratio 32 sailplane (Horten VI), a design for a 6-engine 60 passenger transport aircraft (Ho VIII), and the Ho IX jet fighter, all tailless aft-swept designs.

Some of the most famous tailless airplanes are the Northrop flying wings. At the center right of the figure is a drawing of the XB-35, first flown in 1946, following years of experience with tailless aircraft, beginning with the 1940 flight of the Northrop Model 1 Mockup (N-1M). Bill Sears was closely involved in the development of these aircraft, culminating in the YB-49 jet-powered flying wing bomber. Northrop engineers also had plans for flying wing passenger transports [4], but cancellation of the YB-49 program spelled the end of large scale subsonic aircraft programs for many years. Tailless designs continued to be pursued, especially in the sailplane community as indicated by the success of the Fauvel flying wings (lower right of figure 3). Charles Fauvel patented an unswept untwisted tailless design in 1929 and continued working on this configuration for many years. The sailplane shown, the AV. 45 was first flown in 1960 [5].

More recent work on tailless designs is illustrated in figure 4. The SB-13 (upper left) is the result of work at the Akademische Fliegergruppe Braunschweig, begun in 1982 with first full scale flights in 1988. This 15 m span tailless sailplane has demonstrated good performance, but with many difficulties associated with aeroelasticity and handling [6].

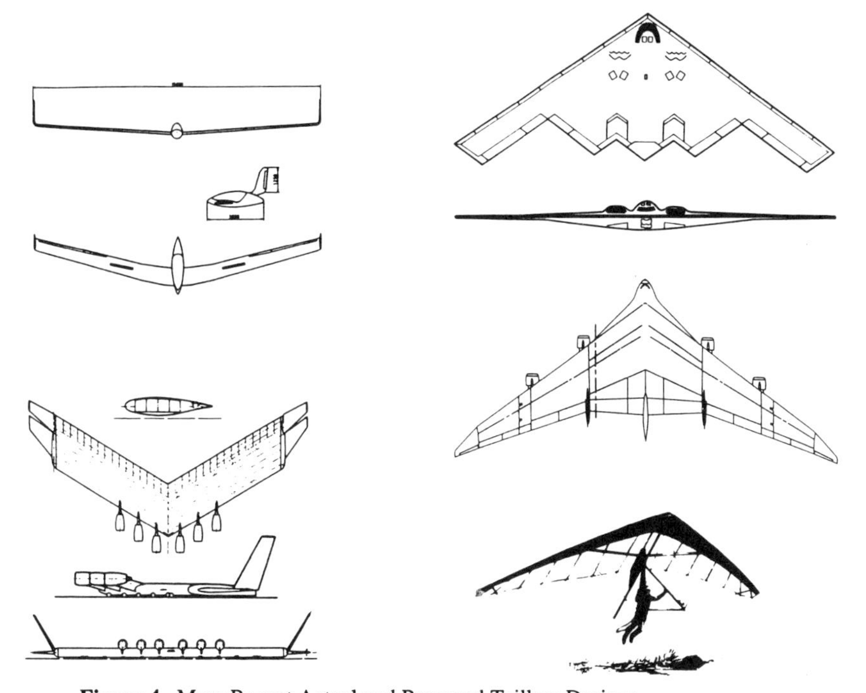

Figure 4. More Recent Actual and Proposed Tailless Designs.

In the lower left portion of figure 4 is a span-loaded cargo airplane, studied by Boeing in the 1970's. The far-sighted design was to have a take-off weight in excess of 2 million pounds, 19% thick airfoil sections, six 93,000 lb thrust engines, and a span of over 400 ft. [7]

The Northrop B-2 design (upper right) was motivated by the need for low radar cross section and represents the first large-scale modern tailless program. It is discussed in more detail in Ref. 8. The B-2 has inspired new looks at the tailless design, such as another Boeing concept for a large capacity transport (center right).

Some have said that the flying wing finally found a niche in the world of hang gliding, evolving quickly from the original design by Francis Rogallo to high aspect ratio sailplanes capable of 300 mile flights where oxygen is used routinely. Indeed, despite several attempts to introduce conventional configurations, tailless designs have dominated hang gliding for more than 20 years [9].

So, there actually are a great number of flying wings, but, except perhaps in hang gliding circles, the conventional design still dominates, and it is widely believed that large performance penalties are associated with this design.

> *"A few airplanes ... have been designed without tails in a misguided attempt to save drag. Actually, without a tail, high-lift devices cannot be used on the wing... The tailless high aspect ratio airplane must have a swept wing... All of these effects account for the virtual disappearance of the tailless configuration..."* , [10]

In order to evaluate the performance advantages or disadvantages associated with tailless aircraft, it is necessary to compare a well-designed tailless aircraft with a similarly-designed conventional airplane. One difficulty with this is that it is very easy to design a very bad tailless aircraft while conventional design wisdom leads to a quite good conventional design. In this paper, such a comparison will not be attempted, but rather some of the approaches to achieving a high performance tailless aircraft will be discussed. Many different types of tailless aircraft exist, and may be categorized by the method used to achieve longitudinal stability and trim. We illustrate three approaches with three aircraft developed over the last decade.

2 LONGITUDINAL STABILITY AND TRIM REQUIREMENTS

One of the basic problems with tailless airplanes involves longitudinal trim.
The requirement is: $C_{m_{cg}} = 0$.

Without a tail: $C_{m_{cg}} = - x_{cg}/c \;\; C_L + C_{m_0} = 0$,
with, x_{cg}/c , the static margin.

For trim, then: $C_{m_0} = sm \;\; C_L$.

If we consider conventional values for static margin, of .1 to .2, the wing must generate values of C_{m_0} that are very large. In fact with just a 5% static margin at an aft c.g., maximum speed condition, many airplanes will fly with a static margin of .35 to .40 at forward c.g. cruise conditions, requiring very large pitching moments about the wing aerodynamic center to trim. The problem is aggravated at higher C_L's as the required C_{m_0} increases linearly with C_L.

There are three approaches to generating these trimming moments without a tail:
1. Use airfoil sections with large, positive values of C_{m_0} (reflexed, self-trimming sections).
2. Rely on three-dimensional effects (sweep and twist) to make the wing C_{m_0} positive.
3. Employ active controls to make the static margin negative and require conventional negative values of wing C_{m_0}.

The following sections illustrate the use of each of these methods in recent tailless aircraft designs, addressing some of the advantages and problems with each approach.

3 TRIM BASED ON SECTION PROPERTIES

The Zanonia seed employs the first approach. Its airfoil section has an obvious reflexed curvature with a positive C_{m_0}. The seed itself is located near the leading edge of the wing to move the c.g. forward of the a.c.. The design is very stable and simple with a sink rate that surprises even an avid model builder. The Zanonia design problem is particularly simple since it need not accommodate a large c.g. range nor a range of lift coefficients. Thus, although the Reynolds number is very small, it is possible to design an airfoil with relatively good performance subject to these constraints. For aircraft, the section design problem is more difficult.

Section Design Issues

In order to obtain sufficient pitching moment from the section itself, tailless aircraft with typical static margins must use highly reflexed sections. Recall that the C_{m_0} needed to trim at a C_L of 1.0 is numerically equal to the static margin, a number of order 0.1. Figure 5 shows a section with a C_{m_0} of +0.1. The aft portion of the section is turned upward by a rather absurd amount and we are forced to sacrifice a great deal of lift that could have been carried by the aft part of the airfoil to obtain this pitching moment; the aft portion of the airfoil actually pushes down. With similar roof-top C_p levels and recovery gradients, a conventional section could produce more than twice the C_L. This is the fundamental compromise involved in achieving trim with positive section C_m. This extreme section gives a static margin of only 10% at $C_L = 1.0$.

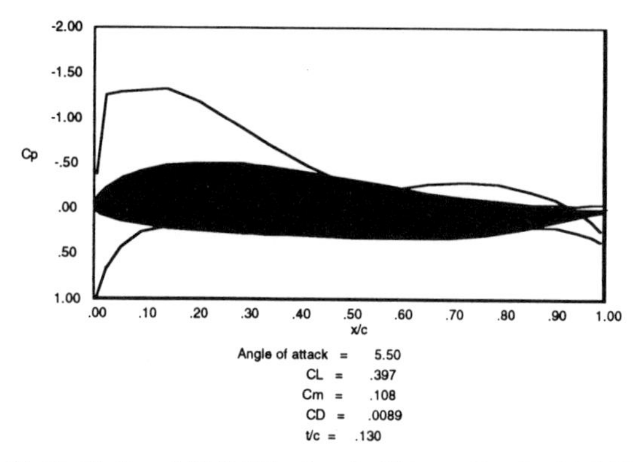

Figure 5. Highly Reflexed Airfoil Section for Trim at C_L =1 with 10% Static Margin.

Although several tailless designs have been flown using conventional sections with negative flap deflections or crude reflex modifications, a more structured approach to positive moment section design can yield considerably better results. By designing a section with a rather conventional upper surface pressure distribution, the less critical lower surface may be used to achieve the required moment with the minimum penalty in section performance. Positive C_p's on the forward lower surface combined with negative C_p's far aft with rather rapid recovery lead to sections with the potential for long runs of natural laminar flow on the lower surface with minimized lift reduction. The section in Figure 6 was designed by R. Liebeck [11] and achieves a respectable $C_{l_{max}}$ of 1.35 measured at a Reynolds number of 250,000, although the C_{m_0} is only +0.015. Modern design methods for airfoils are making it possible to use even this first, simple approach to tailless design without as much penalty as might have been required in the past.

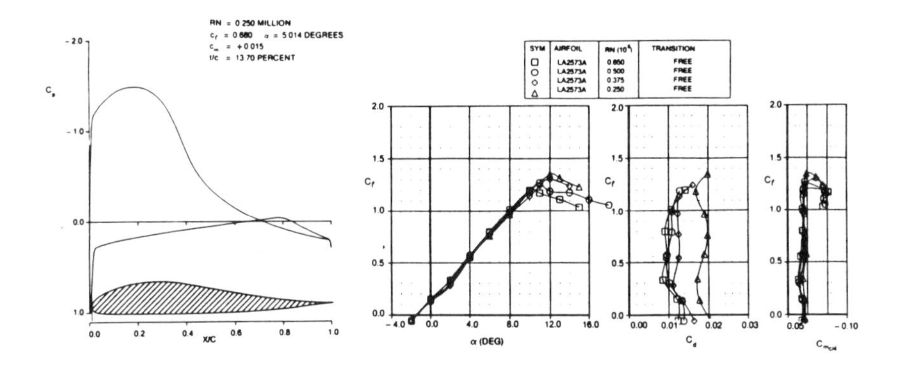

Figure 6. A Carefully-Designed Reflexed Section.

Additional Considerations

Even with well-designed airfoil sections, unswept or "plank"-type tailless aircraft often suffer from a variety of other problems. These include: 1. a very small tail arm and corresponding lack of directional stability and control; 2. a non-minimum-phase control response (large direct lift reduction before the aircraft can pitch up and increase lift); 3. generally poor maximum lift coefficients resulting from "negative" flap deflections and inability to employ conventional high lift devices; 4. very low pitch damping often producing poor handling qualities, including the possibility of tumbling [12].

Example: Pathfinder

One example of this design approach is the Pathfinder aircraft, built by AeroVironment about ten years ago, although details have just been released. This prototype of a high altitude long endurance airplane uses an unswept wing with a 100 ft span and 8 ft chord. Its low wing loading of 0.5 lbs/ft^2 leads to a flight speed of about 13 miles per hour (EAS). It is powered by eight electric engines using about 5KW provided by 30 lbs of Silver-Zinc batteries although it carried sample solar cells and aluminum plates to simulate the eventual power source. The airplane was stable in pitch with trailing edge pitch control surfaces and no vertical surfaces. The distributed engines provide yaw control through differential thrust and result in a very lightweight airframe, as the wing is span-loaded to minimize bending loads. The engines are also located so as to minimize torsion loads. This application of a simple tailless design was, like the Zanonia seed, successful because of its restricted design requirements and good design practice. There was no requirement for a high lift system, no c.g. range, a small C_L range, and the engines naturally provided yaw control.

4 TRIM THROUGH SWEEP AND TWIST

Although a positive pitching moment about the wing's aerodynamic center is required for stable, trimmed flight, each section need not produce a positive moment about its a.c.. By sweeping the wing back and including washout, the wing tips lie behind the a.c. and at zero lift, produce a downward force while the inner portion of the wing is lifting, thus producing a positive C_m at zero lift and hence about the a.c. at all C_L's. This was the idea behind the Horten and Northrop flying wings. The penalties associated with this twist may be large if not done carefully, but they may be negligible if the correct approach to wing design is taken.

It is reported [13] that the Horten's attempted to achieve a distribution of lift with the centroid of lift located at 33% of the semi-span. Indeed, measured lift distributions on the Horten IV suggest that this was achieved (Fig. 7). The result of this distorted lift distribution is a poor span efficiency, estimated in Ref. 13 based on the shape of the span loading, to be about 0.70 at a C_L of 1.0. This location of the lift centroid is required for trimmed flight with the desired static margins and the highly tapered planform shapes used for these sailplanes. The span loading for minimum induced drag subject to a constraint on centroid position is just the problem of minimum induced drag with fixed root bending moment, solved by R.T. Jones and others [14]. The closed form analytic result is:

$$1/e = 9/2\, \pi^2 \eta_c{}^2 - 12\, \pi\, \eta_c + 9$$

where η_c is the position of the centroid in units of the semi-span.

e is maximized for $\eta_c = 4/3\pi = 0.4244$, the centroid of the elliptic loading. When η_c is reduced to 33%, the maximum value of e is 0.72. In addition to its poor span efficiency, this sailplane suffered from a rather low $C_{L_{max}}$ of 1.13.

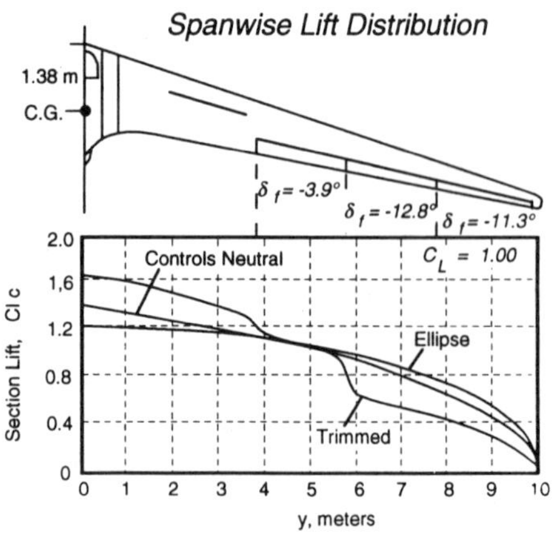

Figure 7. Span Load Distribution for the Horten IV Sailplane from Ref. 13.

To achieve stable trimmed flight without a penalty in span efficiency it is only necessary to create a wing with its aerodynamic center farther outboard (farther aft) than the 42.4% point. Although it is not obvious, computational studies have shown that this is quite possible for wings with sufficiently high aspect ratio, taper ratio, and sweep.

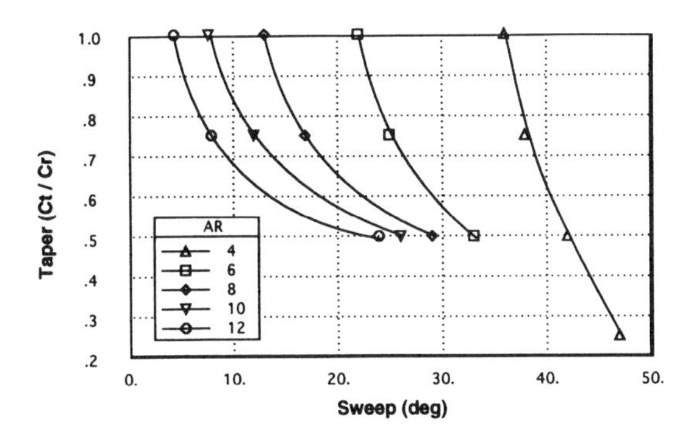

Figure 8. Combinations of Sweep and Taper Required to Move Aerodynamic Center Outboard of 45% Semi-Span. Computations Based on Lifting Suurface Analysis.

This approach may be viewed as illustrated below. The basic lift distribution (at zero lift) has negatively loaded tips to produce the positive C_m needed for trim. This inefficient shape (very non-elliptic) is combined with an additional lift distribution (due to angle of attack) that is also rather non-elliptic because it carries too much lift near the tip. The two distributions combine to form an elliptic distribution. Since the additional loading alone determines the stability, the airplane is stable as well as trimmed with elliptic loading.

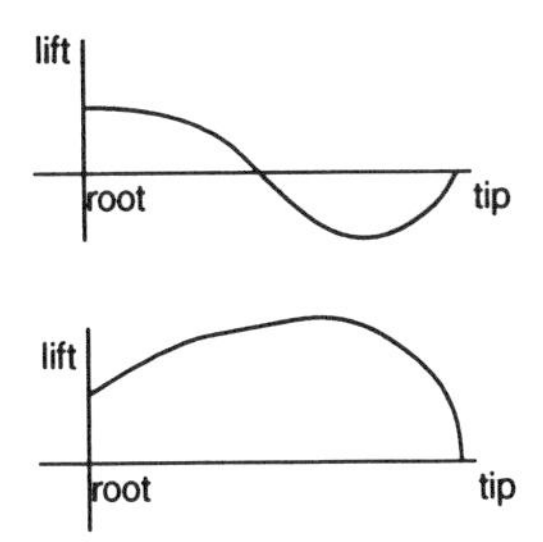

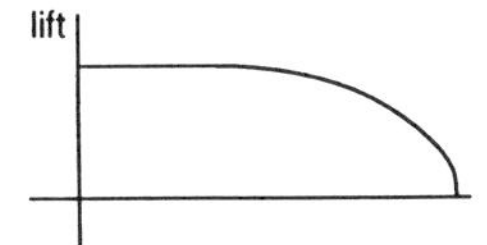

Figure 9. Combining a Very Non-Elliptic Basic Lift Distribution and Additional Loading to Obtain C_{m_0}, Aft a.c., and Elliptic Net Lift.

The aft position of the a.c. also permits the use of rather conventional high-lift devices on this tailless aircraft. Although small inboard flaps were included in early designs by Lippisch and by Buxton in the 1930's, the extent of the surfaces was very limited [2]. The adoption of a planform with outboard a.c. position makes a conventional plain flap of 50% to 60% of the span possible, increasing maximum lift coefficients to values similar to conventional configurations, and provides a desirable camber change for reduced drag and increased lift at low speeds.

Flaps of smaller extent may be used in place of elevons. Although the control authority is less substantial, inboard flaps for pitch control, suggested by Lippisch [15] and R.T. Jones [16], among others, have the following advantages. They avoid the non-minimum phase control of elevons (which reduce airplane lift, to produce pitch up, to increase airplane lift). They do not require control mixing and reduced aileron authority. They eliminate possibility of pitch control reversal, and remove the pitch control surfaces from stall-prone tips.

Example: SWIFT Development

These ideas studied in the design of a foot-launched sailplane, depicted below [17].

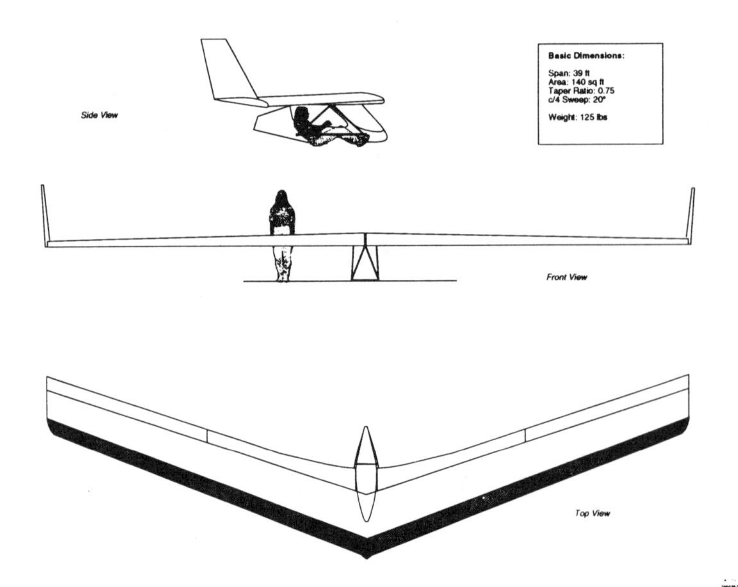

Figure 10. Basic Arrangement of the SWIFT Foot-Launched Sailplane.

Sections with small negative pitching moments were designed with interactive programs and included a great deal of iteration with the structural design. Because the airplane was to have a wing larger than that of a Cessna and yet weigh less than 100 lbs, thick sections were required. Pitching moment constraints along with concerns about surface finish lead to a conservative laminar flow design with a designed laminar run of about 30% of the upper surface, tripped before the start of recovery. The section design was complicated by the low Reynolds numbers (700,000 to 3,000,000), large C_l range, and the required inboard flap and elevon deflections.

Figure 11. Pressure Distributions on SWIFT Airfoil Sections at Two Angles of Attack. Note small negative moment, conservative laminar design.

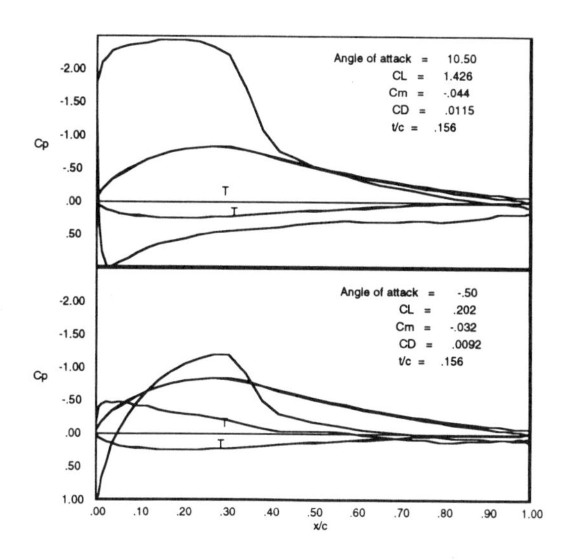

The planform itself was designed based on the previously mentioned ideas about span loading. The detailed sizing, control deflections, and twist distributions were determined by numerical optimization of the design based on a simulated cross-country cruising flight. The large tip chords on the resulting planform not only move the a.c. outboard but increase local Reynolds number (a problem with the Horten sailplanes), and lower local C_l 's reducing the tendency for tip stall and maintaining roll control to high angles of attack. The design's fixed winglets provide directional stability, increase the effective span, move the a.c. farther outboard, and interfere in a favorable way with the ailerons to improve turning coordination.

At this time about a dozen SWIFT's have been built by Bright Star Gliders of Santa Rosa, California. They have won several U.S. and international competitions with a top speed of 80 mph, a stalling speed of 23 mph, and a maximum L/D of about 25.

Aeroelasticity

The SWIFT's moderate aspect ratio, thick sections, and composite design make it quite stiff and aeroelastic problems have not been observed. However, many higher aspect ratio swept wings are strongly affected by aeroelasticity. Wing bending along the elastic axis increases washout in the streamwise direction and reduces the static margin [18]. The effect is often significant (a 10% reduction in static margin for an aspect ratio 20 wing with taper = .5, t/c = .15, C_L = .5 and σ/E = .001). Dynamic instabilities involving the same basic aerodynamic phenomena, coupling wing bending modes and airplane short-period modes, have also been observed—in some cases causing major design revisions [6]. For wings with low torsional stiffness elevon reversal is also possible with even more severe implications than conventional aileron reversal. The possibility of using elevon reversal constructively has also been explored in tests of hang gliders in the late 1970's [19]. In fact, hang gliders represent tailless designs incorporating sweep and twist that have been successful largely because of their exploitation of aeroelastic effects. At high angles of attack hang glider wings have (and need) large twist angles. Twist is reduced due to flexibility at low angles of attack where it would cause a performance penalty. Substantial changes in twist and section camber are produced at low angles of attack using the mechanisms shown in Figure 12. This results in a C_m curve shown below with an increase in stability at low angles of attack to avoid excess speed and reduce the possibility of tumbling.

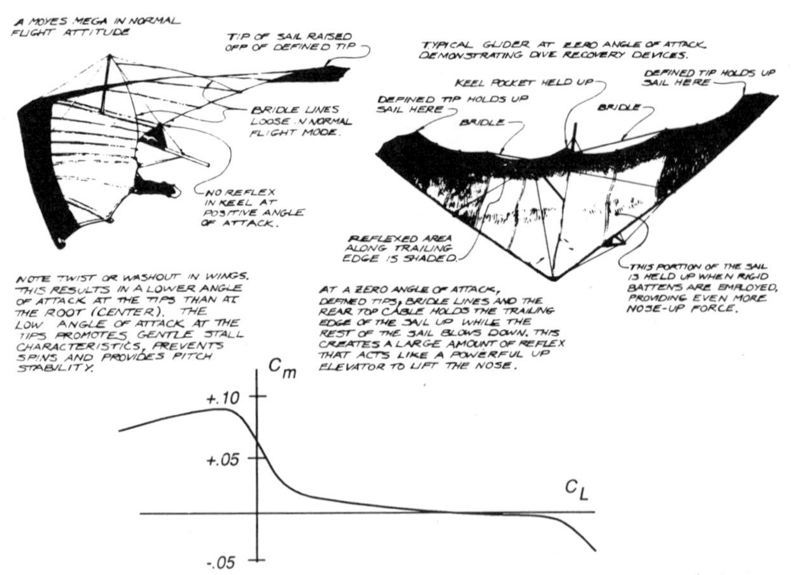

Figure 12. Constructive Use of Aeroeleasticity in Hang Gliders, Drawings by D. Pagen [20]. Below: Typical C_m Curve.

5 UNSTABLE AIRCRAFT AND ACTIVE CONTROL

Basic Approach / Controllability

Some of the difficulties with achieving trim through section design may be eliminated if we discard the requirement for static stability. In this case, as shown in Figure 13, a nose down (negative) moment is required about the section aerodynamic center. With conventional section C_m's of about -0.1, this would require a static margin of -10% (c.g. at 35% chord) at a C_L of 1.0. Trim at lower C_L requires a more aft c.g. position, or less camber—a desirable change. If this level of instability can be tolerated (through the use of active controls) many of the objections to the unswept tailless aircraft are answered: conventional sections and high lift devices may be used.

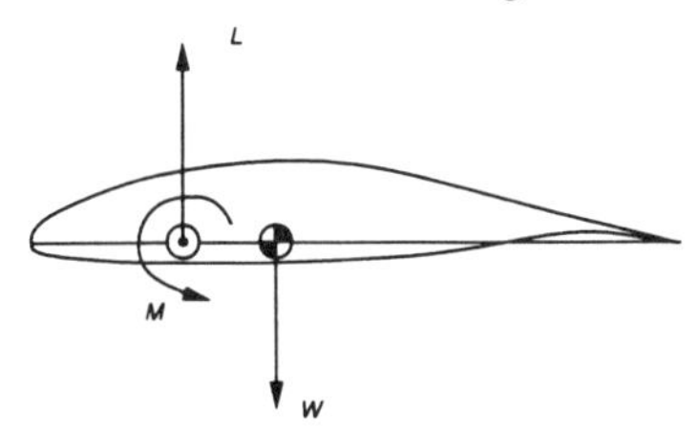

Figure 13. Arrangement of c.g. and a.c. for Unstable Trimmed Section.

If the trailing edge of the wing is the only pitch control, however, moving the c.g. aft reduces control effectiveness and at some point (near 20% unstable) the control effectiveness goes to zero as shown in figure 14. The range of c.g. position between 30% and 35% chord is intriguing, however, and this concept is being explored in a number of on-going projects.

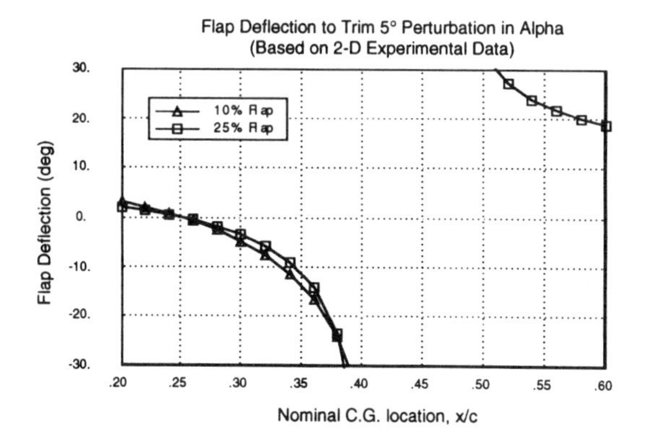

Figure 14. Flap Deflection Required to Trim 5° α Perturbation (based on 2-D experimental data).

To explore the use of active controls in a region in which controllability becomes an issue, Dr. Steve Morris at Stanford constructed an actively-controlled model shown in Figure 15. The aircraft weighed about 20 lbs with a 12 ft wing span and incorporated a custom-built 68020-based computer, high-speed servos, an angle of attack sensor and pitch-rate gyro. Many successful flights were performed with levels of instability ranging from 6.5% to 9%.

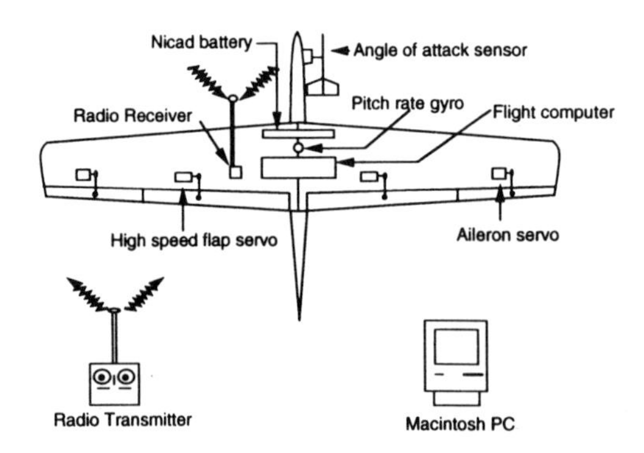

Figure 15. Unstable Flying Wing Model — Hardware Layout.

In-flight data recording (Fig. 16) shows how the closed loop response of the aircraft followed the pilot's commands at the 6.5% stability level. Large variations from the commanded performance, indicative of the reduction in control authority, appear during flights with 9% instability.

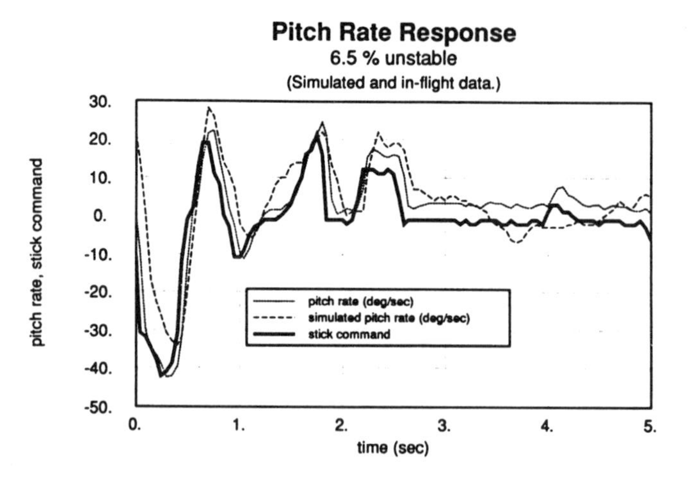

Figure 16. Flight Test Data from Unstable Model Tests.

Example: Oblique All-Wing Airplane

A direct application of this approach was suggested by G.H. Lee in 1961 [21]. The oblique flying wing was based on an idea proposed by R.T. Jones to reduce supersonic wave drag [22]. In this concept, passengers were accommodated inside a large obliquely-swept tailless airplane with the potential for large savings in structural weight (due to span loading and thick sections) and drag. In 1961, active control was not practical for a commercial aircraft and subsequent work focussed on oblique wing-body combinations [23]. However, the all-wing version of the oblique wing concept eliminates much of the undesirable wing-body interference, aeroelastic problems, and weight and drag of the wing-body design, leading to a renewed interest in the oblique all-wing configuration in the last few years. Research on the oblique all-wing aircraft at NASA Ames Research Center, Boeing, Douglas, and Stanford is discussed in Refs. 24-26. Current computations suggest lift to drag ratios that are somewhat higher than conventional designs at Mach 1.6, very high subsonic L/D, a potential for large structural weight savings, and a host of problems associated with passenger accommodations, stability and control.

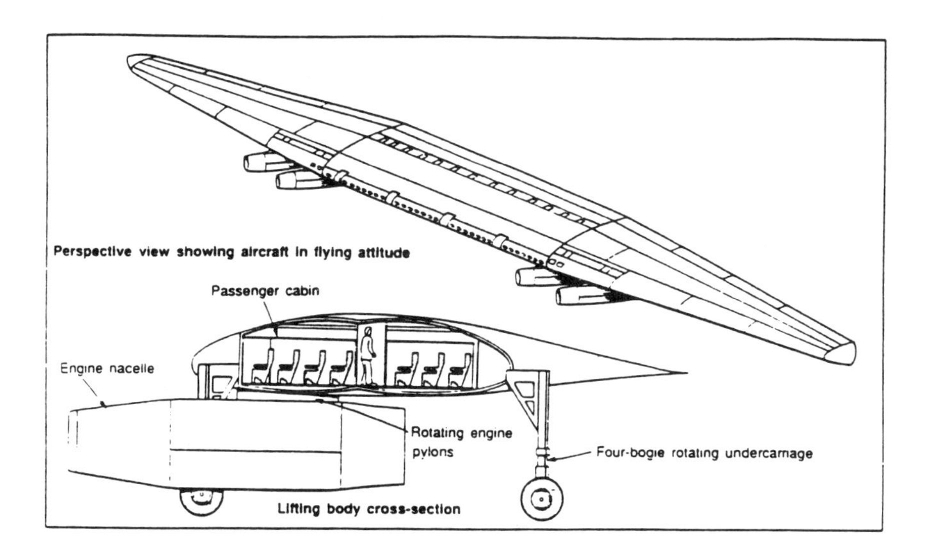

Figure 17. Artist's Concept of an Oblique Flying Wing SST.

Limiting the unswept span to about 400 ft based on airport compatibility considerations leads to an SST design for 450 passengers with an unswept aspect ratio somewhat less than 10, a gross weight of about 900,000 lbs with 4 pivoting engines permitting efficient low speed climb with low sweep angles and efficient cruise up to Mach 1.6 with 68° of sweep.

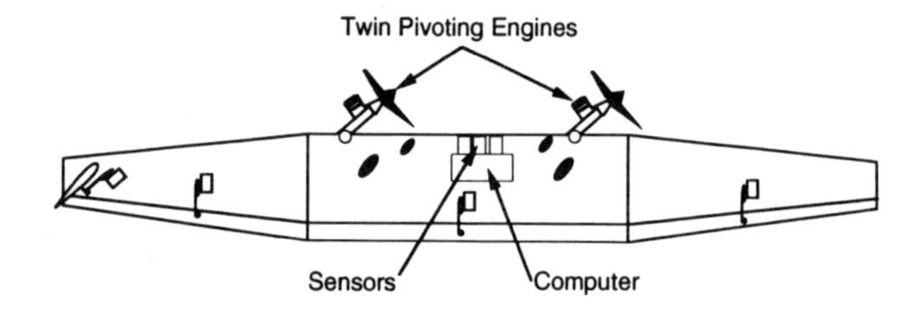

- 20 foot span, 50lb weight
- 7% Statically unstable in pitch
- Onboard computer and sensors for SAS
- Data Recording System

Figure 18. Small-Scale Oblique Wing Flight Research Aircraft.

Recent work at Stanford has focussed on the stability and control of the unstable, low control-authority configuration, starting with a 10 ft span, stable, radio-controlled version of the design. Figure 19 shows the characteristic roots of the equations of motions for the unaugmented model as a function of sweep angle. As the sweep is increased, the model becomes very difficult to fly

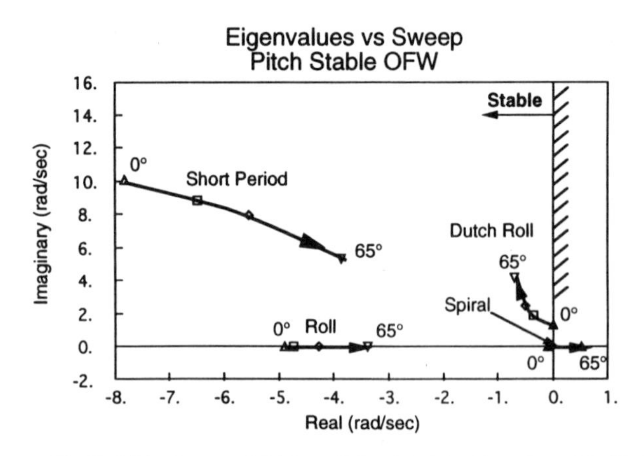

Figure 19. Stability Roots of Oblique Wing Model. Note Unstable "Spiral" Mode.

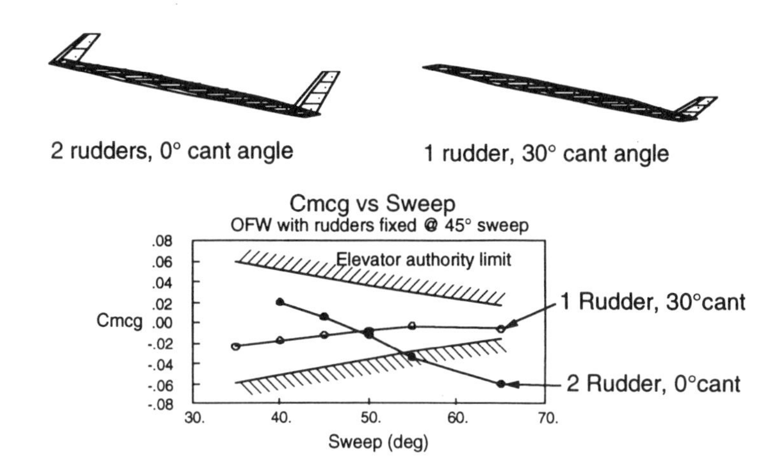

Figure 20. Coupling of Vertical Fins into "Pitch" Motion and Low Control Authority.

because of increased coupling and reduced control authority. Figure 20 shows why the original design with two vertical fins became uncontrollable in pitch (motion about the long axis) and crashed spectacularly, while the subsequent single, canted fin design was flown successfully. Figure 21 shows the unusual asymmetric spiral instability that makes turns to the right more difficult than those to the left. Many additional analyses, a wind tunnel test, and flight of a 20 ft span actively-controlled model are planned in the next year.

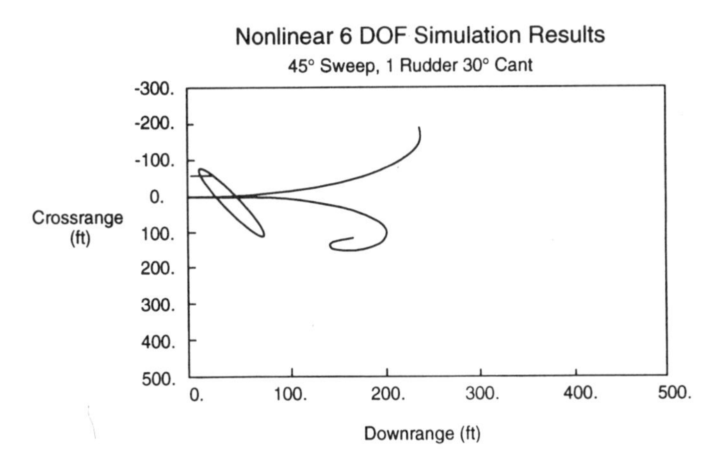

Figure 21. Asymmetric (nonlinear) Spiral Mode Makes Unaugmented Turns Difficult.

One alternative to conventional trailing edge controls that does not suffer from the loss of control effectiveness at aft c.g. positions is that used by birds. Shifting the position of the wing relative to the c.g., either by shifting the pilot weight (hang gliders) or by changing the wing sweep in flight (first suggested by Louis Mouillard in the late 1800's and used by the G.T.R. Hill Pterodactyl, Mk IV [2]) can permit the use of active control without controllability loss. This was the approach taken by AeroVironment in the development of its replica of the largest flying creature, the Quetzelcoatlus northropi. Actively-controlled fore and aft motion of the wings combined with a zero-moment section [11] produced successful gliding and powered flights of this model.

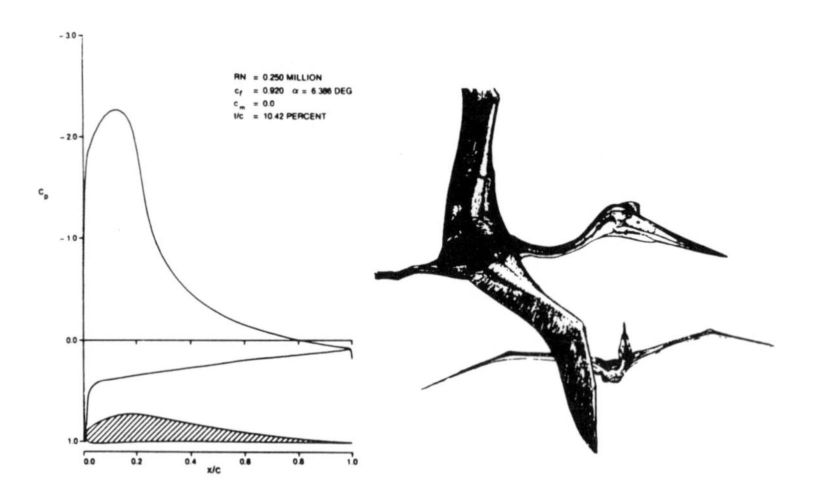

Figure 22. Sections for QN Model Using Active Pitch Control Through Wing Sweep.

Perhaps the most promising compromise for an airplane with active controls involves enhancing the advantages associated with swept tailless aircraft through relaxed stability. This approach, exemplified by the Northrop B-2 and suggested by Professor Sears [27], provides good controllability in pitch, increased pitch damping, and the potential for good lateral handling qualities [28].

6 CONCLUSIONS

The tailless configuration has been considered by airplane designers for one hundred years and has generally been regarded as inferior to the conventional aft-tail design. However, new roles and requirements, such as the need for low radar cross section or very large aircraft, and the growing acceptance of new technologies, such as active controls, change the rules of the game and may make tailless aircraft attractive alternatives for many new applications.

There is also continued interest in stable tailless aircraft and this paper has briefly described how proper combinations of sweep, taper, and twist can lead to stable, trimmed tailless aircraft without aerodynamic penalties. High aspect ratio, moderately tapered wings can be constructed to have very high span efficiencies over a wide range of C_L's and can accommodate reasonably large trailing edge flaps for good maximum lift performance.

ACKNOWLEDGEMENTS

I would like to acknowledge the work of Dr. Steven Morris, currently a post-doctoral student at Stanford, who has been centrally involved in many of the projects discussed here. R.T. Jones has also contributed to many of these ideas through frequent and cherished conversations. Colleagues at NASA Ames Research Center and Boeing are leading many of the projects mentioned here, and I appreciate our many discussions. I would also like to thank Bill Sears, who, in addition to inspiring my interest in aerodynamics, helped bring Richard Shevell to Stanford; Dick , in turn, helped bring me here so that I might pursue passions such as these.

REFERENCES

1. Kroo, I.: Tail Sizing for Fuel Efficient Transports, AIAA 83-2476, October 1983.

2. Weyl, A.R.: Tailless Aircraft and Flying Wings, A Study of Their Evolution and Problems, A Series of 10 Articles in *Aircraft Engineering*, Dec. 1944 - Nov. 1945.

3. Wooldridge, E.T.: *Winged Wonders, The Story of the Flying Wings*, Smithsonian Press, 1983.

4. Maloney, E.: *Northrop Flying Wings*, World War II Publications, 1975.

5. Coates, A.: *Jane's World Sailplanes*, Flying Books 1978.

6. Schweiger, J., Sensburg, O., Berns, H.: Aeroelastic Problems and Structural Design of a Tailless CFC Sailplane, MBB/LKE291/S/Pub 193, Mar. 1985.

7. Whitlow, D., Whitner, P.: Technical and Economic Assessment of Span-Distributed Loading Cargo Aircraft Concepts, NASA CR-144963, Boeing Commercial Airplane Co., 1976.

8. Grellmann, H.: B-2 Aerodynamic Design, AIAA 90-1802, Feb. 1990.

9. Kroo, I.: *Aerodynamics, Aeroelasticity, and Stability of Hang Gliders*, Ph.D. Thesis, Stanford University, 1983.

10. Seckel: *Stability and Control of Aircraft and Helicopters*, Academic Press, 1964.

11. Liebeck, R.: Subsonic Airfoil Design, *Applied Computational Aerodynamics*, P. Henne, ed., Progress in Astronautics and Aeronautics, **125**, AIAA, 1990.

12. Smith, A.M.O.: On the Motion of Tumbling Bodies, *Journal of the Aeronautical Sciences*, **20**, Feb. 1953.

13. Gyorgyfalvy, D.: Performance Analysis of the Horten IV Flying Wing, presented at the 8th OSTIV Congress, Cologne, Germany, June 1960.

14. Jones, R.T.: The Spanwise Distribution of Lift for Minimum Induced Drag of Wings Having a Given Lift and Root Bending Moment, NACA TN 2249, 1950.

15. Lippisch, A.: D.L.V. German Patent Spec. No. 558959, 1930.

16. Jones, R.T.: Notes on the Stability and Control of Tailless Airplanes, NACA TN 837, 1941.

17. Kroo, I., Beckman, E.: Development of the SWIFT—A Tailless Footlaunched Sailplane, *Hang Gliding*, Jan. 1991.

18. Kroo, I.: Aeroelasticity of Very Light Aircraft, *Recent Trends in Aeroelasticity, Structures, and Structural Dynamics*, Univ. of Florida Press, 1986.

19. Private communication with Roy Haggard of Ultralight Products, 1982.

20. Pagen, D.: Hang Glider Technology, A Pictorial Survey, *Hang Gliding*, Feb. 1981.

21. Lee, G.H.: Slewed Wing Supersonics, *The Aeroplane*, March 1961.

22. Jones, R.T.: New Design Goals and a New Shape for the SST, *Astronautics and Aeronautics*, Dec. 1972.

23. Kroo, I.: The Aerodynamic Design of Oblique Wing Aircraft, AIAA 86-2624, 1986.

24. Waters, M., Ardema, M., Roberts, C., Kroo, I.: Structural and Aerodynamic Considerations for an Oblique All-Wing Aircraft, AIAA 92-4220, August 1992.

25. Galloway, T., Gelhausen, P. Moore, M., Waters, M.: Oblique Wing Supersonic Transport Concepts, AIAA 92-4230, August 1992.

26. van der Velden, A., *Aerodynamic Design and Synthesis of the Oblique Flying Wing Supersonic Transport*, Ph.D. Thesis, Stanford University, 1991.

27. Sears, W.: Flying Wing Airplanes: The XB-35/YB-49 Program, AIAA 80-3036, 1980.

28. Morris, S.: Integrated Aerodynamics and Control System Design for Tailless Aircraft, AIAA 92-4604, August 1992.

Response of a Thin Airfoil Encountering a Strong Density Discontinuity

Frank E. Marble

1 INTRODUCTION

Airfoil theory for unsteady motion has been developed extensively, assuming the undisturbed medium to be of uniform density, a restriction accurate for motion in the atmosphere [1,2,3,4,5,6]. In some instances, notably for airfoils comprising fan, compressor and turbine blade rows, the undisturbed medium may carry density variations or "spots," resulting from non-uniformities in temperature or composition, of a size comparable to the blade chord. This condition exists for turbine blades [7,8] immediately downstream of the main burner of a gas turbine engine where the density fluctuations of the order of 50% may occur. Disturbances of a somewhat smaller magnitude arise from the ingestion of hot boundary layers into fans [9] and exhaust into hovercraft. Because these regions of non-uniform density convect with the moving medium, the airfoil experiences a time varying load and moment which we propose to calculate.

Now if the fluid may be treated as incompressible and the velocity disturbances (of the order of angle of attack α) small in comparison with the uniform free stream velocity U, then the density field, assumed known at some time, is given as $\rho\ (\xi - Ut,\ \eta)$. The complicating feature of the problem arises from the vorticity γ, normal to the $\xi - \eta$ plane, generated by the interaction of the convected density field with the pressure field generated by the airfoil. The vorticity satisfies the linearized relation

$$\left(\frac{\partial}{\partial t} + U\frac{\partial}{\partial \xi}\right)\gamma = \frac{1}{\rho^2}\ \mathbf{grad}\, p\ \ \mathrm{x}\ \mathbf{grad}\ \rho\Big|_{\perp} \tag{1}$$

Therefore, if **grad** ρ is large (of zeroth order), the vorticity γ is of the same mathematical order as the pressure field of the airfoil and hence of size comparable to that distributed on the airfoil camber line. Clearly, then, solution of the thin airfoil problem involves determination of the unknown field vorticity as well as the unknown vorticity shed from the trailing edge as a result of the unsteady motion. Note that in this problem, in contrast with the conventional problem of kinematically unsteady airfoils, the time dependence arises solely from the convection of the non-uniform density fluid past the airfoil. We may note also that when the density field is of perturbation order but the pressure field is of zeroth order, a related problem arises which has been examined by Marble & Candel [10] in a particular context, and by Goldstein [11] under more general circumstances.

The case that will be examined in this paper is the response of a plane lifting airfoil to the passage of a strong density discontinuity normal to the direction of uniform motion of the main stream. The vorticity which is created in the manner described by Equation (1) is then concentrated on the density discontinuity, generated at a rate proportional to the pressure gradient, normal to the flow direction, produced by the field of the airfoil. The problem will be formulated through representing the lifting plane airfoil as a sheet of vortex elements whose distribution is determined to satisfy the boundary conditions on the airfoil in the presence of i) the vorticity shed from the trailing edge of the airfoil and ii) the vorticity generated in the free stream by non-uniform density. Thus it will use techniques familiar from conventional formulation of thin airfoil theory but, because of the non-uniform density field, will require the introduction of some innovative features.

We first examine the field generated by a single vortex element in the presence of a strong plane density discontinuity convected by a uniform free stream of velocity U parallel the horizontal axis. A novel and convenient representation of this field is found which greatly simplifies determination of the vorticity distribution on the airfoil. The determination of the distribution of wake vorticity raises another issue because, as the density jump passes over the airfoil, Kelvin's theorem may not be applied in the usual fashion. However, through considering the local impulsive generation of airfoil loading [2,5], the conventional relationship between airfoil circulation and shed vorticity may be recovered for the airfoil in a non-uniform density field.

2 VORTEX ELEMENT NEAR A MOVING DENSITY JUMP

When the fluid density is uniform, an element of vorticity $\gamma(\xi_1)\, d\xi_1$, where $\gamma(\xi_1)$ is the vorticity distribution on the airfoil, has a complex potential

$$w_o(\zeta,t) = \frac{i\gamma(\xi_1,t)d\xi_1}{2\pi}\, ln\,(\zeta-\xi_1) \tag{2}$$

where $\zeta = \xi + i\eta$ is the complex variable in the airfoil plane and ξ_1 is the coordinate of the vorticity element. When, however, the flow has a density jump from ρ_1 on the right to ρ_2 on the left, Figure 1, the potential of the field has a discontinuity at $\xi = \lambda(t)$ but is regular on either side of this discontinuity. Therefore the actual field of the vorticity element requires the addition of a potential $w_1(\zeta, t)$ for $\xi > \lambda$ and a potential $w_2(\zeta, t)$ for $\xi < \lambda$. These potentials must vanish in the far field, produce equal values of $u(\lambda, \eta)$, the ξ-velocity component at $\xi = \lambda$, and equal values of the perturbation pressure on either side of the discontinuity, $p_1(\lambda, \eta) = p_2(\lambda, \eta)$. These pressure perturbations are related to the potentials through the Bernoulli integral,

$\frac{\partial \phi_i}{\partial t} + U\left(\frac{\partial \phi_o}{\partial \xi} + \frac{\partial \phi_1}{\partial \xi}\right) = \frac{p_i}{\rho_i}$ and the condition to be satisfied at $\xi = \lambda(t)$ is

$$\rho_1\left(\frac{\partial \phi_1}{\partial t} + U\frac{\partial \phi_1}{\partial \xi}\right) - \rho_2\left(\frac{\partial \phi_2}{\partial t} + U\frac{\partial \phi_2}{\partial \xi}\right) = -(\rho_1 - \rho_2)U\frac{\partial \phi_0}{\partial \xi} \tag{3}$$

In satisfying this condition it must be kept in mind i) that the potentials depend upon $\lambda(t)$ and that $\frac{d\lambda}{dt} = U$, and ii) that the strength of the vorticity elements representing the airfoil will depend upon time because of the unsteady flow field.

The supplementary potentials w_1 and w_2 are, of course, analytic in the entire ζ-plane although only a portion of the field is used in each case to describe the physical solution. These potentials may be deduced by Fourier methods but it is physically more rewarding to recognize that they should have the nature of a partial image of the actual vortex element. They are, in fact

$$w_1(\zeta,t) = \frac{i\gamma(\xi_1,t)d\xi_1}{2\pi}\left(\frac{\rho_1 - \rho_2}{\rho_1 + \rho_2}\right)\ln\big((\zeta - \lambda) + (\xi_1 - \lambda)\big) \tag{4}$$

$$w_2(\zeta,t) = \frac{i\gamma(\xi_1,t)d\xi_1}{2\pi}\left(\frac{\rho_1 - \rho_2}{\rho_1 + \rho_2}\right)\ln\big((\zeta - \lambda) - (\xi_1 - \lambda)\big) \tag{5}$$

which hold for $\xi_1 > \lambda$, the situation shown in Figure 1. Specifically for $\xi > \lambda$, the complex potential $w_o + w_I$ consists of the actual vortex element plus a vortex at the image point with strength $(\rho_1 - \rho_2) / (\rho_1 + \rho_2)$ times that of the original. On the other hand, when $\xi < \lambda$ the complex potential $w_o + w_2$ consists of the original vortex plus a coincident vortex of strength $(\rho_1 - \rho_2) / (\rho_1 + \rho_2)$ times that of the original.

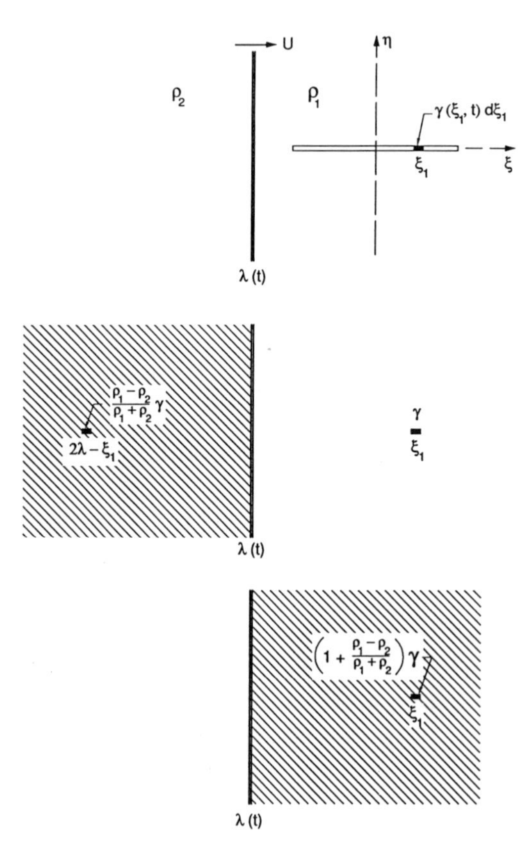

Figure 1. Construction of field induced by a vortex near a moving density jump.

When the density jump has moved downstream of the vortex element, $\lambda > \xi_1$, the corresponding supplementary potentials are

$$w_1(\zeta,t) = -\frac{i\gamma(\xi_1,t)d\xi_1}{2\pi}\left(\frac{\rho_1 - \rho_2}{\rho_1 + \rho_2}\right)\ln\big((\zeta-\lambda)-(\xi_1-\lambda)\big) \tag{6}$$

$$w_2(\zeta,t) = -\frac{i\gamma(\xi_1,t)d\xi_1}{2\pi}\left(\frac{\rho_1-\rho_2}{\rho_1+\rho_2}\right)\ln\big((\zeta-\lambda)+(\xi_1-\lambda)\big) \tag{7}$$

The complementary potentials, Equations (4) & (5), allow calculation of the vorticity distribution on the discontinuity,

$$\gamma_\lambda(\eta) = -v(\lambda+,\eta)+v(\lambda-,\eta) = \Im\frac{d}{d\zeta}(w_1-w_2)_{\xi=\lambda} \tag{8}$$

When the density jump is upstream of the vorticity element, $\lambda < \xi_1$, we find

$$\gamma_\lambda(\eta) = \frac{\gamma(\xi_1,t)d\xi_1}{\pi}\left(\frac{\rho_1-\rho_2}{\rho_1+\rho_2}\right)\frac{\xi_1-\lambda}{(\xi_1-\lambda)^2+\eta^2} \tag{9}$$

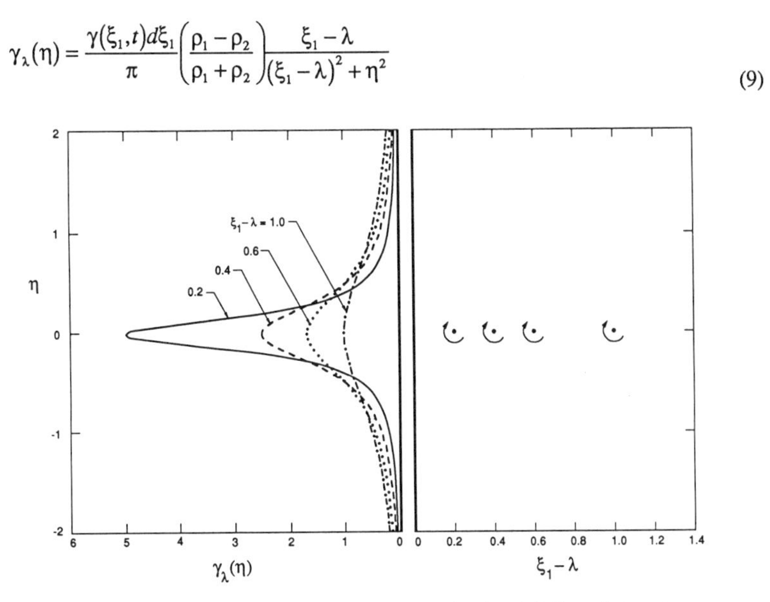

Figure 2. Vorticity induced on density discontinuity by airfoil vorticity element at various positions.

and this vorticity distribution is shown in Figure 2 for several relative locations of the density jump and the vorticity element. As the density jump convects toward the vorticity element $\lambda(\xi_1, t)$, the vorticity distribution $\gamma_\lambda(\eta)$ concentrates nearer the airfoil axis and has a higher maximum value. It is also easily shown, either by application of Kelvin's theorem or by direct integration of Equation (9), that the total vorticity on the density jump is

$$\int_{-\infty}^{\infty} \gamma_\lambda(\eta) d\eta = \frac{\gamma(\xi_1,t)d\xi_1}{2\pi}\left(\frac{\rho_1-\rho_2}{\rho_1+\rho_2}\right) \tag{10}$$

the strength of the "image vortex" we employed to construct the solution. After the density jump has passed downstream of the airfoil vorticity element, the vorticity distributions repeat themselves but with opposite sign, as might be inferred from the appropriate "image vortex" given by Equations (6) & (7).

From the viewpoint of airfoil theory, the important consequence of this calculation is the downwash velocity, $v(\xi, 0)$, which is induced at the plane of the airfoil, that is $-1 \le \xi_1 \le 1$. When the density jump is upstream of the airfoil vortex element, $\xi_1 > \lambda$, the vertical velocity induced by the density jump is

$$v(\xi,0) = -\left(\frac{\rho_1-\rho_2}{\rho_1+\rho_2}\right)\frac{\gamma(\xi_1,t)d\xi_1}{2\pi}\frac{1}{(\xi-\lambda)+(\xi_1-\lambda)} \quad ; \ \xi > \lambda \tag{11}$$

$$v(\xi,0) = -\left(\frac{\rho_1-\rho_2}{\rho_1+\rho_2}\right)\frac{\gamma(\xi_1,t)d\xi_1}{2\pi}\frac{1}{(\xi-\lambda)-(\xi_1-\lambda)} \quad ; \ \xi < \lambda \tag{12}$$

and after the density jump has passed the vorticity element, $\xi_1 < \lambda$, the corresponding downwash velocity is

$$v(\xi,0) = -\left(\frac{\rho_1-\rho_2}{\rho_1+\rho_2}\right)\frac{\gamma(\xi_1,t)d\xi_1}{2\pi}\frac{1}{(\xi-\lambda)+(\xi_1-\lambda)} \quad ; \ \xi < \lambda \tag{13}$$

$$v(\xi,0) = -\left(\frac{\rho_1-\rho_2}{\rho_1+\rho_2}\right)\frac{\gamma(\xi_1,t)d\xi_1}{2\pi}\frac{1}{(\xi-\lambda)-(\xi_1-\lambda)} \quad ; \ \xi > \lambda \tag{14}$$

The downwash velocity induced by the vortex sheet on the density jump is shown in Figure 3 for several positions of the density jump with respect to the vortex element. As is evident from the potentials w_1 and w_2, the downwash is anti-symmetric about the density jump and has maximum absolute value at the density jump of magnitude proportional to $1 / |\xi_1 - \lambda|$.

It is important to note that the foregoing discussion pertains only to vortices that are in relative motion with respect to the density jump. In contrast, when we consider the vortex elements that comprise the wake, the situation is quite different. If we follow a particular vortex element that has been shed into the wake, it i) moves with the free stream velocity U, and ii) it is of constant strength. It therefore induces a steady flow of perturbation order and hence generates a pressure field of second order at the density jump and, as a consequence, influences the vorticity on the discontinuity to only the second order. Thus, the potential associated with a vortex element moving in the wake has a potential continuous across the density jump to the order of our present calculation.

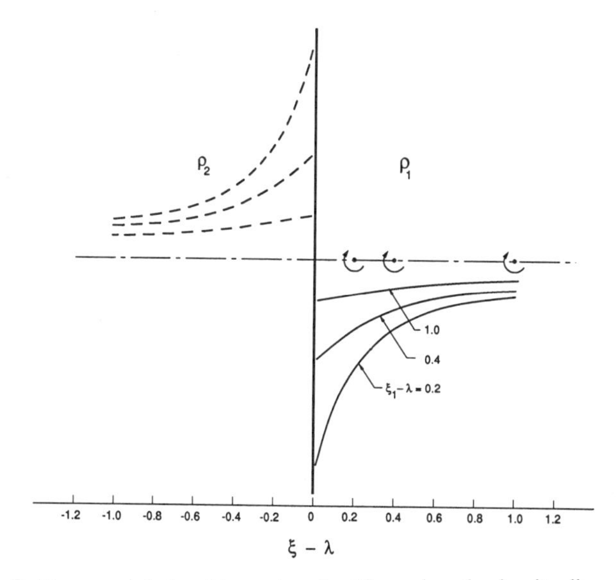

Figure 3. Downwash induced by vortex sheet formed on the density discontinuity.

3 VORTICITY DISTRIBUTION ON A THIN AIRFOIL

The problem of thin airfoil theory is to determine the vorticity distribution along the camber line that provides flow tangential to the given airfoil shape $\eta_1(\xi)$ and regularity at the trailing edge. For steady flow of uniform density about this airfoil shape, the vorticity distribution $\gamma_o(\xi_1)$ is assumed known. For our problem, however, additional components of downwash are induced at the airfoil by i) the vorticity on the density discontinuity and ii) the wake vorticity resulting from the unsteady character as a consequence of the convective motion of the density field. This additional downwash then necessitates a supplementary vorticity distribution $\gamma_1(\xi_1)$ on the airfoil to satisfy the boundary conditions. It is this vorticity distribution that must be found to allow pressure, force and movement on the airfoil to be calculated.

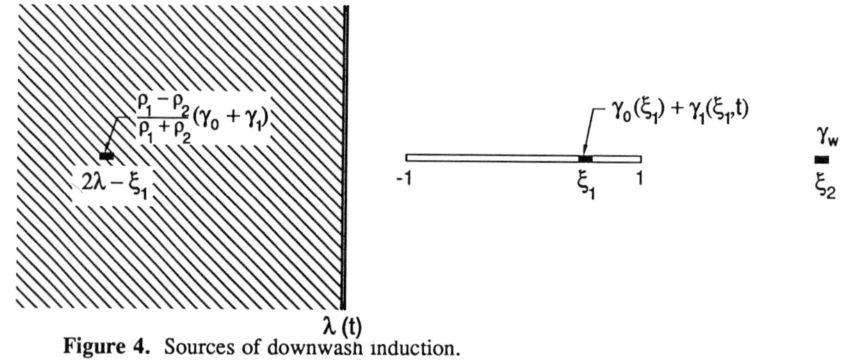

Figure 4. Sources of downwash induction.

The sources of downwash are sketched in Figure 4. At each point ξ on the airfoil, the induced vertical velocity $v(\xi,0)$ must cause the fluid to move tangentially to the airfoil surface, that is

$$v(\xi,0)=U\frac{d\eta_1}{d\xi} \tag{15}$$

But the known vorticity distribution $\gamma_o(\xi_1)$ induces a flow that satisfies the condition given by Equation (15).

$$\frac{1}{2\pi}\int_{-1}^{1}\frac{\gamma_o(\xi_1)}{\xi-\xi_1}d\xi_1=U\frac{d\eta_1}{d\xi} \tag{16}$$

and consequently the remaining induction must give zero vertical velocity on the line segment $-1<\xi\leq 1$.

The remaining induction falls into three categories:

1. The direct induction of the supplementary vorticity distribution $\gamma_1(\xi_1)$ on the airfoil itself

$$\frac{1}{2\pi}\int_{-1}^{1}\frac{\gamma_1(\xi_1,t)}{\xi-\xi_1}d\xi_1 \tag{17}$$

2. The induction of the vorticity distribution on the density jump resulting from the airfoil vorticity $\gamma_o(\xi_1)+\gamma_1(\xi_1,t)$. Using the technique introduced in Section 2, this is

$$\frac{1}{2\pi}\left(\frac{\rho_1-\rho_2}{\rho_1+\rho_2}\right)\int_{-1}^{\lambda}\frac{\gamma_o+\gamma_1}{\xi-2\lambda+\xi_1}d\xi_1-\frac{1}{2\pi}\left(\frac{\rho_1-\rho_2}{\rho_1+\rho_2}\right)\int_{\lambda}^{1}\frac{\gamma_o+\gamma_1}{\xi-\xi_1}d\xi_1 \tag{18}$$

for values of $\xi<\lambda$, and

$$\frac{1}{2\pi}\left(\frac{\rho_1-\rho_2}{\rho_1+\rho_2}\right)\int_{-1}^{\lambda}\frac{\gamma_o+\gamma_1}{\xi-\xi_1}d\xi_1-\frac{1}{2\pi}\left(\frac{\rho_1-\rho_2}{\rho_1+\rho_2}\right)\int_{\lambda}^{1}\frac{\gamma_o+\gamma_1}{\xi-2\lambda+\xi_1}d\xi_1 \tag{19}$$

for values of $\lambda>\xi$.

3. The induction of the wake vorticity

$$\frac{1}{2\pi}\int_{1}^{\infty}\frac{\gamma_w}{\xi-\xi_2}d\xi_2 \tag{20}$$

The expressions in Equations (18), (19) and written explicitly for the position $\lambda(t)$ of the density jump lying between the leading and trailing edges of the airfoil. When the density jump is upstream of the leading edge of the airfoil, only the second term in each expression survives; similarly, when the density jump is downstream of the trailing edge, only the first term remains in each expression.

For points ξ on the airfoil where $\lambda<\xi\leq 1$, the sum of these parts vanishes at each point ξ of the range.

$$\int_{-1}^{1}\frac{\gamma_1}{\xi-\xi_1}d\xi_1+\left(\frac{\rho_1-\rho_2}{\rho_1+\rho_2}\right)\left(\int_{-1}^{\lambda}\frac{\gamma_1+\gamma_o}{\xi-\xi_1}d\xi_1-\int_{\lambda}^{1}\frac{\gamma_1+\gamma_o}{\xi-2\lambda+\xi_1}d\xi_1\right)+\int_{1}^{\infty}\frac{\gamma_w}{\xi-\xi_2}d\xi_2=0 \tag{21}$$

A corresponding integral equation, utilizing Equation (18) rather then Equation (19), holds at each value of ξ ;in the range $-1<\xi<\lambda$. When the wake vorticity γ_w is functionally related to the supplementary vorticity $\gamma_1(\xi_1, t)$, these relations constitute an integral equation determining γ_1 as a function of position on the airfoil and time.

4 VORTICITY DISTRIBUTION IN THE WAKE

The strength and distribution of the wake vorticity is determined by the vorticity $\gamma'_w(\tau)$ shed at any time τ, and by the fact that, in the linearized approximation, this vorticity element is transported with free stream velocity U along the horizontal axis. Then the vorticity in the wake is just

$$\gamma_w\left(t - \frac{\xi_2 - 1}{U}\right) = \gamma'_w(\tau) \tag{22}$$

that is, the wake vorticity at a position $\xi_2 > 1$ at time t is equal to that shed from the trailing edge at the earlier time

$$\tau = t - \frac{\xi_2 - 1}{U} \tag{23}$$

The usual argument to demonstrate that the shed vorticity element $\gamma'_w(t)\, U\, dt$ is given by the negative of the change of total circulation about the airfoil in the same period

$$\gamma'_w(t) U\, dt = -\frac{d\Gamma}{dt} dt \tag{24}$$

involves the application of Kelvin's theorem to a contour encompassing the entire airfoil. And although this familiar argument may be carried out when the density jump is either upstream or downstream of the airfoil, the presence of the discontinuous potential raises complications as the density jump passes over the airfoil. Instead, we shall employ the concept of the impulsive generation of lift described by Burgers [2] which was used by Kármán & Sears [5] as the basis of their unsteady airfoil theory.

An impulsive load δI applied along an element $\delta\xi$ of the horizontal axis creates a vortex pair, each vortex having a circulation Γ, according to the relation $\delta I = \rho\Gamma\delta\xi$ where ρ is density of the fluid in which impulse is applied. This vortex pair then drifts downstream with the free stream velocity U. A succession of impulses applied to the same element $\delta\xi$ produces a corresponding succession of vortex pairs drifting downstream along the ξ axis. Now, let us choose the time intervals between these impulses to be $\delta t = \delta\xi/U$ and the impulses to be of the same magnitude

$$\delta I = \int_0^{\delta t} f dt$$

f being the equivalent steady load on the element. Then each impulse generates a vortex at the downstream edge of the element equal and opposite to the vortex transported from the leading edge and which was generated by the previous impulse. As a consequence there remains only the vortex of circulation Γ at the upstream edge of the element and a vortex of strength $-\Gamma$, the one generated at the downstream edge of the element by the initial impulse, drifting far downstream. The former constitutes the "bound vortex" representing the lift on the element (more appropriately designated the "regenerate" vortex) while the latter constitutes the "starting vortex." This configuration constitutes the basis for the Kármán-Sears [5] formulation of the theory of airfoils in non-uniform motion.

The complete airfoil consists of a distribution of such elements, each with its peculiar load which may vary with time, and each producing its succession of vortex pairs. Figure 5 illustrates this representation by an airfoil of three elements. At the start of the motion, $t = 0$, each element produces a vortex pair according to its load and each of these vortex pairs moves downstream with the undisturbed velocity U. A second set of impulses is applied at a time $\delta\xi/U$ later, the resulting set of vortex pairs moves correspondingly and each successive set of impulses contributes its own set of vortex pairs. The residual vortices moving away from the trailing edge constitute the vortex wake.

Now, if we choose a point A between two elements of the airfoil, the field to the left of this point consists of a collection of vortex pairs. Furthermore, if we move with the undisturbed fluid, this region continues to consist of vortex pairs, more being generated at each successive impulse. As a consequence, the circulation in the region does not change with time in spite of the fact that the loading of the airfoil may be changing.

This situation may be interpreted as shown also in Figure 5. As the contour moves with the fluid, the time derivative of the circulation Γ about this contour vanishes. Furthermore, any contour that moves with the fluid across the airfoil has a circulation that does not change, simply because any loading changes that occur create vortex pairs which remain within the contour in question.

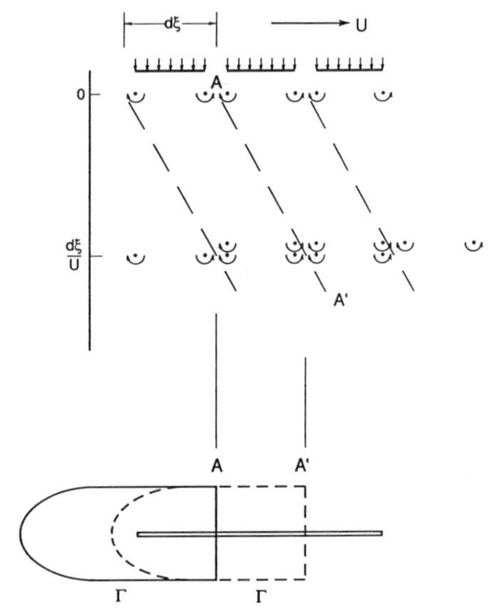

Figure 5. Generation of lift by successive impulses – conservation of vortex pairs.

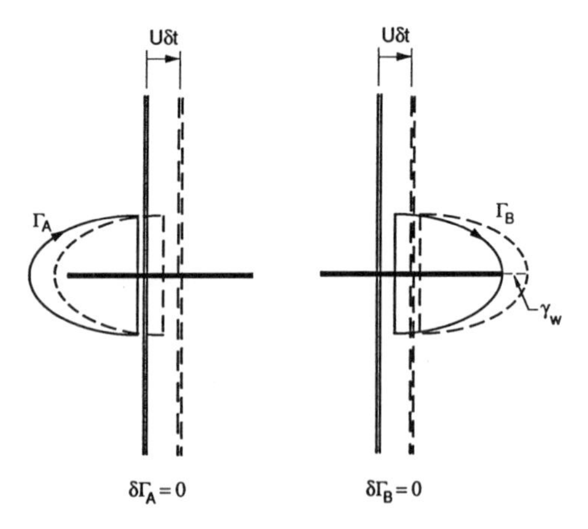

Figure 6. Relation of shed vorticity to airfoil circulation.

These observations are particularly useful in determining the shed vorticity as the density discontinuity is passing over the airfoil. Referring to Figure 6, consider the contour including the portion of the airfoil to the left of the discontinuity, lying entirely within the fluid of density ρ_2. Our argument regarding the elementary impulses holds and, as a consequence, the circulation Γ_A does not change with time. On the other hand, consider the contour which contains the remaining portion of the airfoil and lying entirely within the fluid of density ρ_1. Again, as we move with the fluid, the circulation Γ_B about this contour does not change. Hence, as it moves downstream, the contour covers an element $U\delta t$ of the ξ-axis comprising a vorticity element $\gamma'_w(t)U\,\delta t$ of the wake vorticity. Now in the time δt

$$\delta\Gamma_A + \delta\Gamma_B = \delta\int_{-1}^{1}\gamma_1(\xi_1,t)d\xi_1 + \gamma'_w(t)U\delta t = 0$$

so that the shed wake velocity is

$$\gamma'_w(t) = -\frac{1}{U}\frac{d}{dt}\int_{-1}^{1}\gamma_1(\xi_1,t)d\xi_1 \tag{25}$$

The usual relationship between airfoil circulation and shed vorticity holds also as the density jump passes over the airfoil.

Equation (25), together with the relation given in Equation (23), allows the wake vorticity distribution $\gamma_w(\xi_1, t)$ to be expressed in terms of $\gamma_1(\xi_1, t)$ so that Equation (22) becomes an integral equation for $\gamma_1(\xi_1, t)$.

5 PRESSURE DISTRIBUTION, LIFT AND MOMENT

If we denote by u_+, p_+ and u_-, p_- the velocity and pressure perturbations on the upper and lower surfaces of the airfoil, then the linearized momentum equation gives

$$\frac{\partial u}{\partial t} + U\frac{\partial u}{\partial \xi} = -\frac{1}{\rho}\frac{\partial p}{\partial \xi} \tag{26}$$

where the density may be ρ_1 or ρ_2 depending upon the region in which Equation (26) is applied. Now the vorticity representing the airfoil, $\gamma(\xi, t)$, is equal to $u_+ - u_-$ and $p_+ - p_- = \Delta p$ is the airfoil load per unit area in the downward direction. By writing Equation (26) for the upper and lower surfaces and subtracting the results we obtain the following relation between the vorticity and the airfoil loading

$$\frac{\partial \gamma}{\partial t}+U\frac{\partial \gamma}{\partial \xi}=-\frac{1}{\rho}\frac{\partial \Delta p}{\partial \xi} \tag{27}$$

When the fluid density is uniform this may be integrated from the trailing edge, $\xi = 1$, at which point $\gamma=0$ and $\Delta p=0$. Then

$$\rho\int_{\xi}^{1}\frac{\partial \gamma}{\partial t}d\xi-\rho U\gamma(\xi,t)=\Delta p(\xi,t) \tag{28}$$

and when the density jump is upstream of the leading edge of the airfoil, this holds for $-1 < \xi \leq 1$ and $\rho = \rho_1$. The lift on the airfoil is just

$$L=\int_{-1}^{1}(-\Delta p)d\xi=\rho_1 U\int_{-1}^{1}\gamma(\xi)d\xi-\rho_1\frac{d}{dt}\int_{-1}^{1}d\xi\int_{\xi}^{1}\gamma(\xi_1)d\xi_1$$

$$=\rho_1 U\Gamma-\rho_1\frac{d}{dt}\int_{-1}^{1}(1+\xi)\gamma(\xi)d\xi \tag{29}$$

Likewise, after the density jump has passed downstream of the trailing edge of the airfoil, the lift is given by an expression identical with Equation (29) with the exception that ρ_1 is replaced by ρ_2. In each case, $\gamma(\xi, t) = \gamma_o(\xi) + \gamma_1(\xi, t)$ and consequently

$$\Gamma(t)=\int_{-1}^{1}\gamma_o(\xi)d\xi+\int_{-1}^{1}\gamma_1(\xi,t)d\xi=\Gamma_o+\Gamma_1(t) \tag{30}$$

When the density jump lies between the leading and trailing edges of the airfoil, Equation (28) holds with $\rho=\rho_1$ so long as $\xi>\lambda$. For the range $\lambda>\xi>-1$, Equation (27) must be integrated from $\xi=\lambda$ to the left, which gives

$$\rho_2\int_{\xi}^{\lambda}\frac{\partial \gamma}{\partial t}d\xi+\rho_2 U\big(\gamma(\lambda)-\gamma(\xi)\big)=-\big(\Delta p(\lambda)-\Delta p(\xi)\big) \tag{31}$$

Now the conditions at the density jump require that the pressure and the horizontal velocity components are continuous. As a consequence, $\Delta p(\xi)$ and $\lambda\,(\xi)$ are continuous at $\xi = \lambda$, so that when we write the loading integral, Equation (28), for the range $\lambda < \xi \leq 1$,

$$\rho_1 \int_\lambda^1 \frac{\partial \gamma}{\partial t} d\xi - \rho_1 U \gamma(\lambda) = \Delta p(\lambda) \tag{32}$$

where the values of $\gamma\,(\lambda)$ and $\Delta p\,(\lambda)$ are identical with those occurring in Equation (32).

Note further that because $d\lambda/dt = U$, Equation (32) may be written

$$\rho_1 \int_\lambda^1 \frac{\partial \gamma}{\partial t} d\xi - \rho_1 U \gamma(\lambda) = \rho_1 \frac{d}{dt} \int_{\lambda(t)}^1 \gamma(\xi, t) d\xi = \Delta p(\lambda) \tag{33}$$

and similarly for Equation (31)

$$\rho_2 \frac{d}{dt} \int_\xi^\lambda \gamma(\xi, t) d\xi - \rho_2 U \gamma = -\Delta p(\lambda) + \Delta p(\xi) \tag{34}$$

The value of the pressure loading $\Delta p(\lambda)$ at the discontinuity may be eliminated between Equation (33) and Equation (34) to give the loading $\Delta p(\xi)$ in the range $-1 < \xi < \lambda$

$$\Delta p(\xi) = \rho_2 U \gamma(\xi) + \rho_1 \frac{d}{dt} \int_\lambda^1 \gamma(\xi) d\xi + \rho_2 \frac{d}{dt} \int_\xi^{\lambda 1} \gamma(\xi) d\xi \tag{35}$$

Together with the relation valid for the range $\lambda < \xi < 1$

$$\Delta p(\xi) = -\rho_1 U \gamma(\xi) + \rho_1 \frac{d}{dt} \int_\xi^1 \gamma(\xi) d\xi \tag{36}$$

the loading is then determined for all points of the airfoil.

The lift may now be computed, using Equations (35) & (36), during passage of the discontinuity over the airfoil

$$L = \int_{-1}^{1}(-\Delta p)d\xi = \rho_2 U\int_{-1}^{\lambda}\gamma(\xi)d\xi + \rho_1 U\int_{\lambda}^{1}\gamma(\xi)d\xi$$

$$-\int_{-1}^{\lambda}d\xi\left\{\rho_1\frac{d}{dt}\int_{\lambda}^{1}\gamma(\xi_1)d\xi_1 + \rho_2\frac{d}{dt}\int_{\xi}^{\lambda}\gamma(\xi_1)d\xi_1\right\} - \int_{\lambda}^{1}d\xi\left\{\rho_1\frac{d}{dt}\int_{\xi}^{1}\gamma(\xi_1)d\xi_1\right\}$$

which, after partial integration, gives

$$L = \rho_2 U\int_{-1}^{\lambda}\gamma(\xi)d\xi + \rho_1 U\int_{\lambda}^{1}\gamma(\xi)d\xi - \rho_2\frac{d}{dt}\int_{-1}^{\lambda}(1+\xi)\gamma(\xi)d\xi - \rho_1\frac{d}{dt}\int_{\lambda}^{1}(1+\xi)\gamma(\xi)d\xi \tag{37}$$

Likewise, the counter-clockwise moment about the mid-chord point is

$$M = \rho_2 U\int_{-1}^{\lambda}\gamma(\xi)\xi d\xi + \rho_1 U\int_{\lambda}^{1}\gamma(\xi)\xi d\xi$$

$$-\rho_2\frac{d}{dt}\int_{-1}^{\lambda}\frac{1}{2}(\xi^2-1)\gamma(\xi)d\xi - \rho_1\frac{d}{dt}\int_{\lambda}^{1}\frac{1}{2}(\xi^2-1)\gamma(\xi)d\xi \tag{38}$$

6 APPLICATION TO THE FLAT PLATE AIRFOIL

A flat plate airfoil of chord 2 at a small angle of attack α has a vorticity distribution

$$\gamma_o(\xi) = 2U\alpha\sqrt{\frac{1-\xi}{1+\xi}} \tag{39}$$

for steady motion in a uniform medium. As a consequence, the integral equation for $\gamma_1(\xi, t)$, Equation (20), becomes, for $\lambda < -1$ ahead if the leading edge of the airfoil,

$$\int_{-1}^{1}\frac{\gamma_1(\xi_1,t)}{\xi-\xi_1}d\xi_1 - \frac{\rho_1-\rho_2}{\rho_1+\rho_2}\int_{-1}^{1}\frac{\gamma_1(\xi_1,t)}{\xi-2\lambda+\xi_1}d\xi_1 + \int_{1}^{\infty}\frac{\gamma_w(t-(\xi_2-1)/U)}{\xi-\xi_2}d\xi_2$$

$$= \alpha\left(\frac{\rho_1-\rho_2}{\rho_1+\rho_2}\right)\left(1-\sqrt{\frac{\xi-2\lambda+1}{\xi-2\lambda-1}}\right) \tag{40}$$

where

$$\gamma_w(t) = -\frac{1}{U}\frac{d}{dt}\int_{-1}^{1}\gamma_1(\xi_1,\tau)d\xi_1 \tag{41}$$

Corresponding expressions are easily obtained for the density jump over the airfoil, $-1 < \lambda \leq 1$, and for the density jump downstream of the airfoil, $1 < \lambda$. Although we may proceed analytically when either $\lambda < -1$ or $\lambda < 1$, the situation is more complex when the discontinuity is passing over the airfoil and consequently a straightforward numerical calculation for $\gamma_1(\xi, t)$ is chosen. Examples of the vorticity distribution on the airfoil, vorticity distribution on the density jump, and pressure distribution on the airfoil are shown in Figures 7 and 8 for values of $\lambda = -1.2$ and $\lambda = -0.8$ respectively.

As the density jump approaches the leading edge of the airfoil, Figure 7a shows that a distribution of positive vorticity is induced on the density jump when $\lambda = -1.2$. The sign of the distribution follows directly from Equation (9) because the vorticity distribution $\gamma_1(\xi_1, t)$ is dominated by the initial loading $\gamma_o(\xi_1)$, Equation (36). As a consequence, the density jump induces a downwash at the leading edge, producing the negative vorticity γ_1 shown in Figure 7b. The resulting pressure coefficient perturbation (the negative of the loading), Figure 7c, shows a strong unloading of the leading edge corresponding to the action of a downward gust. It is important to notice here that, in contrast to the effect of the sharp edge gust [5], the density discontinuity has a downwash field that precedes its arrival.

The situation changes when the density jump lies within the airfoil chord, shown in Figure 8. The vorticity distribution on the discontinuity, Figure 8a, shows the effect of strong leading edge vorticity γ_o which now lies to the left of the density jump, largely negating the effect of the vorticity of the same sign lying to the right. Figure 8b shows a comparably small induced vorticity on the airfoil. The significant pressure coefficients in Figure 8c arise largely from the local temporal variation of vorticity, the time-differentiated term of Equations (35) & (36). The pressure distribution of Figure 8c implies a significant adverse pressure gradient induced on the lower surface.

Now because the position of the airfoil and the velocity of the undisturbed flow are constant during the encounter with the density jump, the potential and the vorticity distribution $\gamma_o(\xi)$ given in Equation (39) are independent of time. However, the pressure coefficient induced by this vorticity is time dependent because, specifically, the variable position $\lambda(t)$ of the discontinuity

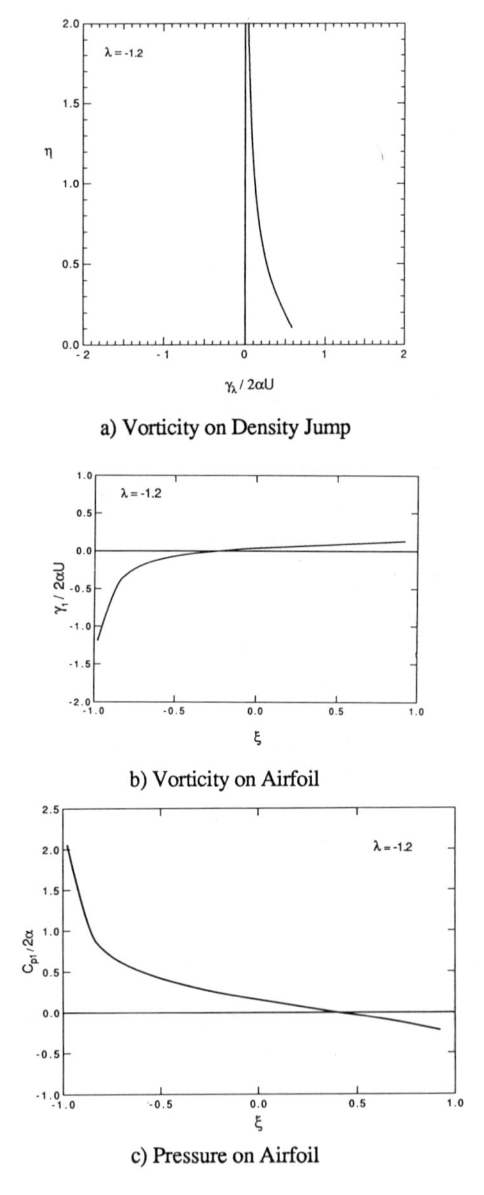

Figure 7. Flow induced on plane airfoil by the density discontinuity, λ=-1.2.

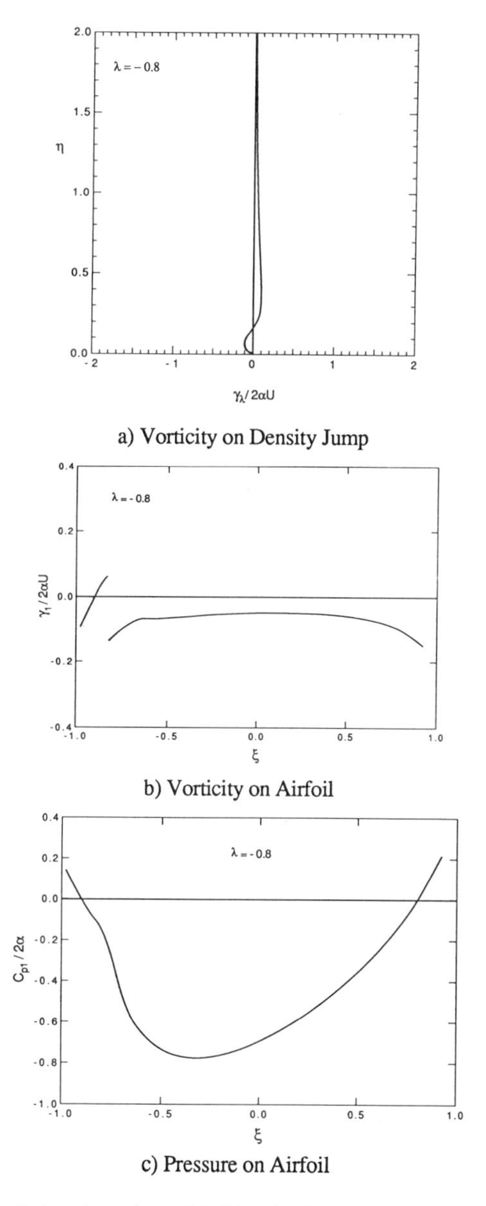

a) Vorticity on Density Jump

b) Vorticity on Airfoil

c) Pressure on Airfoil

Figure 8. Flow induced on plane airfoil by the density discontinuity, λ=-0.8.

appears in the limits of the integrals of Equation (35). When the discontinuity is ahead of the airfoil, that is $\lambda < -1$,

$$C_{po} \equiv \frac{\Delta p}{\frac{1}{2}\rho_1 U^2} = -4\alpha\sqrt{\frac{1-\xi}{1+\xi}} \tag{42}$$

and after it has passed downstream, $\lambda < 1$,

$$C_{po} = -4\alpha\frac{\rho_2}{\rho_1}\sqrt{\frac{1-\xi}{1+\xi}} \tag{43}$$

During the process of passage, however, the pressure ahead of the discontinuity, $-1 < \xi < \lambda$,

$$C_{po} = -4\alpha\frac{\rho_2}{\rho_1}\sqrt{\frac{1-\xi}{1+\xi}} - 4\alpha\left(1-\frac{\rho_2}{\rho_1}\right)\sqrt{\frac{1-\lambda}{1+\lambda}} \tag{44}$$

and downstream of the discontinuity, $\lambda < \xi \le 1$,

$$C_{po} = -4\alpha\sqrt{\frac{1-\xi}{1+\xi}} \tag{45}$$

which we note is unchanged from its value when the discontinuity was far upstream of the airfoil. Figure 9 shows $C_{po}(\xi)$ for several values of λ during the period the discontinuity is passing over the airfoil. The C_{po} distribution for $\lambda = -1$ holds for all $\lambda < -1$ and the C_{po} distribution for $\lambda = 1$, equal to ρ_2/ρ_1 times the C_{po} distribution for $\lambda = -1$, holds for all $\lambda > 1$ after the discontinuity has left the trailing edge. The changes in the C_{po} distribution take place from the leading edge and progress following the density jump. These values of C_{po} are, generally, larger than the values of C_{pI} and we may anticipate that the forces on the airfoil will to a considerable extent be dominated by C_{po} and thus confined to the period when the density jump is actually passing over the airfoil.

The lift coefficient C_{Lo} resulting from the pressure coefficient C_{po} is easily calculated, remembering that the airfoil chord is equal to 2,

$$\begin{aligned} C_{Lo} &= -\frac{1}{2}\int_{-1}^{1} C_{po}\,d\xi = 2\alpha\left\{\frac{\rho_2}{\rho_1}\int_{-1}^{\lambda}\sqrt{\frac{1-\xi}{1+\xi}}\,d\xi + \left(1-\frac{\rho_2}{\rho_1}\right)\sqrt{\frac{1-\lambda}{1+\lambda}}\int_{-1}^{\lambda} d\xi + \int_{\lambda}^{1}\sqrt{\frac{1-\xi}{1+\xi}}\,d\xi\right\} \\ &= 2\pi\alpha\left\{1-\left(1-\frac{\rho_2}{\rho_1}\right)\left(1-\frac{1}{\pi}\cos^{-1}\lambda\right)\right\} \end{aligned} \tag{46}$$

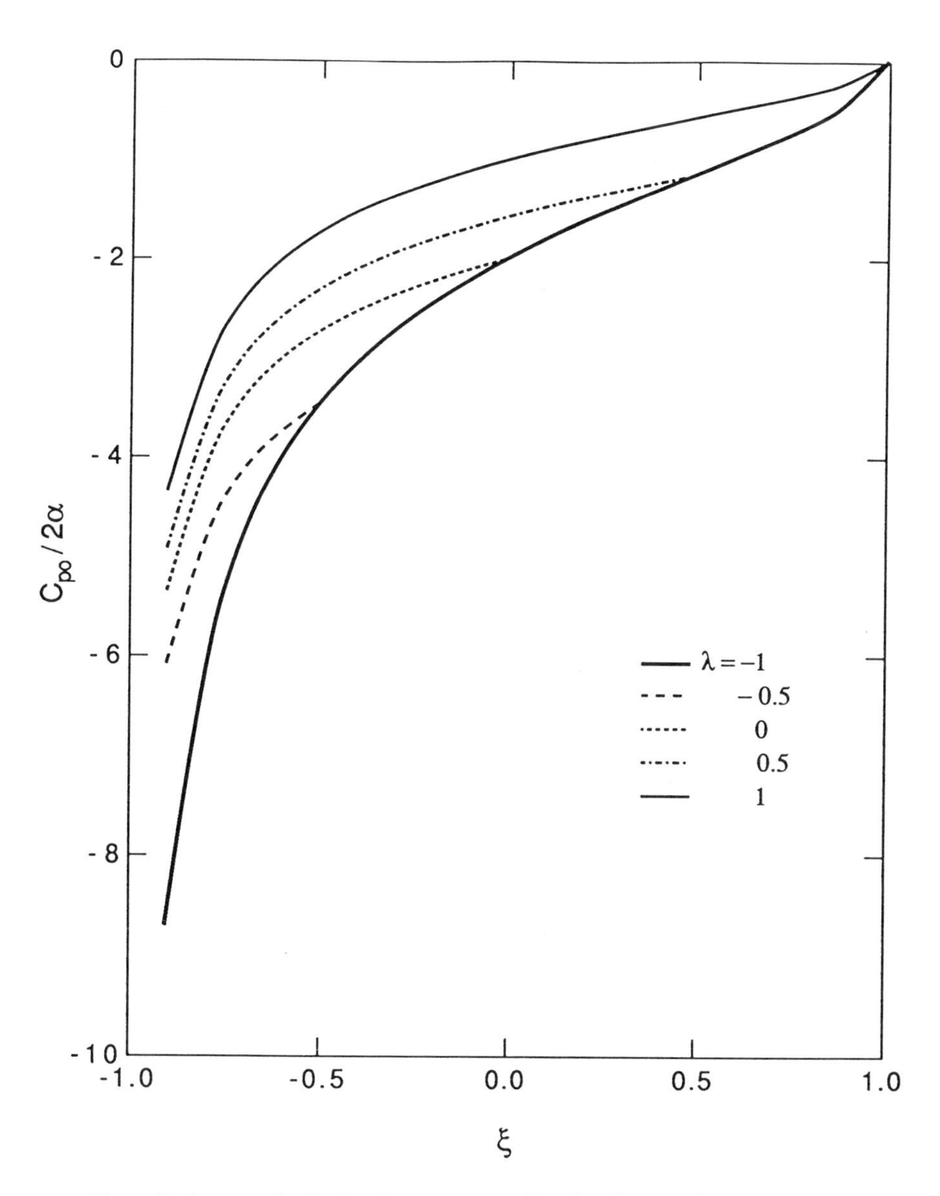

Figure 9. Pressure distribution due to passage of density discontinuity over the airfoil.

This is shown as the broken line in Figure 10.

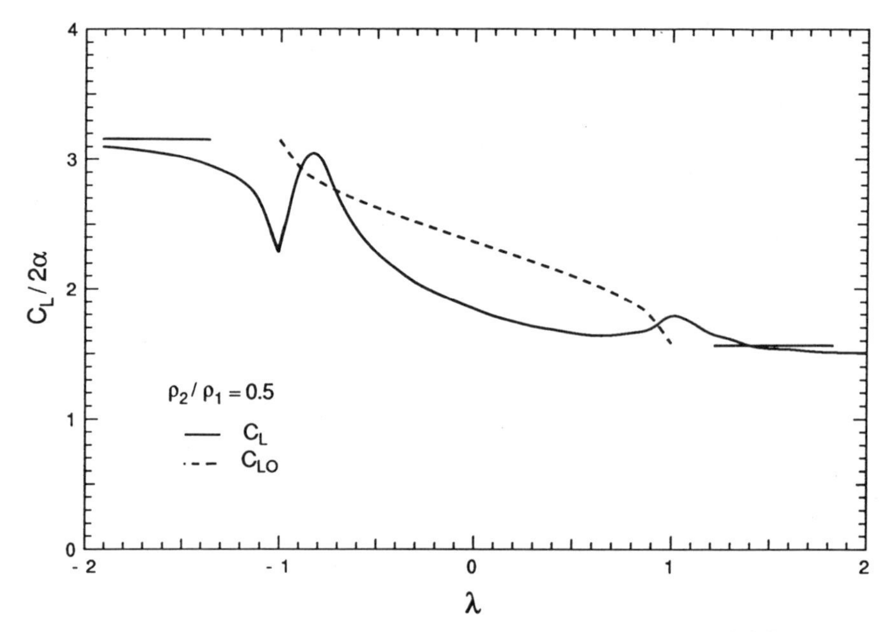

Figure 10. Lift coefficient during passage of density discontinuity over airfoil.

The clockwise moment about the airfoil midpoint is also easily calculated, again recalling that the airfoil chord is 2,

$$C_{Mo} = \frac{1}{4}\int_{-1}^{1} C_{po}\xi d\xi = \frac{1}{4}(2\alpha)\left\{\frac{\rho_2}{\rho_1}\sqrt{\frac{1-\xi}{1+\xi}}\xi d\xi + \left(1-\frac{\rho_2}{\rho_1}\right)\sqrt{\frac{1-\lambda}{1+\lambda}}\int_{-1}^{\lambda}\xi d\xi + \int_{\lambda}^{1}\sqrt{\frac{1-\xi}{1+\xi}}\xi d\xi\right\}$$

which may be evaluated as

$$C_{Mo} = \frac{\pi}{4}\alpha\left\{1-\left(1-\frac{\rho_2}{\rho_1}\right)\left(1-\frac{1}{\pi}cos^{-1}\lambda + \frac{\lambda}{\pi}\sqrt{1-\lambda^2}\right)\right\} \tag{47}$$

This result appears as the broken line in Figure 11.

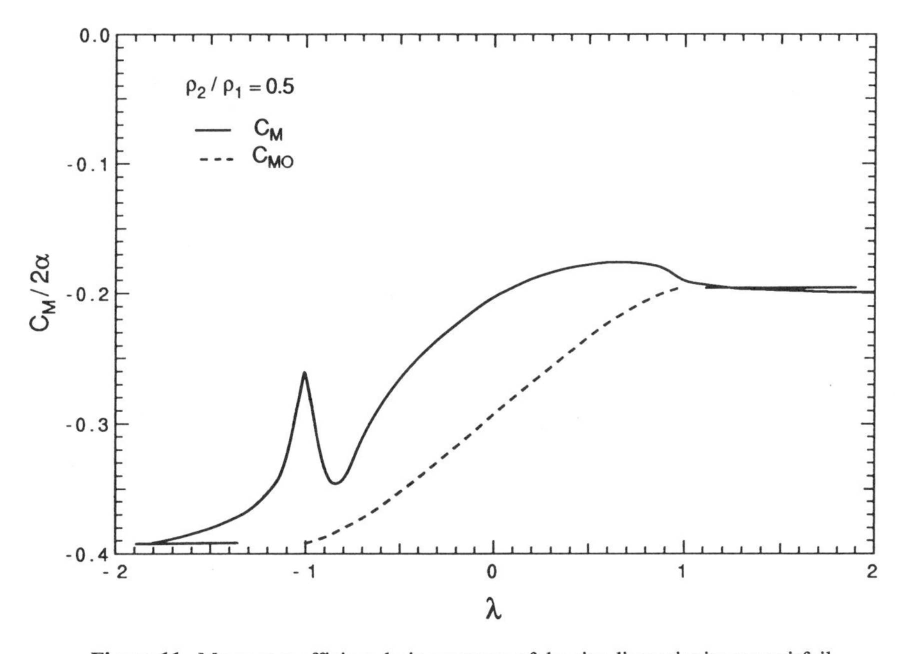

Figure 11. Moment coefficient during passage of density discontinuity over airfoil.

The complete solutions for the lift and moment coefficients shown in Figures 10 and 11 were obtained by numerically solving Equation (40) & (41) for the $\gamma_1(\xi, t)$, the additional vorticity distribution induced by the density jump. Integration was begun with the density jump 2 chord lengths upstream, $\lambda = -4$, and continued until the density jump had passed 2 chord lengths downstream, $\lambda = 4$. At $\lambda = -4$ the values of γ_1 were negligible and the vorticity shed from the trailing edge correspondingly small. As the discontinuity moved downstream, the successive distributions $\gamma_1(\xi, t)$ allowed determination of the γ_w values to be used at each succeeding value of λ.

The unique features of the problem are most readily described with reference to the lift coefficient shown in Figure 10. As the density jump approaches within about one half chord of the leading edge, the downwash, induced as illustrated in figures 2 and 3, generates a reduction of the airfoil vorticity distribution, Figure 7b, and the consequent lift reduction, Figure 9, for $\lambda < -1$. As the discontinuity passes the leading edge, the flow following the density jump constitutes a strong upwash, creating a situation resembling the sharp-edge gust. In fact, the cusp appearing at

$\lambda = -1$ is related to the lift growth for a sharp-edged gust shown in Figure 8 of Kármán & Sears [5]. The subsequent rise in C_L overshoots the value of C_{Lo} to a degree that evidently exceeds that shown simply because the number of intervals used to describe the airfoil surface was not sufficient to resolve the peak. The moment coefficient, Figure 11, reflects these events in local loading. The response as the density jump passes off the trailing edge is relatively mild, largely because the values of $\gamma_o(\xi)$ are small in this region and consequently the near-field reflected vorticity is small.

Initially, it seems surprising that the variations in lift and moment are confined so closely to the period when the density discontinuity is passing directly over the airfoil, particularly so in view of our experience with other examples of unsteady airfoil behavior. Significant response to a sharp-edge gust, for example, continues until the gust has passed 5 to 10 chord lengths downstream. The difference lies in the fact that in the present case the flow field is kinematically identical when the discontinuity is far upstream and when it has passed far downstream and, in particular, the circulation about the airfoil is the same in both cases. Consequently, the shed vortex sheet contains zero net vorticity and behaves much like a vortex pair with a spacing of the order of the airfoil chord whose induced field dies out as $1/\lambda^2$ rather than as $1/\lambda$.

REFERENCES

1 Glauert, H.: The Force and Moment on an Oscillating Aerofoil, British A.R.C., R. & M. No. 1242, 1929.

2 Burgers, J. M.: Problems of Non-Uniform Motion, Section A, Ch. 5, V. II, *Aerodynamic Theory,* W. F. Durand ed., pp. 280-310, Springer Publishing, 1935.

3 Theodorsen, Th.: General Theory of Aerodynamic Instability and the Mechanism of Flutter, N.A.C.A. Technical Report No. 496, Washington, DC, 1935.

4 Kussner, H. G.: Zusammenfassender Bericht uber den instationaren Auftrieb von Flugeln, *Luftfahrtforschung,* Bd 13, pp. 410-424, 1936.

5 Kármán, Th. v. and Sears, W. R.: Airfoil Theory for Non-Uniform Motion, *J. Aero. Sci.*, **5**, pp. 379-390, 1938.

6 Marble, F. E.: Response of a Nozzle to an Entropy Disturbance – Example of Thermodynamically Unsteady Aerodynamics, *Unsteady Aerodynamics,* V. II, Symposium Proceedings, University of Arizona, Tucson, ed. by R. B. Kinney, 1975.

7 Giles, M. B.and Krouthen, B.: Numerical Investigation of Hot Streaks in Turbines, Paper AIAA-88-3015, 24th Joint Propulsion Conference, Boston, 1988.

8 Wortman, A.: Unsteady Flow Phenomena Causing Weapons Fire – Aircraft Engine Inlet Interference Problems, Theory and Experiments, *Unsteady Aerodynamics,* V. I, Symposium Proceedings, University of Arizona, Tucson, ed. by R. B. Kinney, 1975.

9 Marble, F. E. and Candel, S. M.: Acoustic Disturbance from Gas Nonuniformities Convected through a Nozzle, *J. of Sound and Vibration,* **55** (2), pp. 225-243, 1977.

10 Goldstein, M. E.: Unsteady Vortical and Entropic Distortions of Potential Flows Round Arbitrary Obstacles, *J. of Fluid Mechanics,* **89**, Pt 2, pp. 433-468, 1978.

Section III

ADAPTIVE WALL WIND TUNNELS

Operation of the Adaptive-Wall Wind Tunnel of TsAGI, Moscow

V. M. Neyland
A. V. Semenov
O. K. Semenova

1 INTRODUCTION

The history of the development of the adaptive-wall wind tunnel is well known, primarily due to the studies of Wolf and Goodyer [1]. They discovered publications describing the experiments carried out in such a facility as early as in the pre-war years. However, the second and true birth of the concept of the adaptive boundaries on the flow is closely related to the names of Sears [2] and Ferri and Baronti [3]. Sears is considered a father of adaptive walls and has based his idea on a well-known principle of solid stream lines. Therefore, the first technical implementation of this concept was also realized in a most obvious way, i.e. flexible adaptive walls. Since that time this type of self-adaptive flow boundaries remains the most popular in the world, though none of these facilities were used for industrial purposes.

For many years Russia was carrying out its studies on its own, repeating, or sometimes anticipating, the achievements of Western states. The first non-classified papers in this direction [4] were published soon after the pioneering works [2,3] appeared, so their independent development and implementation is obvious. However, the idea of wall boundary conditions controlled by segmented gas blowing-suction, though proposed in [4], has been thoroughly investigated by computations and theoretically but was never applied in practice.

Instead, at the end of the 1960's TsAGI began to develop a project of a large industrial wind tunnel, T-128, with perforated adaptive panels on the test section walls. The project had been completed by 1976, and the facility itself was put into service in 1982. The new configuration of test section walls was aimed at minimisation of harmful effect of flow boundaries on the flow past a model by optimum adjustment of local perforation. This adjustment should be performed automatically by the controlling computer. During years of its operation certain advantages (rather significant) as well as drawbacks (which can be eliminated) were found out. The present paper discusses some experimental results obtained in T-128 and gives a brief description of the facility itself.

SYMBOLS:

M	- Mach number
Re	- Reynolds number
C_p	- dimensionless pressure coefficient
f	- porosity coefficient = (S orifices) / (S wall)
x	- coordinate along the tunnel axis
R	- mathematical porosity parameter
α	- angle of attack
β	- side slip angle
C_l	- lift

2 T-128 WIND TUNNEL

T-128 wind tunnel, Figure 1, is a closed-circuit facility with variable density (0.3 - 4 atm); the Mach number range is 0.2 - 1.7, and the dimensions of the text section are 2.75x2.75m. Reynolds number range of T-128 is compared with the Re numbers of other facilities of analogous size in Figure 2. The facility is equipped with five interchangeable test sections designed for various experiments:

No.1: - tests of full models installed on the tail sting;
No.2: - tests of the models installed on the struts for determining the sting effect;
- tests of half-models;
No.3: - test of airfoil models;
No.4: - simulation of jets;
- optical/physical studies (is not completed yet);
No.5: - aeroelasticity tests.

Interchangeable test sections allow enhancing the productivity of the facility because the time required for the model installation in the tunnel circuit is reduced considerably.

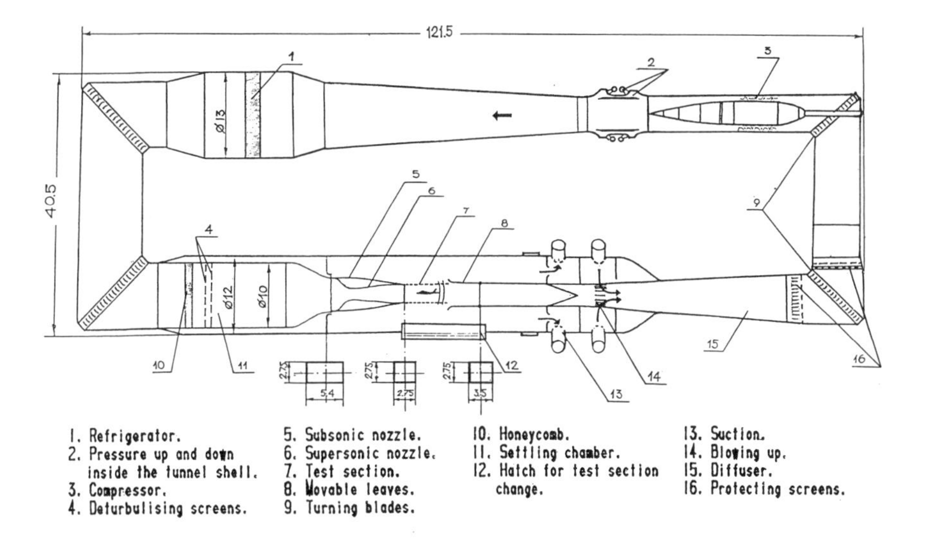

Figure 1. Schematic of tunnel circuit.

All the test sections are equipped with perforated wall panels of different configuration. The present paper describes the tests of complete models in test section No.1 and tests of half-models in test section No.2. The walls of either test section consist of 128 independent panels, the porosity of which may vary from 0 to 18%. Porosity level is varied by overlapping the perforation holes by external plates moving along the fixed frame, Figure 3. Schematic of the wall panels as well as conventional locations of models are presented in Figures 3-6.

When the facility became operational no one had clear understanding of what should be the algorithm for wall porosity control. Investigations showed that the classical scheme of Sears (matching the flow in the tunnel with the imaginary flow outside the circuit) was not practical because of the lack of super-fast techniques of computing 3-D transonic flows which allow to control the experiment in the on-line mode. On the other hand, it was technically difficult to measure the flow velocity component normal to the wall. For these purposes a non-intrusive method of measurements is desirable for transonic regimes, but at present this problem remains unsolved for the facilities of this size.

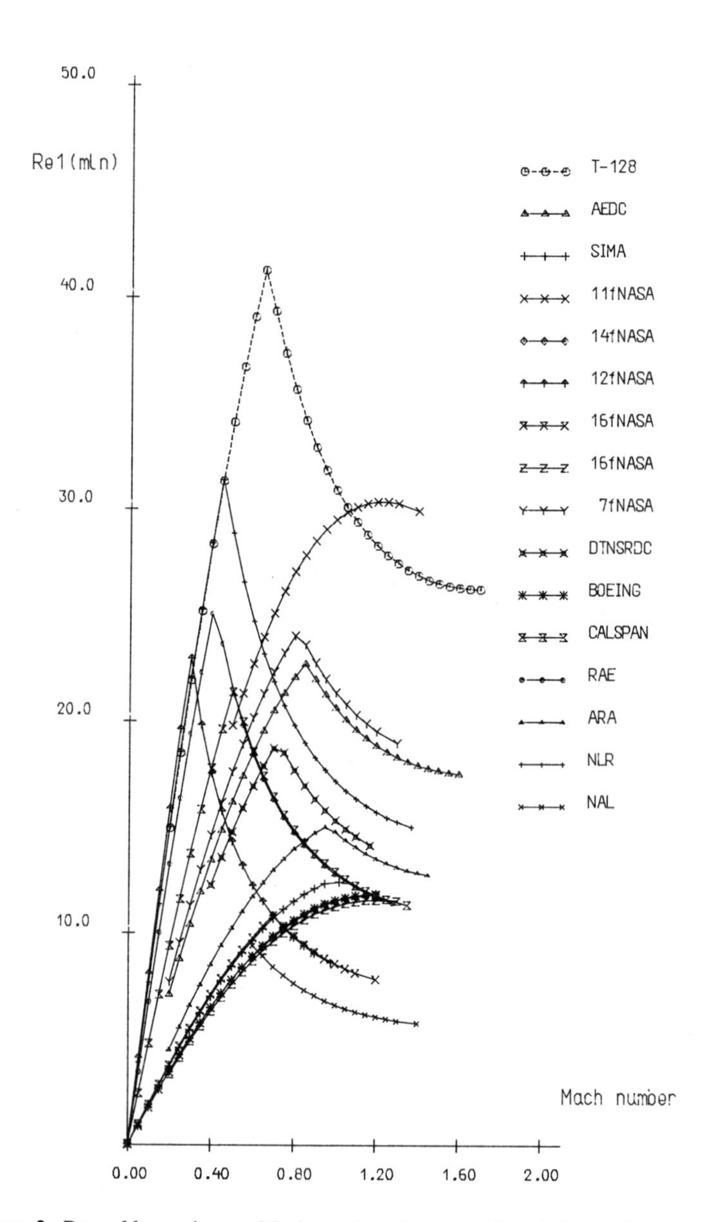

Figure 2. Reynolds number vs. Mach number of comparable wind tunnels.

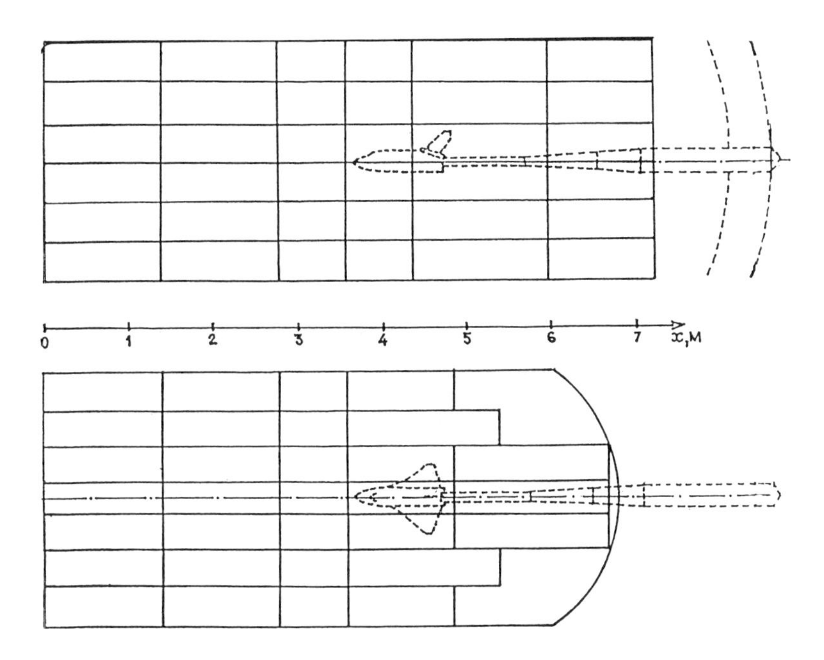

Walls surface segmentation of test section №1
(each rectangle correspons to independent unit).

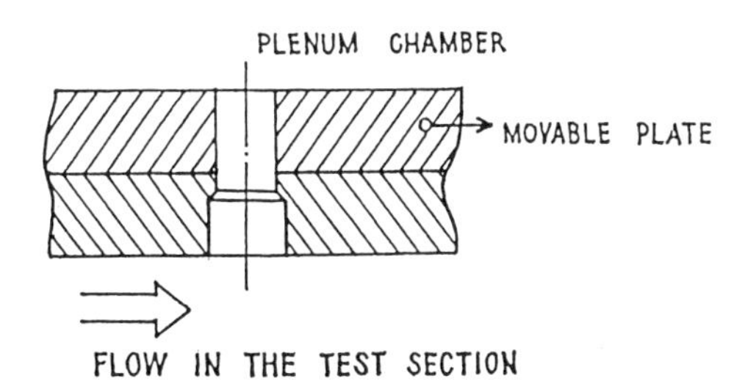

Porosity variation scheme.

Figure 3. Wall surface segmentation and porosity variation scheme.

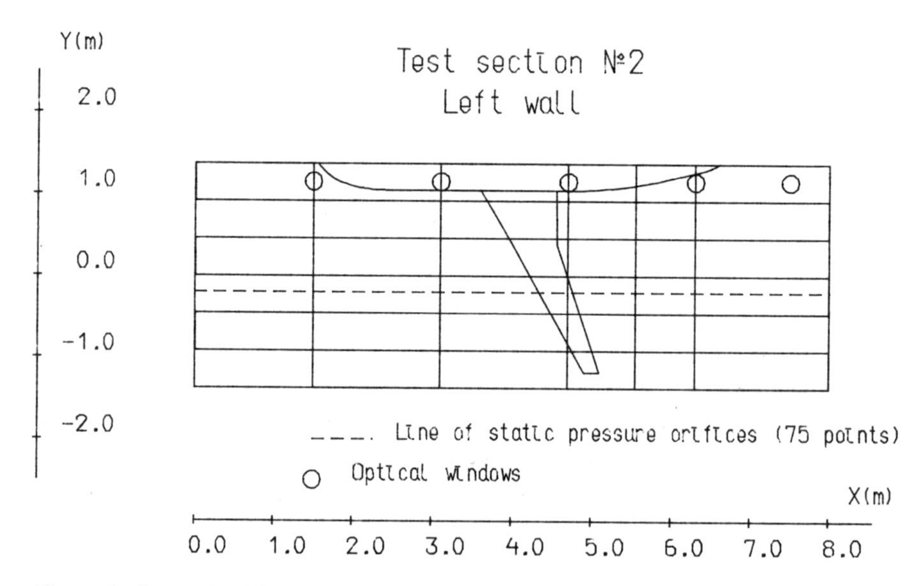

Figure 4. Conventional location of model in Section 2 (right wall).

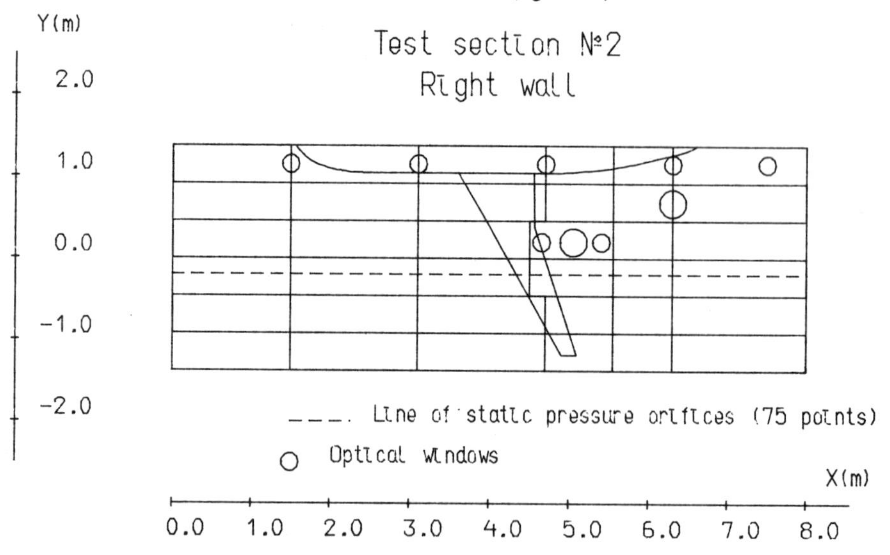

Figure 5. Conventional location of model in Section 2 (left wall).

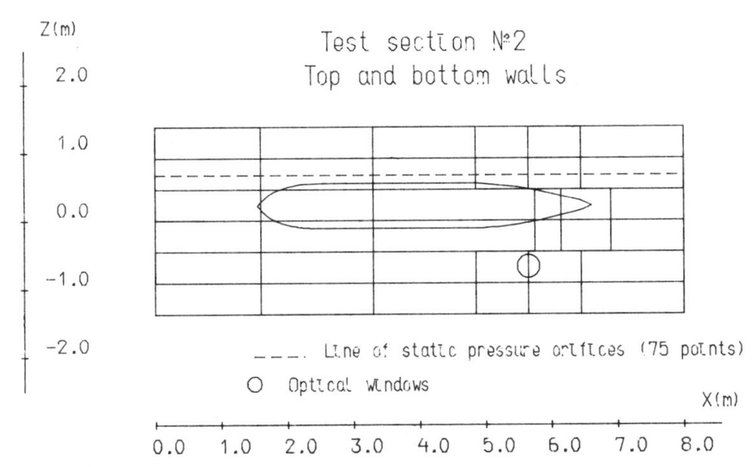

Figure 6. Conventional location of model in Section 2 (top/bottom walls).

To cope with these difficulties, a simplified approach was adopted which was based on the following considerations. It is known that in a purely subsonic potential flow the internal region is fully governed by a single velocity component given on its boundaries. In this case, maintaining the given pressure distribution on the boundary, corresponding to free stream conditions around a model, shall ensure interference-free tests. There are two factors that distinguish the tunnel conditions from the above consideration: viscous layer near the walls and local supersonic regions in the flow.

An account of the first factor may be taken with the help of traditional (though rather complicated) methods. As for the second factor, it deteriorates the assumption of harmonicity of the functions defining the flow in the internal region. So, generally speaking, the above consideration becomes inapplicable, and the problem of errors induced by the local supersonic regions is still waiting in-depth analysis if the wall adaptation procedure is based on the given pressure distribution.

A very simplified version of boundary conditions adaptation for wall interference elimination looks as follows, Figure 7. The viscosity effect correction is applied to the measured wall pressure, and the resultant figure is compared with the designed value of free stream conditions near the model. The value of the mismatch permits us to correct the local wall porosity.

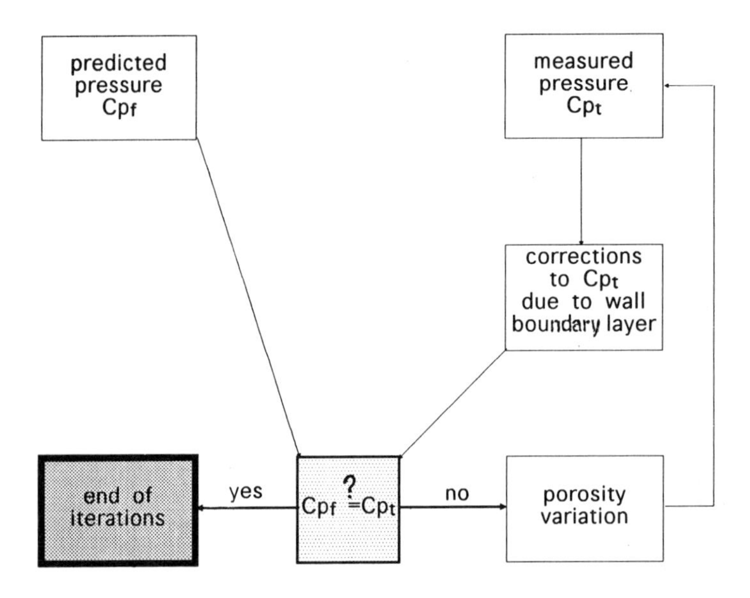

Figure 7. Algorithm for wall interference elimination.

In fact, the practical application of these ideas even in such an over-simplified version causes a lot of problems. First of all, to calculate the reference far field one should choose the proper representation of the model. In subsonic flows a set of singularities is quite adequate. Proper representation is validated by comparison of the computed pressure with the measured pressure distribution on the walls at the given wall boundary conditions, Figures 8 & 9.

In transonic flows at low angles of attack and side-slip angles the far field is governed by the transonic area rule. For high-lift models one may use the transonic lifting line theory. However, one should bear in mind that both cases correspond to the test section size being much larger than the size of a model. As a rule, in wind tunnel conditions the width of the test section is close to the length of the model. To predict the reference field for these conditions, the calculations of a flow past a real configuration should be done by the known numerical methods.

Re-calculation of pressure measured on the tunnel walls to the edge of the potential flow core is done by computation of the boundary layer development on a porous surface at the measured Cp(x) [5]. Fortunately, in the majority of cases the optimum porosity level of the walls is very small, so the corrections applied to Cp due to mass flow through the wall are small.

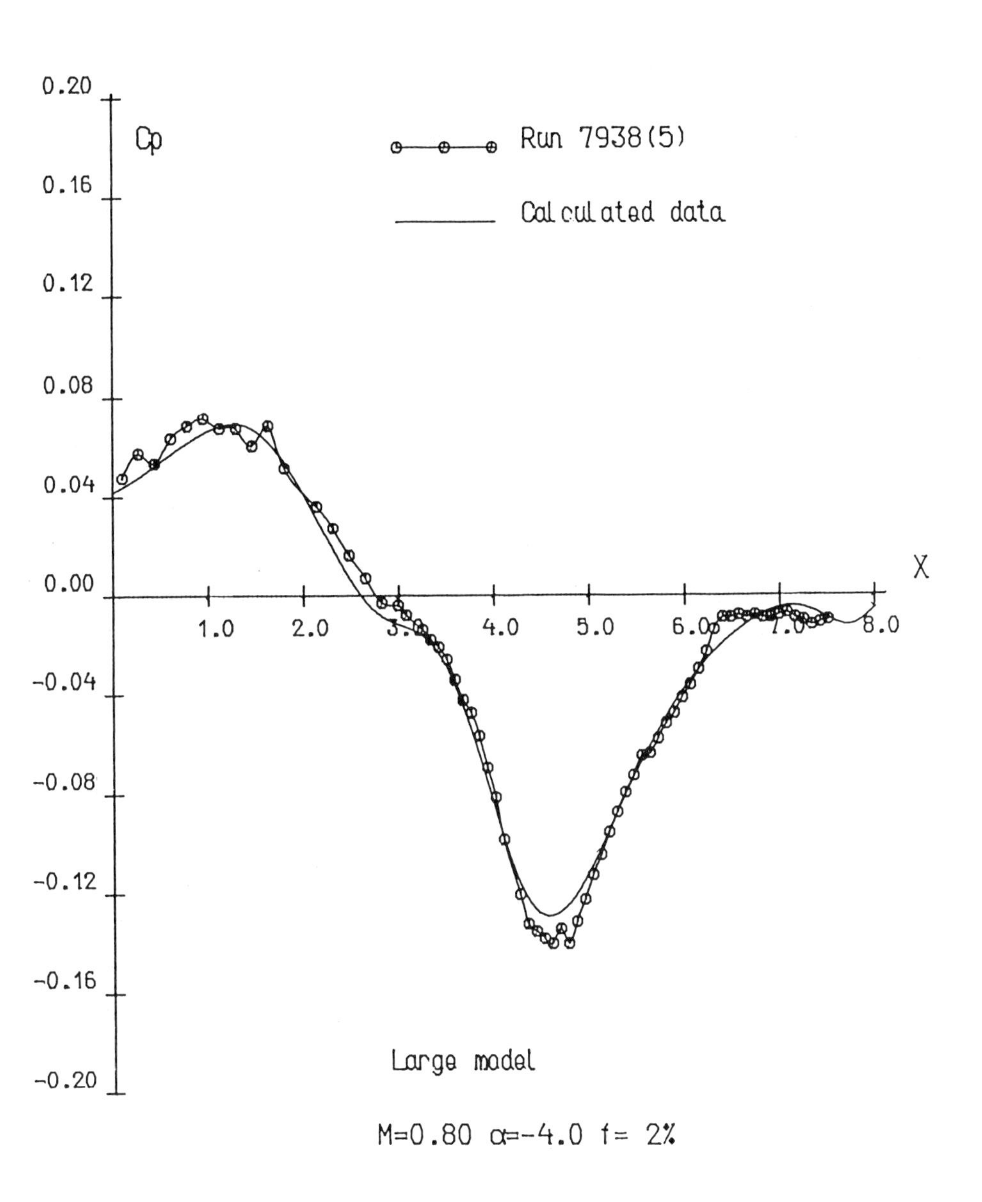

Figure 8. Pressure distribution along left wall.

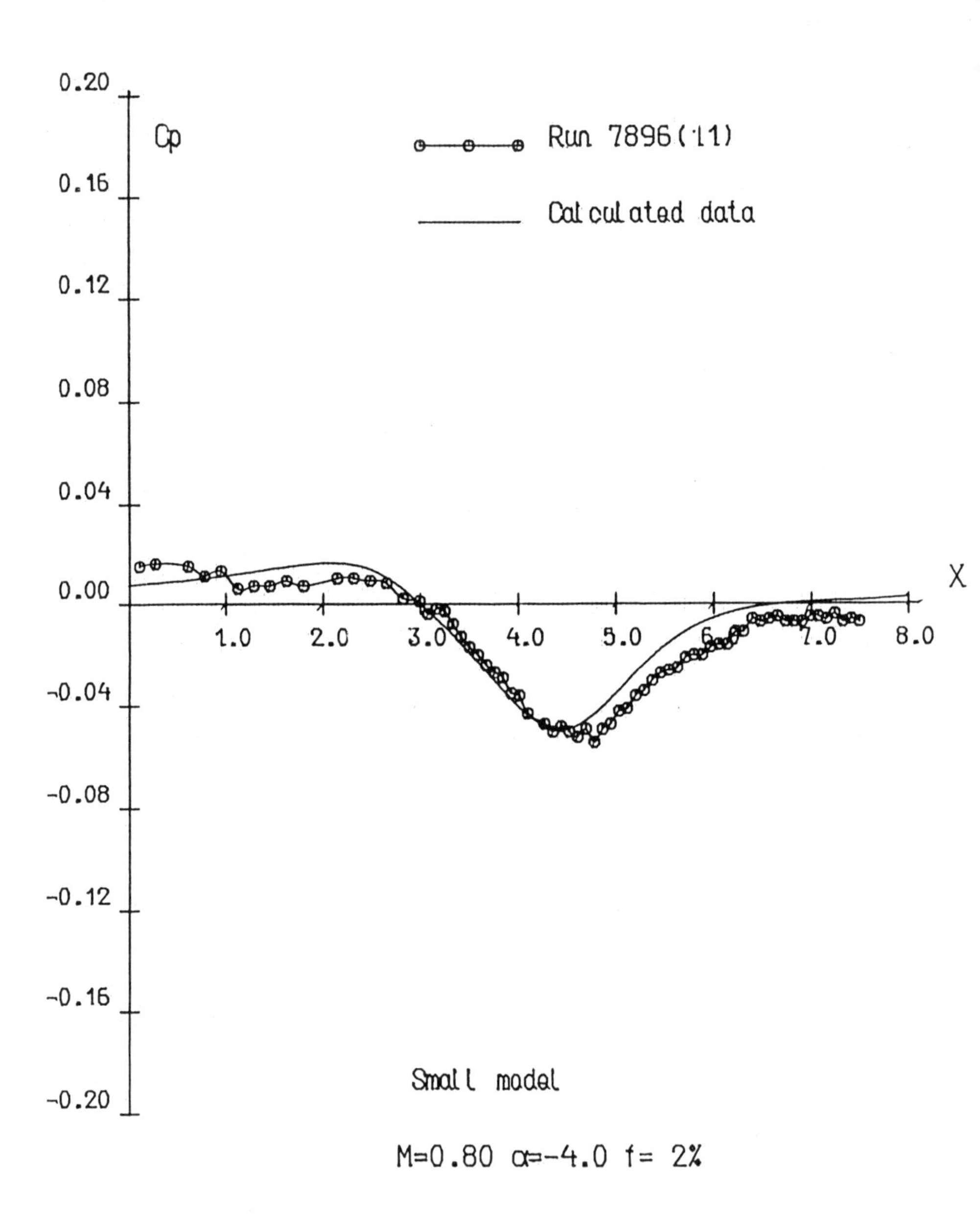

Figure 9. Pressure distribution along left wall.

The adaptation procedure presented in Figure 7 is an iterative process, the convergence of which depends on the choice of the initial approximation. At transonic regimes transonic theory [6] can help us to understand what is the initial porosity level. In accordance with this theory $f_{opt} \sim \sqrt{1-M^2}$, i.e., the porosity is close to zero at M ~ 1. It will be shown below that this conclusion is proved experimentally.

3 OPERATIONS AND TEST RESULTS

Initially, the above technology was used for testing the Buran-Energia models of different sizes at M ~ 1 [7,8] (Schematic of the model and the facility is given in Figures 10 & 11). The adaptation was essentially facilitated due to axial symmetry of the flow in the vicinity of the walls at a certain porosity level. Figure 12 shows pressure distribution on a side wall and the floor at different f's with the carrier model installed in the tunnel. Obviously, at small f's there is no difference between these two walls; also no effect of α variation from 0 to 4° is observed. With the growth of f the difference between the walls becomes more pronounced, though insensitivity to the behaviour of α is preserved.

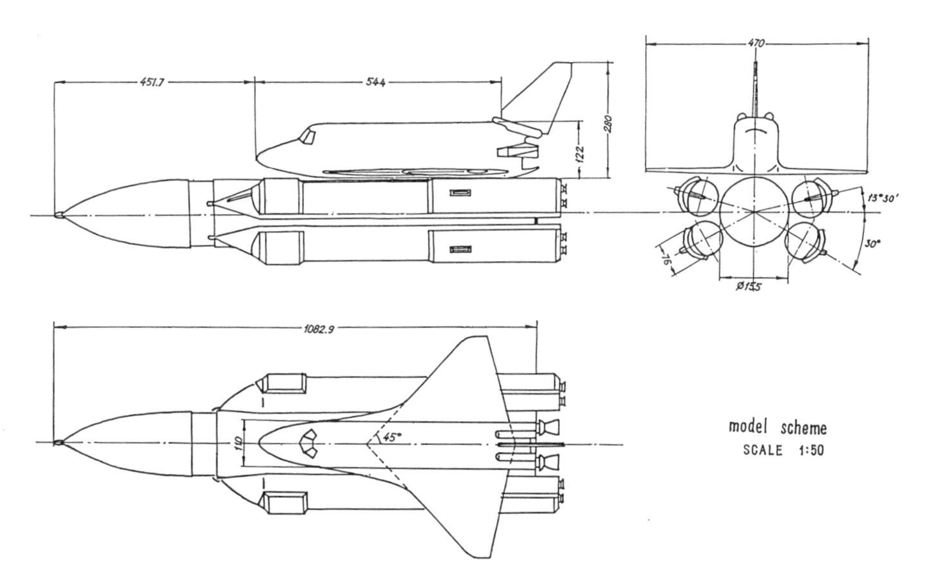

Figure 10. Schematic of Buran-Energia Model.

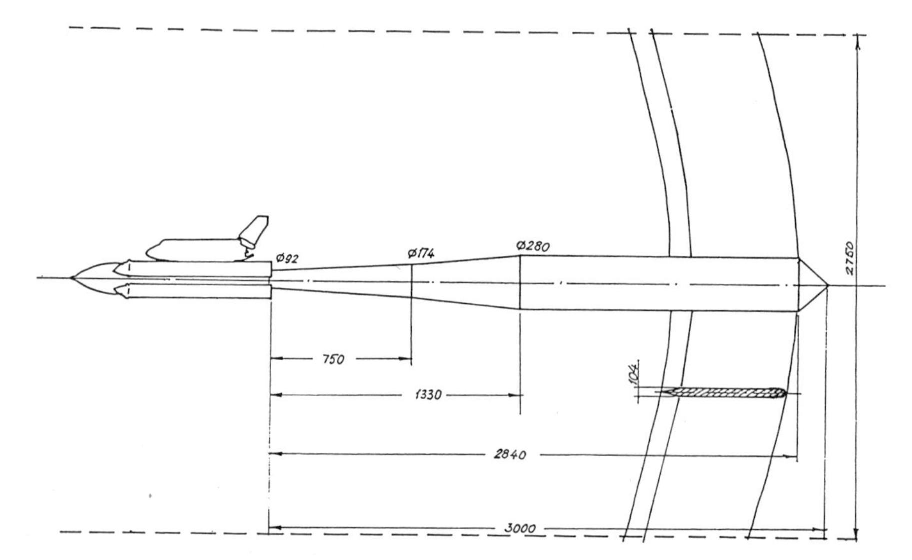

Figure 11. Model of scale 1:50 installation in test section No. 1 of T-128.

The response of Cp is analogous for the models of other sizes, too, Figure 13.

Such behaviour of the flow field near the walls gave us the indication to look for the optimum distribution of f within the axisymmetric piece-wise functions and to perform the adaptation procedure semi-automatically. Figures 14 & 15 present the results where Cp(x) at optimum f(x) is compared with free stream profiles. Obviously, the agreement of the curves is quite good.

As for the comparison of all the characteristics of the models of different scales, the relatively accurate similarity here was observed only for the dependence of $C_l(\alpha)$ of the Buran model in combination with the carrier. The drag value is completely distorted by the tail/sting interference, and the levels of distortion differ for different models, so it is impossible to make any comparisons for the models of different scales.

Figure 16 gives the comparison of the dependence $C_l(\alpha=0)$of the Buran models of different sizes in combination with the carrier. One can see that the models 1:120 and 1:50 fit well due to the adaptation procedure; the scale of C_l variation due to the wall adaptation is shown, too. It should be pointed out that this parameter was of key importance for the first flight of the system as it determines the limit load on the supports mounting the orbiter on the carrier at the stage of launch. Experimental data made it possible to determine the value with a great accuracy.

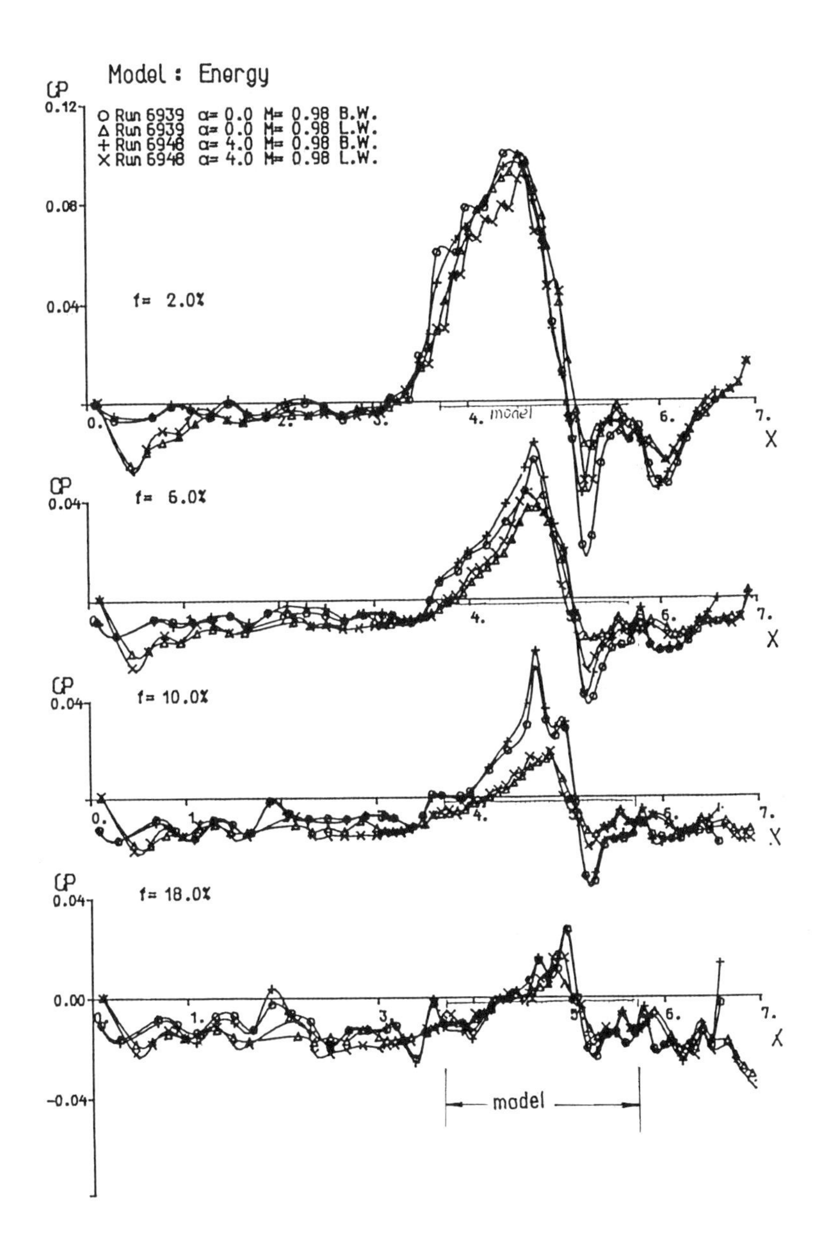

Figure 12. Pressure distribution along bottom and left walls.

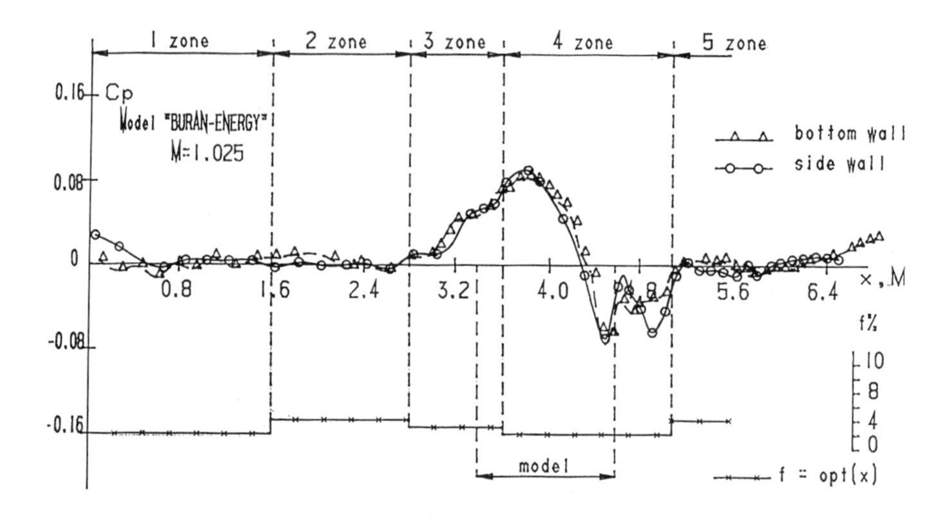

Figure 13. Pressure and porosity distribution along bottom and side walls, $\alpha = \beta = 0^{o}$.

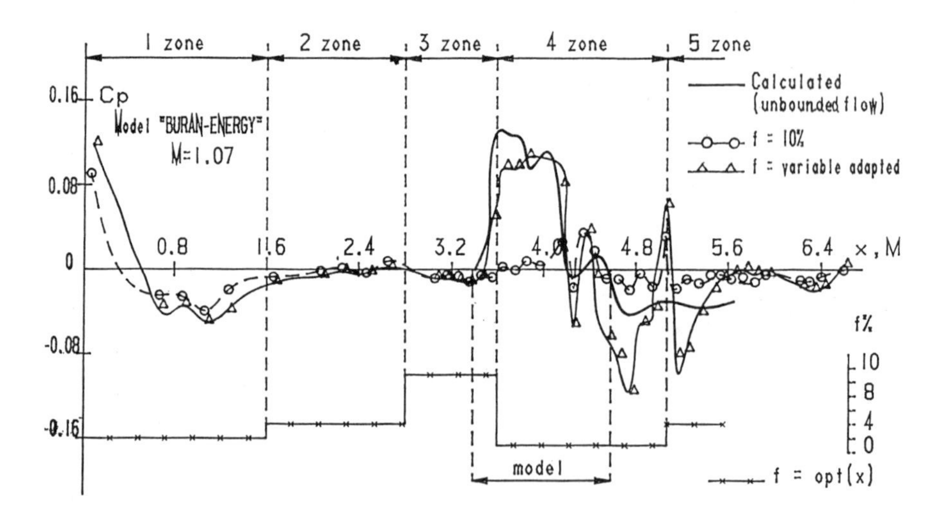

Figure 14. Pressure and porosity distribution along top wall.

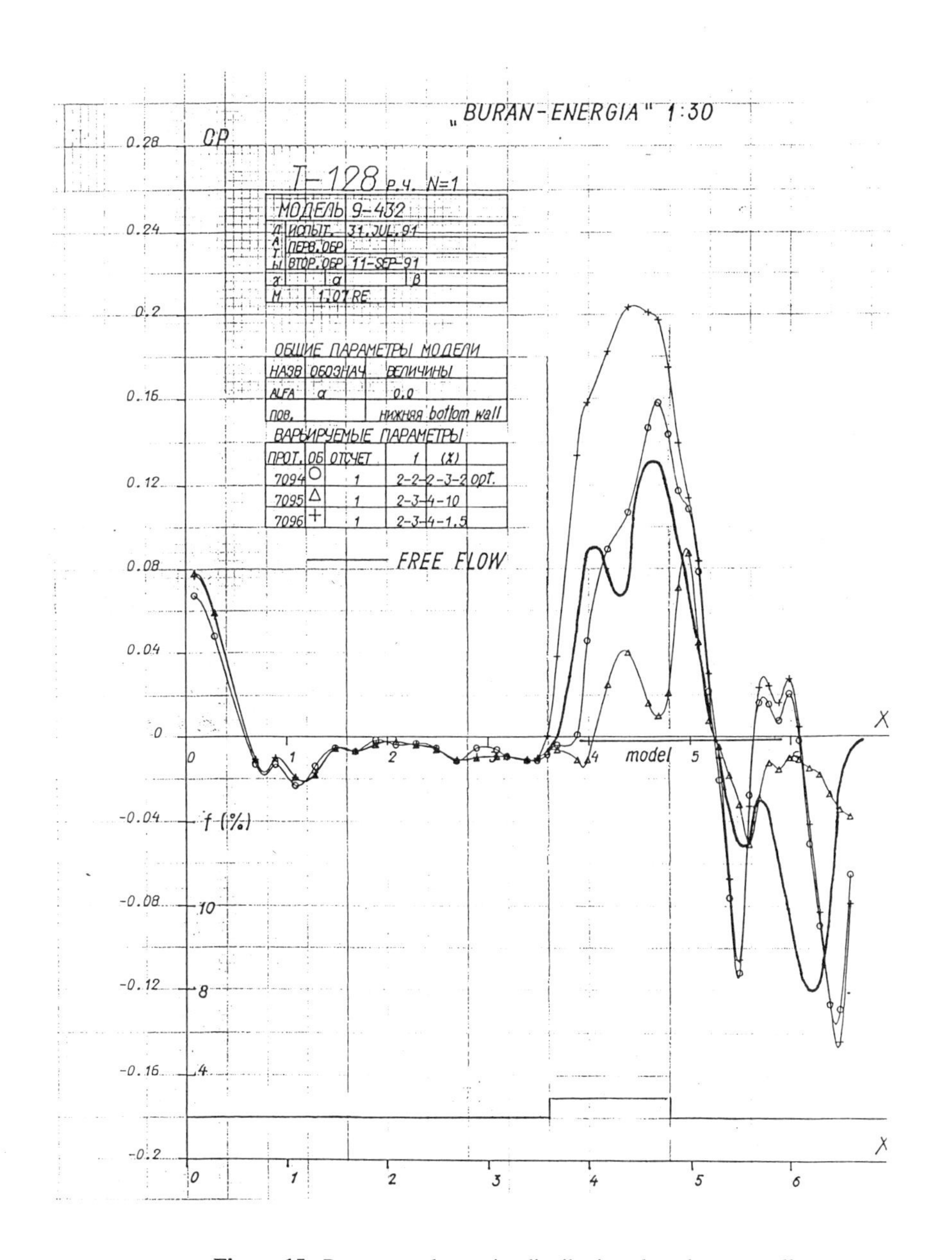

Figure 15. Pressure and porosity distribution along bottom wall.

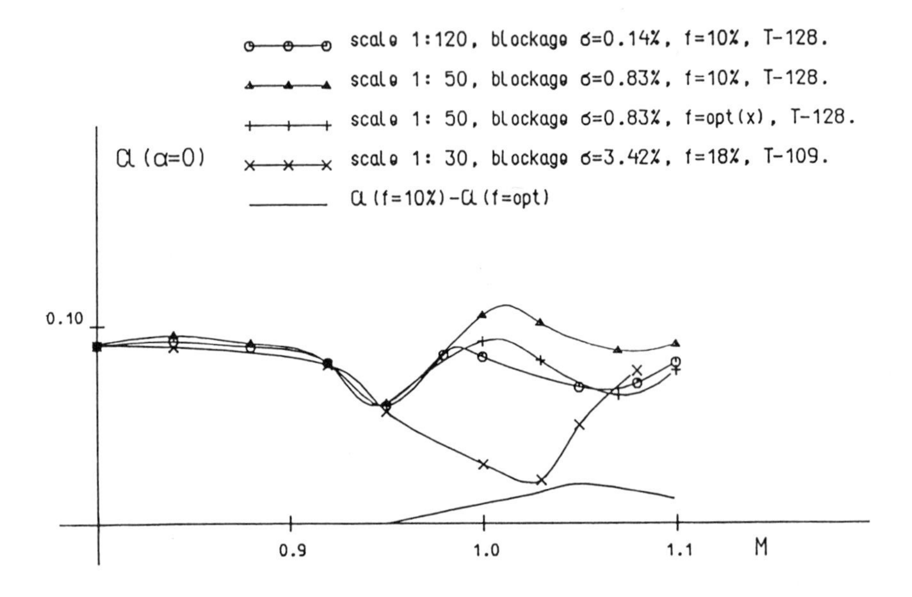

Figure 16. Lift of the Buran-Energia model in complete configuration at $\alpha = \beta = 0^{o}$.

In test section No.2 of T-128 the two geometrically-similar models of a civil aircraft were tested which differ in their size by a factor of 1.75 (Schematic of the large model installation is shown in Figures 4-6). Blockage ratio was 1.03% and 3.16%, respectively. The models were installed on the external 5-component strain gage balance on the test section ceiling. Simultaneously with load measurements, the pressure distributions were measured at eight wing sections and along the four walls of the test section. All the experiments were carried out with fixed transition line on the wing (5% from the leading edge). For the large model tests the wall adaptation was done at M=0.88, Re=7 x 10^6, α=0 and 4°. The interference free pressure field was determined by the linear subsonic theory. Accuracy of the model represented by a set of singularities is obvious from Figure 8.

The wall porosity was adjusted to the optimum value by maintaining the measured $Cp_t(x)$ close to the predicted value of $Cp_f(x)$; the procedure was simulated on a computer prior to the experiment. The value of $Cp_t(x)$ was calculated also within the linear subsonic theory for the given piece-wise wall porosity. To find $f_{opt}(x)$ the technique of coordinate-by-coordinate retrieval was used. The resultant table of the functions f_{opt} for each of the four walls and their dependencies versus M and α were input in the computer controlling the porosity of the walls.

Ideally, the same procedure of coordinate-by-coordinate retrieval should be repeated in the wind tunnel using measured values for $Cp_t(x)$. In this case the predicted distribution of $f_{opt}(x)$ could be used as the initial approximation. Unfortunately, due to the lack of high-speed pressure measurement system and insufficient power of the controlling computer this process could not be performed in real time. Therefore in the author's opinion, the obtained results, though rather optimistic, cannot be regarded as the final goal.

Figure 17 shows the wall pressure distribution obtained for the two constant porosity values: standard level of 10% and the optimum one shown at the bottom. These data are compared with computation corresponding to infinite flow (solid line). One can see that we managed to achieve rather good agreement of pressure profiles on the left wall where the pressure amplitude was maximum, and, consequently, its contribution to the interference effect should be maximum, too. On the right wall Cp is close to 0 and is not sensitive to the porosity variations. On the floor the result of adaptation was worse than on the left wall. It seems that the linear subsonic theory used for simulating the tunnel conditions in the course of adaptation exaggerates the value of local porosity (remember, that by the transonic theory, f_{opt} must be very small). At the same time, as the wing tip is located close to the floor (about 16cm), the local supersonic region is capable of touching the wall, thus, the applicability of subsonic theory in this case seems doubtful. Perhaps this is the proper direction for improving the wall adaptation procedure in the on-line mode which was discussed above.

Pressure distributions across the wing sections are shown in Figures 18a & b. Three cases are compared: large model with standard porosity; large model and adaptive perforation; small model and standard porosity. Data spread for three repetitive experiments is shown. Obviously, in the majority of sections considered the wall adaptation resulted in considerable agreement of pressure profiles of the large and of the small models. The only exclusion was the lower surface of tip section. Possible explanations of this phenomenon are presented above.

When examining the effect of adaptation on the total characteristics of the models and comparing the results obtained for the large and for the small models one should bear in mind that the most appropriate parameter to be examined is $C_l(\alpha)$. It should be pointed out that due to the interaction of the ceiling boundary layer with the model body, the similarity of drag properties of the small and the large models cannot be maintained. Comparison of the dependence of $C_l(\alpha)$ for the case analogous to Figure 18 is shown in Figure 19.

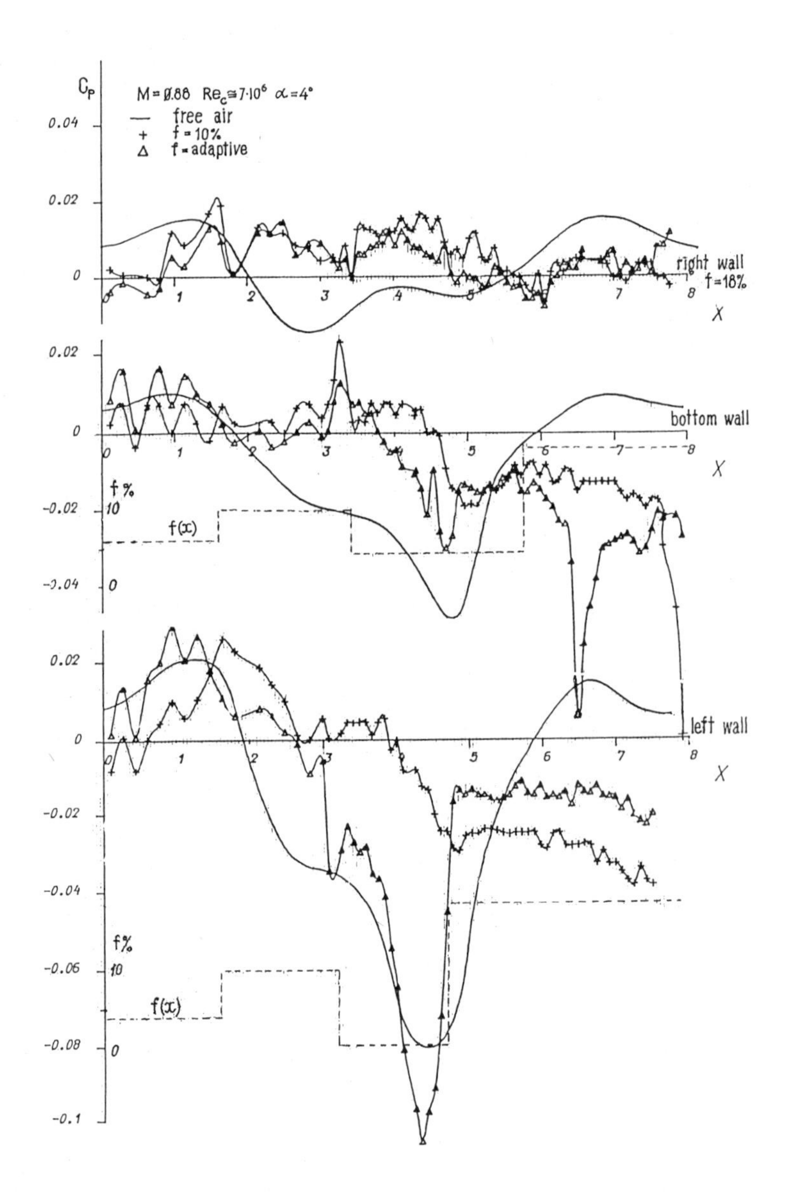

Figure 17. Pressure and porosity distribution along walls.

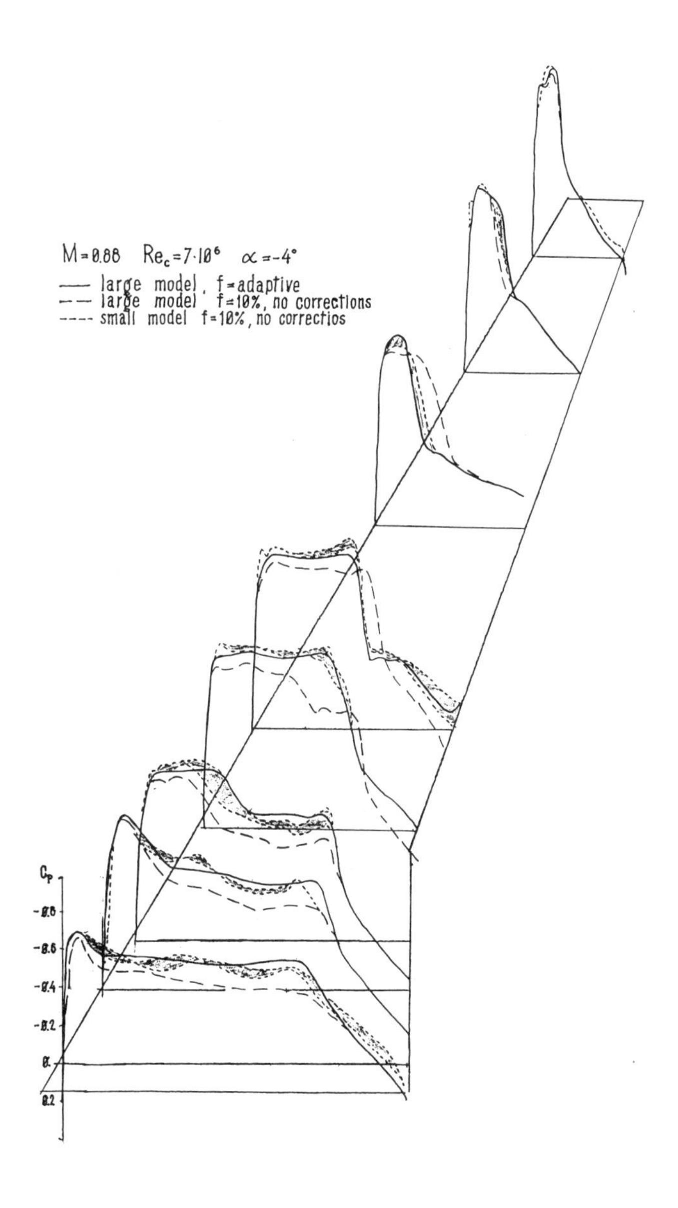

Figure 18a. Pressure distribution over upper surface of model wing.

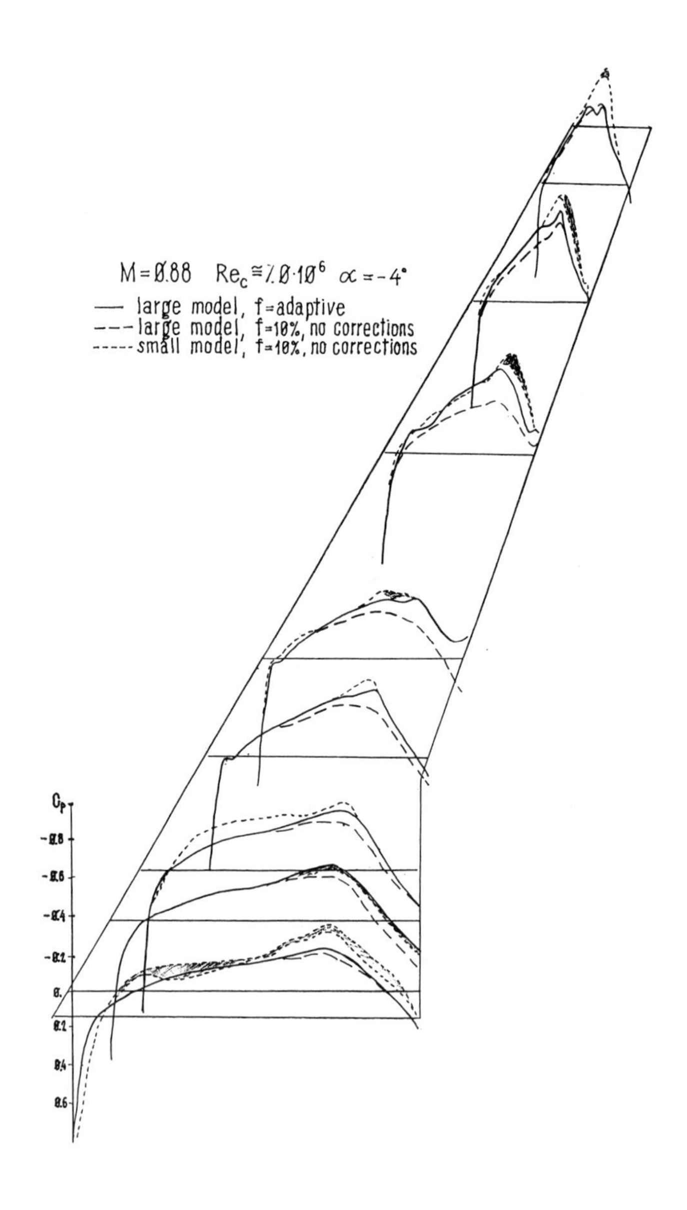

Figure 18b. Pressure distribution over lower surface of model wing.

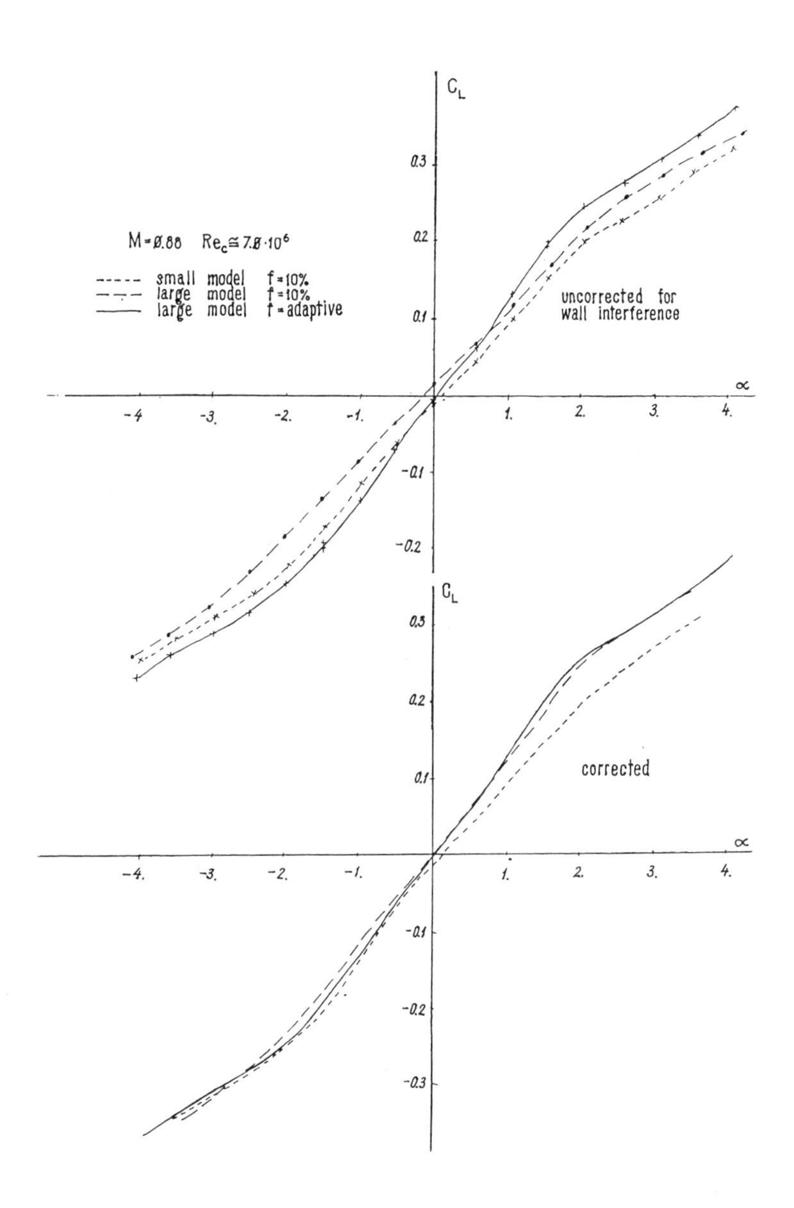

Figure 19. Lift curves.

Three curves present three cases: the large model and standard porosity (10%); the same model and wall adaptation; the small model and standard porosity (10%). All the three curves are evidently different from each other. Then, the data on both models at f=10% were corrected using the linear subsonic theory (the wall interference correction). The results are shown in Figure 18, where one can see that at $\alpha<0$ all three curves become closer, and at $\alpha>0$ the dependence of $C_l(\alpha)$ for the small model differs from the curve characterizing the large model.

It should also be noted that the comparison of slopes of $C_l(\alpha)$'s obtained in T-128 with the results for the small model obtained in BTWT facility of the Boeing Company are somewhat different. The reason of this discrepancy requires further investigation.

The authors are grateful to the Boeing Company, which kindly allowed use of the experimental data on two half-models tested in the T-128 facility of TsAGI; the authors also thank their colleagues A.R. Gorbushin and A.I. Ivanov who were most helpful in this research.

REFERENCES

1 *Adaptive Wall Newsletter,* No. 6, Nov., 1987.

2 Sears, W. R.: Self Correcting Wind Tunnels, *Aeronautic J.,* **78**, Feb.-Mar., 1974.

3 Ferri A. and Baronti, P.: A Method for Transonic Wind Tunnel Corrections, *AIAA J.*, **11**, No. 1, pp.63-66, Jan. 1973.

4 Sichev, V. V. and Fonarev, A. S.: Interference-Free Wind Tunnels for Transonic Research, Science Notes of TsAGI, No. 5, 1975.

5 Ivanov, A. I.: An Experimental Study of Gas Flow Near the Perforated Walls of a Transonic Wind Tunnel, *Fluid Mechanics - Soviet Research,* **17**, No. 4, Jul.-Aug. 1988.

6 Neyland, V. M.: Optimum Porosity of Wind Tunnel Walls at Small Supersonic Speeds, Mechanics of Fluid and Gas, *Soviet Academy of Science News,* No. 4, 1989.

7 Neyland, V. M.: Adaptive Wall Wind Tunnels with Adjustable Permeability – Experience of Exploitation and Possibilities of Development, *Proceedings of International Congress on Adaptive Walls,* China, 1991.

8 Neyland, V. M., Semenov, A. V., Semenova O. K.,Glazkov, S. A., Ivanov, A. I. and Khozyaenko, N. N.: Testing Technique Features of the Experiments in the Wind Tunnel with Adaptive Perforated Walls, Preprint TsAGI, No. 47, 1991.

Adaptive-Wall Wind-Tunnel Research at Ames Research Center: A Retrospective

Edward T. Schairer

1 SUMMARY

This paper reviews adaptive-wall wind-tunnel research conducted at Ames Research Center between 1978 and 1988. This research focused on developing ways to apply the concept of adaptive walls in transonic test sections with ventilated walls. In the approach pursued at Ames, local mass flow through slotted test-section walls was controlled by adjusting the pressures in compartments of a segmented plenum; the flow measurements required to test compatibility of the wind-tunnel flow with free-air boundary conditions were made using a laser velocimeter; empirical influence coefficients were used to predict how pressure changes in the plenum would affect the flow. Both two- and three-dimensional proof-of-concept experiments were conducted in a small indraft wind tunnel. Subsequently, a two-dimensional test section was demonstrated in the Ames 2x2 ft. Transonic Wind Tunnel – a wind tunnel representative of a production facility. The 2x2 ft. tests showed large reductions of wall interference; however, the time required to make the necessary flow measurements and plenum pressure adjustments far exceeded what would be acceptable for production testing. This was primarily because the imaginary surface where free-air compatibility was tested had to be separated from the test section walls to avoid the complex, viscous flow adjacent to the walls.

It is unlikely that the Ames approach to adaptive walls will be applied in production wind tunnels. The two biggest unresolved problems are how to quickly and accurately make the necessary flow measurements and how to predict the effects of wall adjustments. In contrast, flexible-wall technology for two-dimensional testing is ready for application in production wind tunnels. However, neither the ventilated- nor flexible-wall approach has been shown to be a technically viable and cost-effective solution to the three-dimensional adaptive-wall problem.

SYMBOLS

c	model chord length
C_L	lift coefficient
C_p	pressure coefficient
M	freestream Mach number
p_∞	freestream static pressure
u	measured axial perturbation velocity
u_f	free-air axial perturbation velocity
U	freestream axial velocity
w	measured vertical velocity
w_f	free-air vertical velocity
x,ξ	axial distance from model leading edge, positive downstream
z,η	vertical distance from tunnel centerline, positive up
α	angle of attack, deg
β	$\sqrt{1-M^2}$

2 INTRODUCTION

Between 1978 and 1988 researchers at Ames Research Center investigated ways to apply self-streamlining or adaptive walls in transonic wind tunnels with ventilated test section walls. As a major operator of large, high-speed wind tunnels, Ames could not ignore an emerging technology that promised to eliminate wall interference – a major source of uncertainty in transonic tests. Since wall-interference considerations usually restrict the size of models that can be tested in conventional test sections, adaptive-wall technology also had the potential to allow larger models to be tested in wind tunnels of a given size, thereby making more efficient use of the wind-tunnel airstream and increasing the useful Reynolds-number capability of existing wind tunnels.

Wall interference refers to the distortion that test-section walls produce in flow past a model. In general, the walls impose a boundary condition on the flow that is incompatible with free-air boundary conditions. For flows that are governed by linear equations, and when the model is small compared to the tunnel dimensions, it is often possible to separate wall effects and interpret them as slight adjustments or "corrections" to the freestream vector. In transonic wind tunnels, however, it is much more difficult to correct for wall effects because transonic flows are nonlinear, and thus wall effects cannot be neatly separated, and because flow boundary conditions at the ventilated walls are complex and uncertain. If wall effects become too large, wind-tunnel flows become "uncorrectable" in the sense that it is not possible to relate them to any free-air flow.

In principle, wall interference can be eliminated by adjusting the walls so that flow at the walls, or on some other surface surrounding the model, is consistent with free-air boundary conditions. The most intuitive example is a test section with flexible walls in which the walls are adjusted to conform to the shape of a free-air streamtube. Naturally, the free-air wall shape will be different for each flow condition. Flexible-wall test sections were first developed in England [1,2,3] in the late 1930's and in Germany [4] during World War II as solutions to the choking problem (i.e., massive blockage interference) at transonic speeds. Since World War II, test sections with ventilated walls (slotted or perforated) have been universally adopted for transonic testing.

In the early 1970's, Sears and several others independently recognized that wall-settings corresponding to free-air flow can be determined from flow measurements at or near the test section walls without any information about the model. This greatly simplified and systematized the problem of determining the proper wall adjustments compared to the semi-empirical procedures employed by the British in their wartime experiments. In addition, by 1970 computer and automation technology was available that would allow the necessary calculations to be performed "online" and the walls to be quickly and automatically adjusted. These developments made the concept of an adjustable-, or "adaptive-", wall wind tunnel far more practical than it was during World War II.

According to the Sears algorithm for adjusting test-section walls [5], an imaginary surface surrounding the model divides infinite flow into two regions: an inner region that includes the model and is "solved" by the wind tunnel, and an outer region that extends to infinity and is solved mathematically, Figure 1. The goal is to force compatibility of the two solutions everywhere on the surface by adjusting the test section walls. The first step in the algorithm is to measure the distributions of two flow quantities upon this surface. Sears recognized that in free air these two distributions are uniquely related. Therefore, the next step is to apply one distribution as a boundary condition and compute the corresponding free-air flow in the outer region. This "outer-flow solution" includes the distribution of the second quantity on the surface, which can then be compared to the measured distribution of that quantity. Differences between the computed and measured distributions of the second quantity are used as a basis for adjusting the test section walls, and the procedure is repeated until satisfactory convergence is achieved. An important virtue of this procedure is that the outer-flow solution can often be computed by solving linear equations even when flow in the inner region is quite complex and nonlinear.

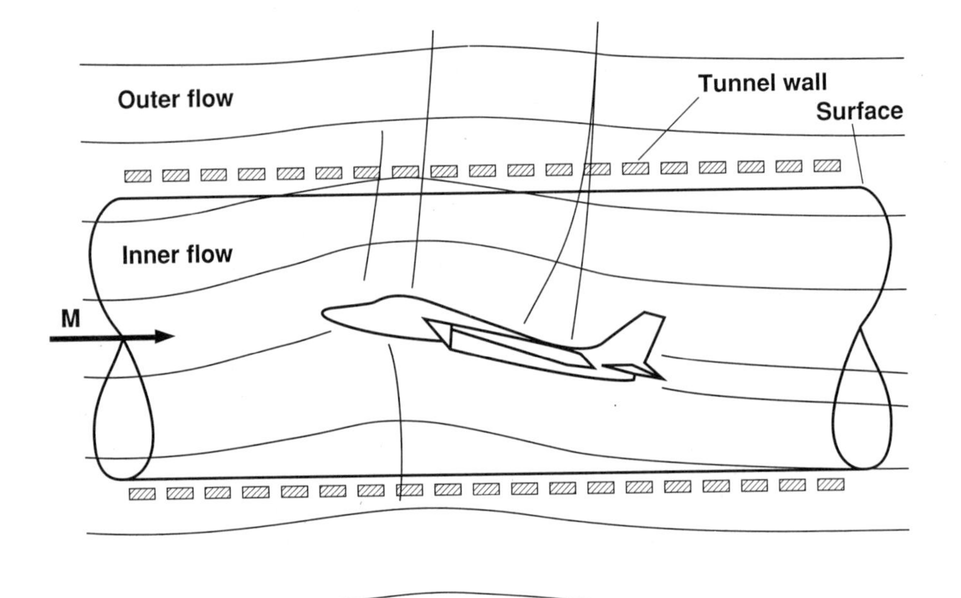

Figure 1. Adaptive wall concept [5].

By the time work at Ames began, two adaptive-wall concepts had been demonstrated for two-dimensional airfoil flows. Test sections with flexible upper and lower walls had been independently developed by European groups at the University of Southampton [6,7], Office National d'Etudes et de Recherches Aerospatiales (ONERA) [8], and the Technische Universitat (TU) Berlin [9]. Ventilated test sections with local control of airflow through the walls had been demonstrated by U.S. groups at Calspan [10,11,12] and Arnold Engineering Development Center (AEDC) [13]. At Ames itself, an independent research group was developing a two-dimensional, flexible-wall test section [14]; however, it was not designed to be "self-streamlining" in the sense of Sears' concept.

Research at Ames focused on the ventilated-wall approach primarily because it seemed to represent a smaller departure from the design of existing transonic test sections at Ames. The Ames Unitary-Plan 11x11 ft. Transonic Wind Tunnel (TWT), the 14 ft. TWT, and the 2x2 ft. TWT all have test sections that are vented through longitudinally slotted walls to a surrounding plenum. Another reason for choosing ventilated rather than flexible walls was that, in principle, ventilated walls can be designed to provide both streamwise and cross-stream adjustments needed for three-dimensional testing. Providing the same capability with flexible walls would require compound wall curvature. Furthermore, the ventilated-wall approach would preserve the proven shock-wave cancellation properties of ventilated walls – an important consideration for transonic testing. In contrast, the complex interaction between a model shock wave and wall boundary layer was a major unresolved uncertainty and potential liability of the flexible-wall approach. Finally, unlike perforated walls or impermeable walls, slotted walls can be designed with windows between slots, thus allowing access for optical flow diagnostic techniques.

Based on these considerations, the wall configuration we chose for all our adaptive-wall research consisted of straight, rigid walls with longitudinal slots that vented the test section to a plenum. The plenum was divided into compartments, each of which was connected through proportional valves to reservoirs of high- and low-pressure air.

One of the principal difficulties in applying adaptive walls in ventilated test sections is making the necessary flow measurements. Flow at the measurement surface must be inviscid, as it would be if the model were indeed in free air. In test sections with flexible walls, inviscid conditions at the edge of the (attached) wall boundary layer can be inferred from measurements at the wall. Thus, the walls themselves can serve as the boundary between inner and outer flows. The wall shapes and the pressure distributions upon them – both of which are easily measured – are sufficient to test for wall interference. In contrast, in test sections with ventilated walls, flow conditions at the edge of the complex, viscous region near the wall generally cannot be determined from conditions at the wall itself. Therefore, flow conditions must be directly measured along an imaginary boundary in the inviscid flow away from the wall.

Beginning with the first proof-of-concept experiments, we chose laser velocimetry (LV) to make these measurements. LV is accurate, non-intrusive, and allows flexibility in choosing and changing measurement locations. We decided not to use flow-angle probes or volumetric techniques for inferring inviscid flow conditions from local mass flow through the walls because of the problems Calspan had with these methods [10]. We also felt that pressure pipes that were eventually used at Calspan and AEDC [15] would be too intrusive and that data from them would be difficult to interpret when a shock wave impinged upon them.

In our preliminary adaptive-wall experiments we employed a unique variation of Sears' method for testing for free-air compatibility. This method, developed by Davis [16], requires measurements of only one flow quantity. The measurements, however, must be made along two boundaries, one closer to the model and one more distant. Davis showed how in free air the velocity distributions at the two boundaries are uniquely related, thereby establishing a test for free-air compatibility. This method simplified the flow-measurement problem since it allowed us to use a one-component LV.

Another major difficulty in applying adaptive-walls in ventilated test sections is determining the wall adjustments needed to produce the desired flow changes at the measurement boundary. Again, the difficulty arises because the measurement boundary is separated from the wall. We dealt with this problem by measuring influence coefficients. These were defined as the effect of pressure changes in each plenum compartment on the velocities at the measurement surface. We linearized the problem both by assuming a linear relationship between pressure and velocity changes and by assuming that the effects of pressure changes in individual compartments on velocities at the control surface could be superimposed. By this method, the plenum pressure changes needed to produce desired velocity changes at control points were computed by multiplying the vector of velocity changes by the inverse of the influence matrix.

We conducted two- and three-dimensional proof-of-concept experiments in a small indraft wind tunnel. In the first experiment, a two-dimensional airfoil was tested in transonic flow [17]. In a subsequent experiment [18], the test section was modified to allow cross-stream adjustments of the top and bottom walls, and a sidewall-mounted semi-span wing was tested at high subsonic speeds. The three-dimensional test was supported by numerical simulations of the same configuration using a linear panel code [19]. The results of this work encouraged us to design a two-dimensional adaptive-wall test section for the Ames 2x2 ft. TWT [20] – a wind tunnel similar in design to the 11x11 ft. TWT and thus representative of a production facility. We also began design studies for a 3x3 ft. three-dimensional adaptive-wall "demonstrator." At this time, management considered the 11x11 ft. TWT as a candidate for eventual conversion to adaptive walls.

While the test section for the 2x2 ft. TWT was being designed and built, further two-dimensional experiments were conducted by Celik of Stanford University in the Ames indraft wind tunnel [21]. These experiments explored one-step convergence schemes and the use of sidewall pressure measurements as an alternative to LV for assessing wall interference. Meanwhile, new ways to process flow measurements obtained in adaptive-wall wind tunnels were being developed. For example, Davis described the "generalized adaptive-wall wind tunnel" in which the walls could be adjusted to simulate a wind tunnel of arbitrary size [22]. Schairer [23] showed how the flow measurements required by Davis' one-component interference assessment algorithm could be used to estimate corrections for wall interference. A byproduct of this analysis was a method for computing free-air conditions at the measurement surface in one step (i.e., from one set of flow measurements) rather than iteratively [24]. In addition, simple, zonal methods were developed to account for nonlinear effects in the assessment of two-dimensional wall interference [25].

Before the test section for the 2x2 ft. TWT was completed, a reorganization at Ames moved the adaptive-wall research group out of the division with responsibility for operating Ames' major transonic wind tunnels, including the 2x2 ft. TWT. Although work on the 2x2 ft. TWT test section continued, plans to build a three-dimensional demonstrator were abandoned. Adaptive-wall research at Ames ended at the conclusion of abbreviated initial experiments in the new 2x2 ft. test section [26,27].

This paper describes the adaptive-wall test sections that were developed at Ames and summarizes the results of experiments that were conducted in them. The paper concludes with a discussion of the future of adaptive-walls for transonic testing.

3 TWO-DIMENSIONAL TESTS IN THE 25x13cm INDRAFT WIND TUNNEL

We conducted our first experiments in a small indraft wind tunnel, Figure 2, that was connected through a valve to a large vacuum sphere of the Unitary-Plan Wind Tunnel complex. The test section was 25 cm wide, 13 cm high, and about 74 cm long, Figure 3. The top and bottom walls were slotted, and the sidewalls were plexiglass. Separate upper and lower plenums were each divided by spanwise partitions into ten compartments. Compartments nearest the model, where longitudinal flow gradients were expected to be large, were smaller than more distant compartments. Mach number was set by changing the height of an adjustable sonic throat immediately downstream of the test section.

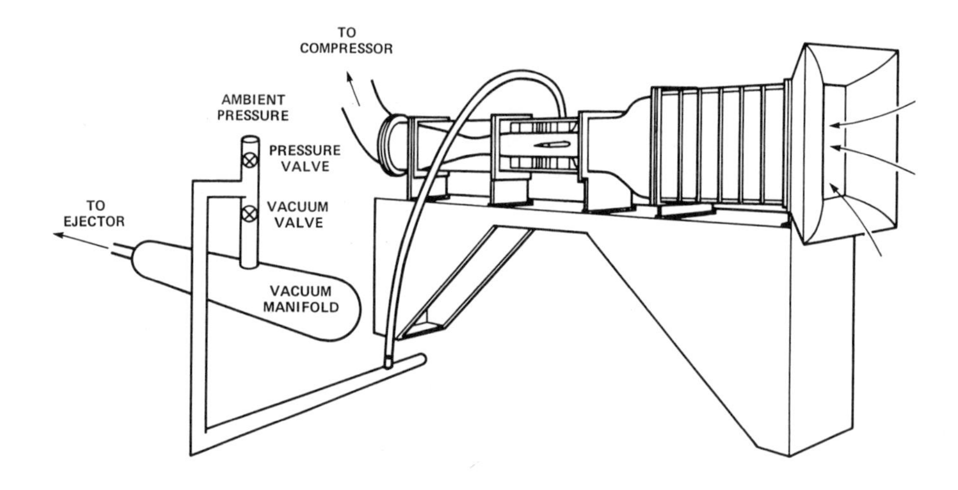

Figure 2. Schematic of indraft wind tunnel showing one channel of the plenum pressure system.

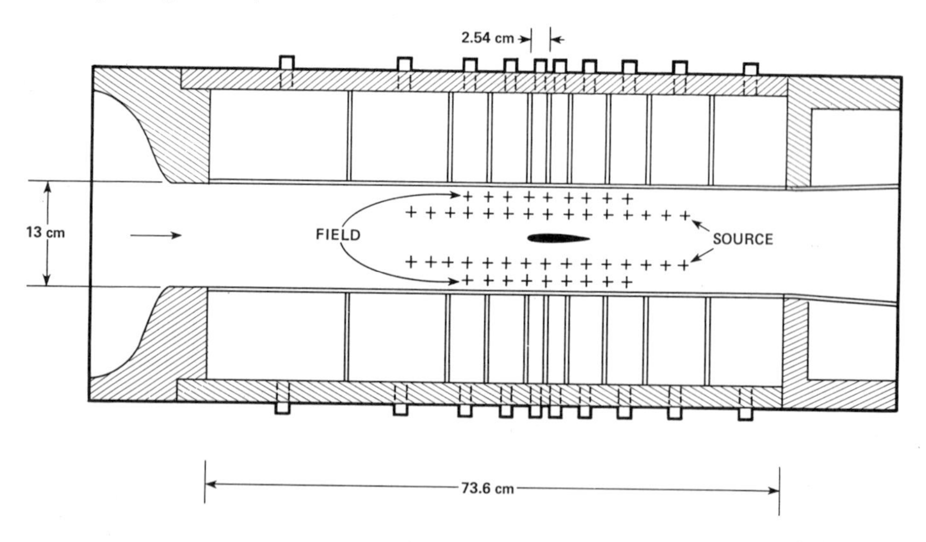

Figure 3. Test section for two dimensional tests in the 25x13cm indraft wind tunnel.

Each plenum compartment was connected by Tygon tubing, PVC pipe, and manually operated plastic ball valves to low- and high-pressure air reservoirs. The low-pressure reservoir was a manifold that was pumped down by an ejector that could operate continuously for about 30 min. This limited the length of the adaptive-wall runs. Since static pressure in the test section was less than atmospheric pressure, the ambient atmosphere was used as the high-pressure reservoir. The plumbing connections for a single compartment are illustrated schematically in Figure 2. Pressures in the plenum compartments were measured by scanivalve pressure transducers and were displayed on a water manometer board. Plenum pressure adjustments were made by manually adjusting the appropriate ball valve while observing the pressure change on the manometer board.

A one-component, forward-scatter laser velocimeter was used to measure vertical velocity (upwash) distributions at two levels, 0.4 and 0.667 chords, above and below the model, Figure 3. The laser and all of the beam-generating optics were mounted on a fixed optics table. The transmitting and collecting lenses were supported by identical two-dimensional positioning platforms on opposite sides of the test section. Light from the laser was directed to the transmitting lens by mirrors mounted on the platforms. Since the wind tunnel was located in a dusty environment, there was no need to artificially seed the flow. Figure 4 is a photograph of the experimental set-up showing the test section, laser velocimeter, and the plenum pressure control valves.

Figure 4. Photograph of the 25x13cm indraft wind tunnel.

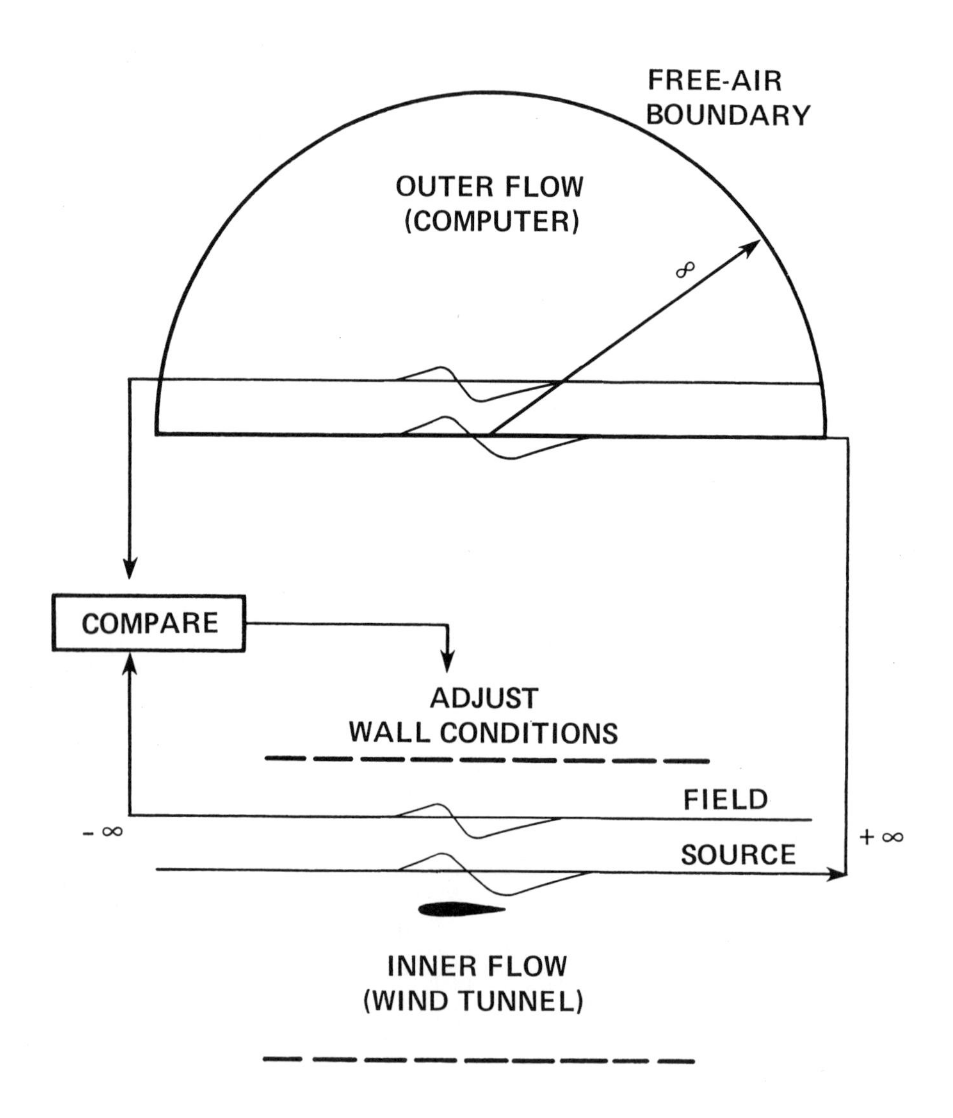

Figure 5. The one component interference assessment algorithm.

Figure 5 illustrates schematically the wall-setting algorithm. First, upwash distributions were measured at the closer ("source,") and more distant ("field,") levels between the model and the test section walls. The measured upwash distribution was applied as a boundary condition at the source level and free-air conditions were applied at infinity to compute the free-air upwash distribution at the field level. For flow that can be described by linear theory, this computation simply involves evaluating the integral [16]:

$$w_f(x,z_2) = \frac{\beta|z_2 - z_1|}{\pi} \int_{-\infty}^{\infty} \frac{w_f(\xi, z_1)}{(\xi - x)^2 + \beta^2 (z_2 - z_1)^2} d\xi \tag{1}$$

This "outer-flow solution" was then compared to the measured upwash distribution at the field level. Differences between the measured and computed distributions were used with an influence coefficient matrix to compute necessary pressure changes in the plenum compartments. After the plenum pressures were adjusted, the algorithm was repeated as necessary. All aspects of the algorithm were automatically controlled by a minicomputer except that the plenum pressure control valves were adjusted by hand, one at a time.

The model was a 3-in (7.62 cm) chord NACA 0012 airfoil. The tunnel blockage ratio was 7%, and the height-to-chord ratio was 1.7. The model was tested at $\alpha = 0°$ and 2° and at Mach numbers up to 0.80. The model boundary layer was tripped at 0.10c. For cases at $\alpha = 0°$ we assumed that the flow was symmetrical about the tunnel centerline. Therefore, for these cases we measured upwashes only above the model and symmetrically applied pressure changes to upper and lower plenum compartments. Tests could not be run at Mach numbers greater than 0.80 because condensation in the test section saturated the LV photodetector.

Figure 6 illustrates representative results from these experiments. It shows how wall adjustments changed upwash distributions at the field level and pressure distributions on the model for the case $M = 0.80$, $\alpha = 0°$. The figure also includes experimental "free-air" pressure distributions for a 6-in chord NACA 0012 airfoil measured at nearly the same conditions in the Calspan 8-ft. wind tunnel [28]. The effect of the wall adjustments was to improve the agreement between the measured and theoretical upwash distributions and to move the model shock wave upstream to a location much closer to its free-air position.

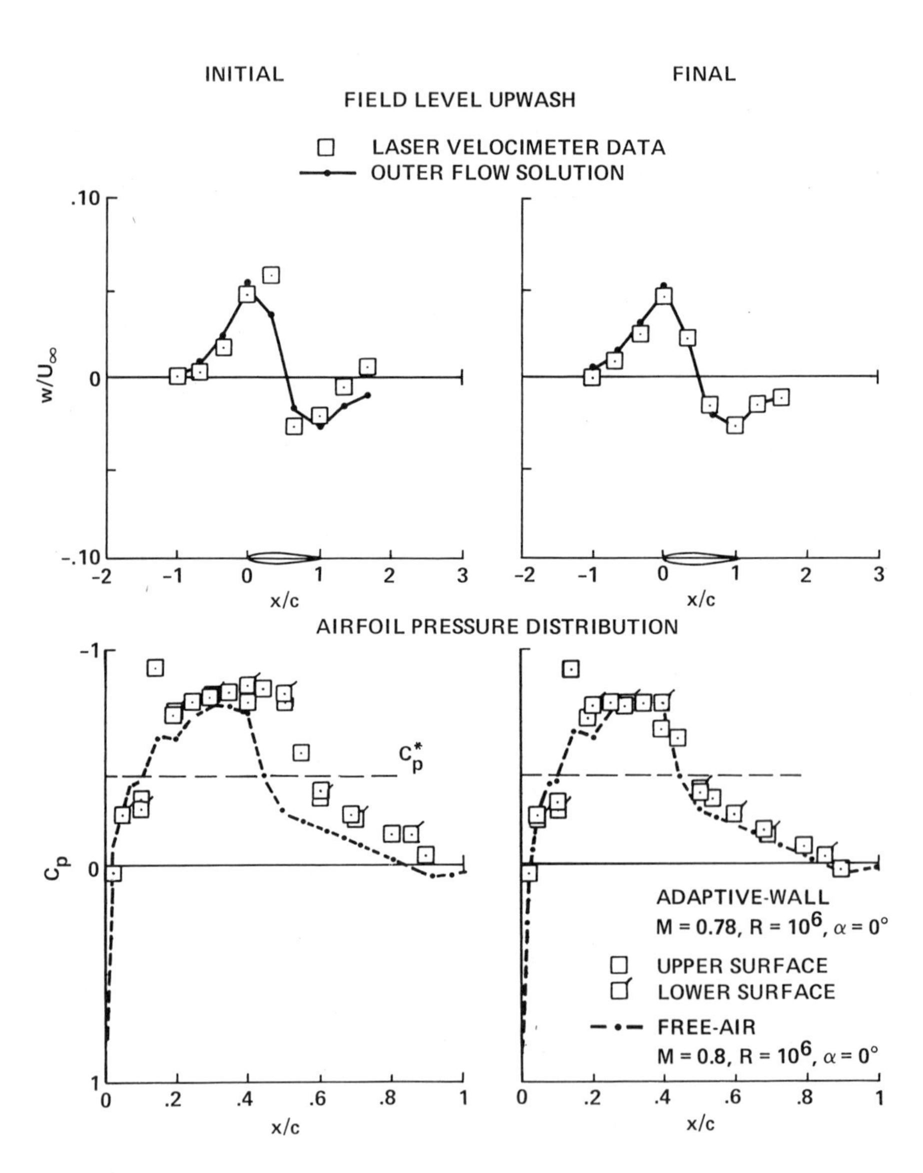

Figure 6. The effect of wall adjustments on upwash distribution at the field level and on the model pressure distribution ($M = 0.8$, $\alpha = 0^o$).

These first experiments, though imperfect, demonstrated that our approach to adaptive walls was technically feasible for two-dimensional flows. In particular, the one-component method for assessing interference was effective, and laser velocimetry showed promise as a means for making the necessary flow measurements. The weak link in the procedure was the method for determining the plenum pressure changes. The magnitudes of upwash changes at the field level were not accurately predicted by the influence coefficients; however, the signs of the velocity changes at the control points were usually as expected. Finally, it was tedious and time-consuming to manually adjust the plenum control valves. This difficulty could be avoided in a facility equipped with automatic valves.

4 THREE-DIMENSIONAL TESTS IN THE 25x13cm INDRAFT WIND TUNNEL

We next applied our approach to adaptive walls in a three-dimensional wing-on-a-wall experiment, Figure 7. The two-dimensional test section was modified by subdividing each upper and lower plenum compartment into three cross-stream compartments in which pressures could be independently adjusted. The number of streamwise compartments was reduced from 10 to 6, so there were 18 compartments above the test section and 18 below it. The plenum-pressure control system was expanded from 20 channels to 36.

The model was a tapered, unswept wing that was mounted to one sidewall window. It was supported by a six-component force and moment balance. The model extended two-thirds of the way across the test section and produced a blockage of 2.6%. The sidewall opposite the model was solid plexiglass and could not be adjusted.

Wall interference was assessed using a three-dimensional version of Davis' one-component algorithm. What were lines or levels in two dimensions became rectangular surfaces in three dimensions. In contrast to the two-dimensional case, the free-air relationship between upwash distributions at the source and field surfaces could not be expressed analytically, in closed form. Therefore, the outer flow problem, which was assumed to be governed by linear equations, was solved by finite differences. Figure 8 illustrates the finite domain and the grid upon which the outer-flow solution was approximated. Free-air boundary conditions on the outer faces of the domain were estimated using a horseshoe vortex at the model station.

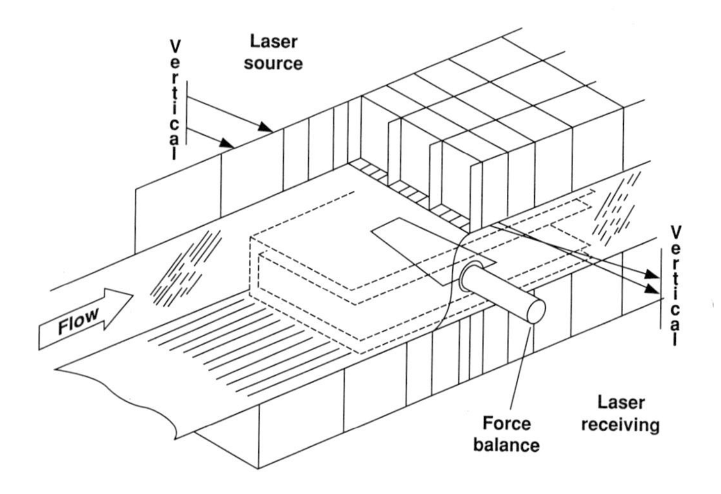

Figure 7. Schematic of the three-dimensional adaptive-wall experiment in the 25x13cm indraft wind tunnel.

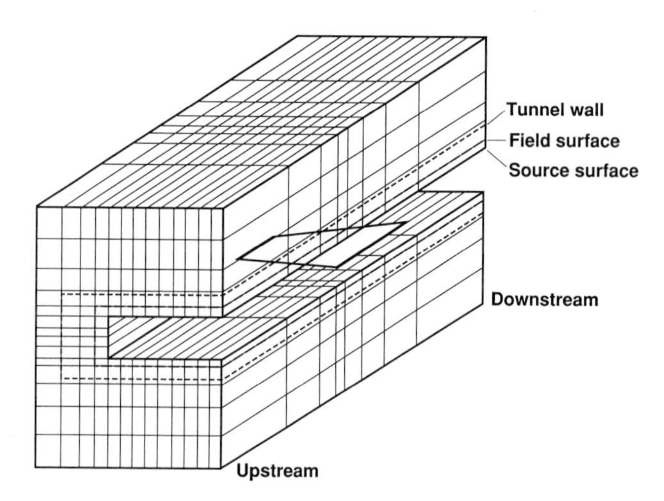

Figure 8. Domain for the three-dimensional adaptive-wall experiment in the 25x13cm indraft wind tunnel.

As in the two-dimensional experiment, laser velocimetry was used to measure upwash distributions at the measurement surfaces. The positioning platforms for the laser velocimeter were modified to add spanwise positioning capability and to allow faster scanning between measurement locations. Upwash was measured at 49 control points on each surface – at seven points at each of seven streamwise stations, Figure 9. Control points on the upper and lower faces of the field surface were centered with respect to the plenum compartments. Boundary conditions between measurements points at the source surface were estimated by linear interpolation.

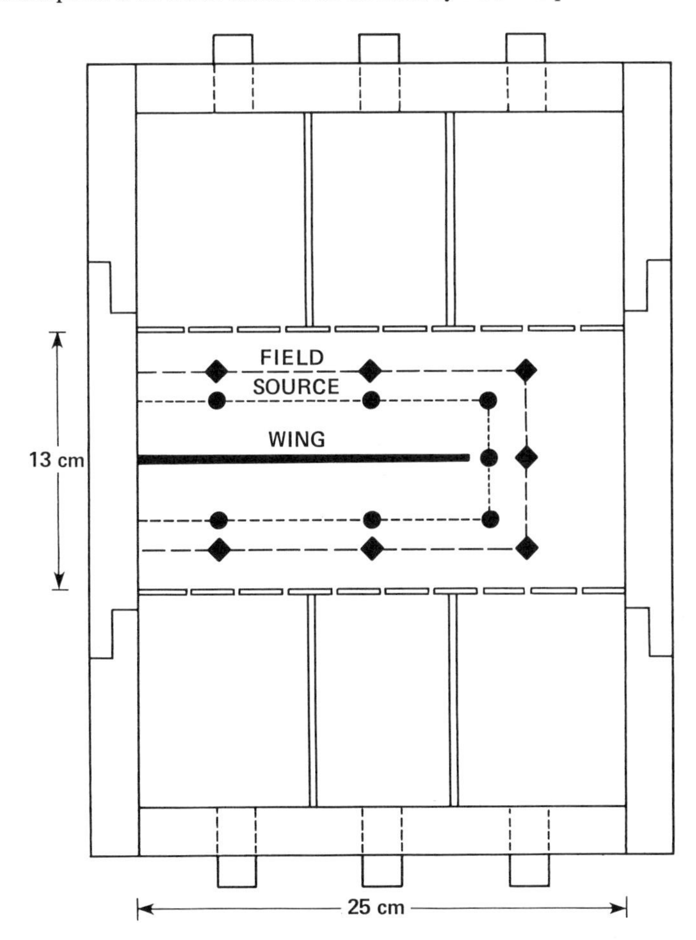

Figure 9. LV measurement points for the three-dimensional experiment.

The accuracy of the outer-flow solution was tested by applying it to the upwash field of a horseshoe vortex in free air. Figure 10 compares the actual free-air upwashes at the field surface with two outer-flow solutions: one where upwashes due to the vortex were "measured" at every grid point at the source surface, and the other where upwashes were only "measured" at the points where they were actually measured in the experiments, Figure 9, with data at intermediate grid points determined by linear interpolation. The accuracy of the solution computed from more complete boundary data is quite good except at stations immediately outboard and in the plane of the wing, where spanwise velocity gradients are very high. The accuracy of the solution computed from interpolated boundary conditions was good at the most inboard station; however, the error became progressively worse at the more outboard stations.

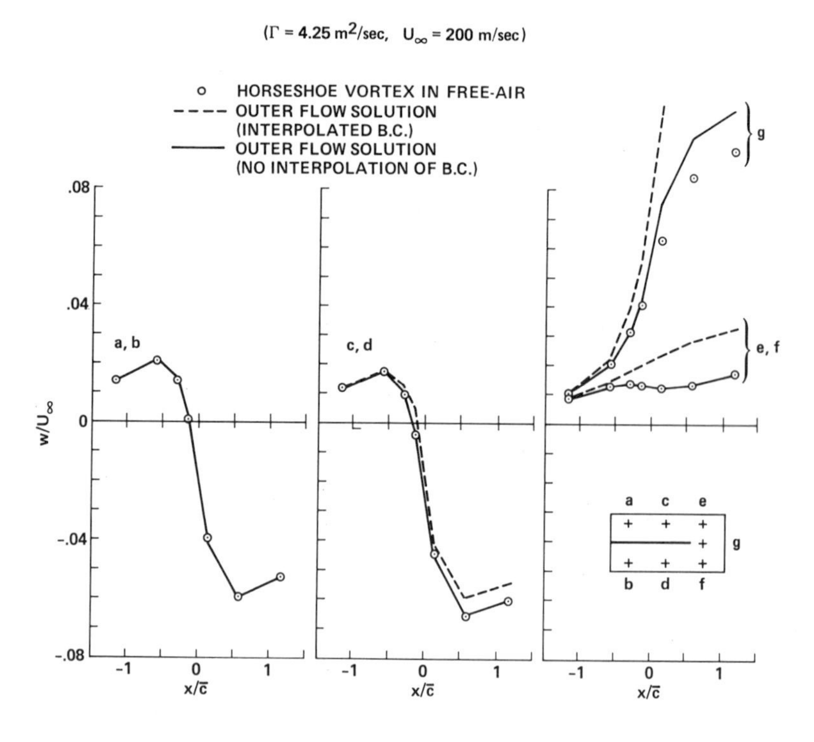

Figure 10. Three-dimensional outer-flow solver applied to a horseshoe vortex in free air.

Figures 11 and 12 illustrate typical results from the three-dimensional experiment. For the case $\alpha = 5.3°$ and $M = 0.60$, measured upwashes at the field level are compared to the outer-flow solutions before and after the walls were adapted. Although the wall adjustments reduced differences between the measured and computed upwashes, significant differences persisted. The most important reason for this was that upwash changes at the field surface could not be accurately predicted using linear influence coefficients. In addition, the low-pressure air system could not produce sufficient suction in several plenum compartments. Figure 13 shows how the wall adjustments changed the lift coefficient of the model. The figure includes "free-air" data for a model of the same geometry tested in a large wind tunnel [29]. As expected, the wall adjustments decreased the lift coefficient; however, the final lift coefficient was still well above the free-air value.

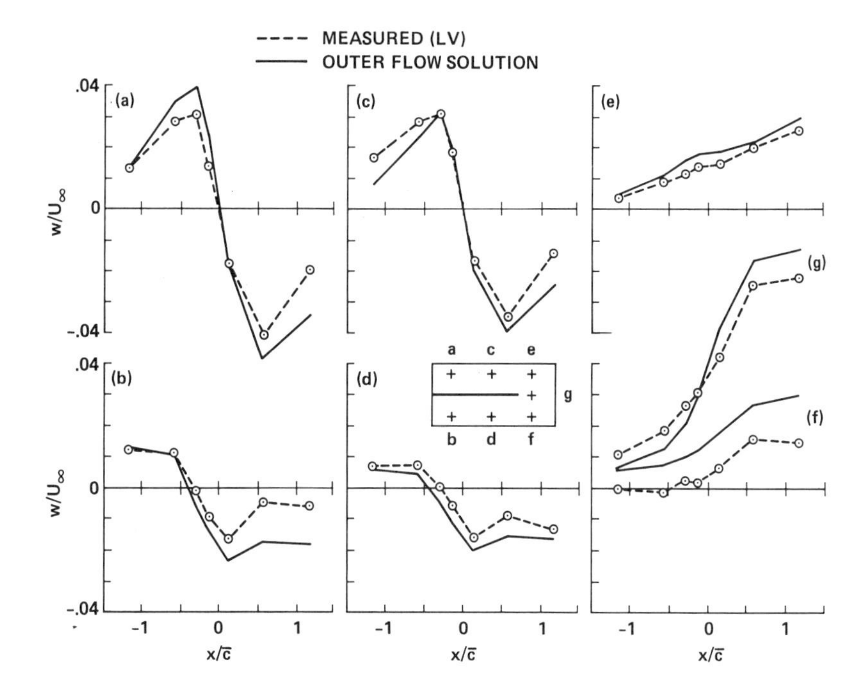

Figure 11. Comparison of the measured upwashes at the field surface with the outer-flow solution before the walls were adjusted ($M = 0.60$, $\alpha = 5.3°$).

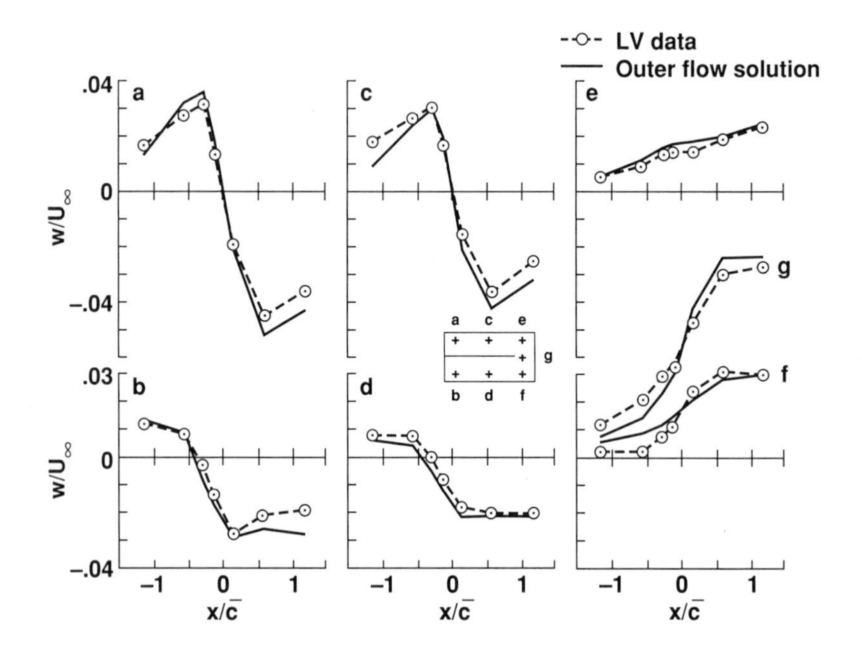

Figure 12. Comparison of measured upwashes at the field surface with the outer-flow solution after the walls were adjusted ($M = 0.60$, $\alpha = 5.3°$).

These experiments established that it is substantially more difficult to apply our approach to adaptive-walls in three dimensions than in two. In particular, LV is not suitable for making the necessary flow measurements. Measurements are required at too many points and the points do not all lie in the same plane. This leads to excessively long LV data-acquisition times and complex scanning mechanisms. In addition, high-power lasers (e.g., 15-watt) would probably be required in larger, production facilities because, in forward scatter (which must be used in transonic flow), light must be collected from across the test section. Optical access would also be a problem if the sidewalls, as well as the top and bottom walls, were adaptive. These experiments did not establish whether or not adaptive sidewalls would be necessary in a production facility. These and other issues were to be resolved in the 3x3 ft. demonstrator, which was never built.

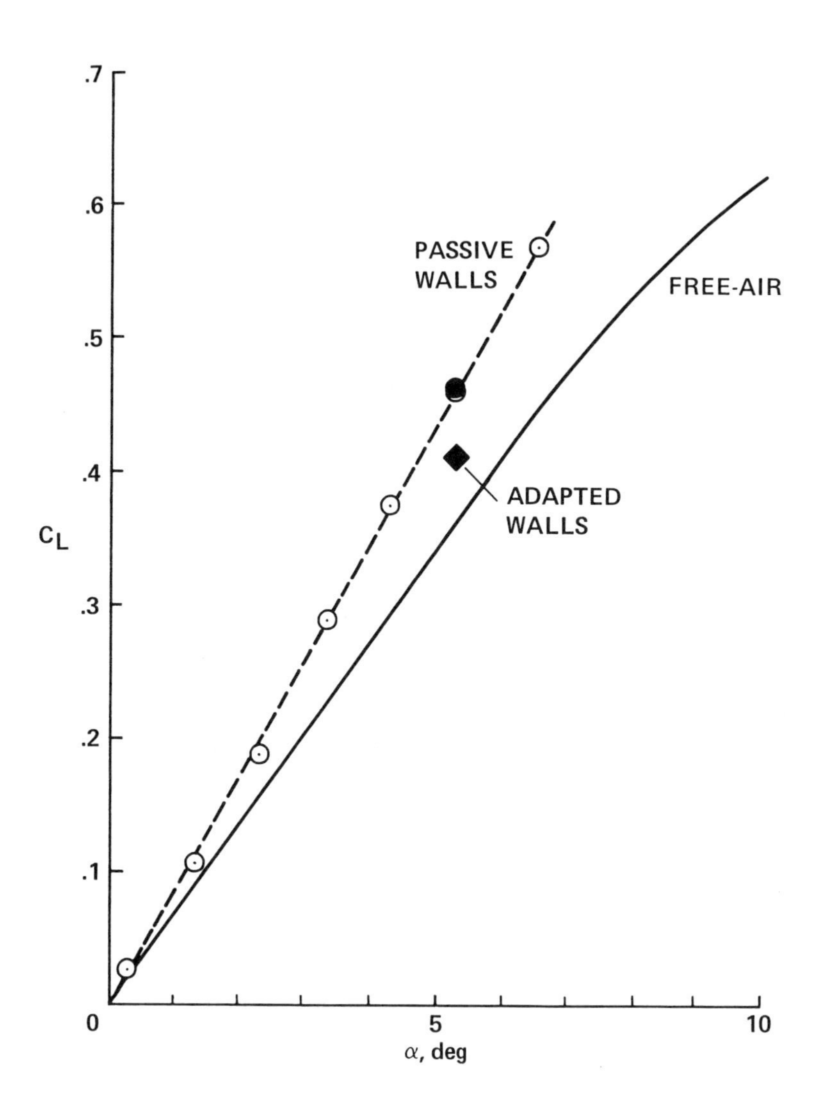

Figure 13. Effect of wall adjustments on the model lift coefficient ($M = 0.60$).

5 FURTHER TWO-DIMENSIONAL TESTS IN THE 25x13cm WIND TUNNEL

Additional two-dimensional experiments were conducted by Stanford researchers in the Ames indraft wind tunnel. These tests were designed to explore ways to speed up the adaptive-wall algorithm. In particular, they demonstrated how side-wall pressure measurements along axial lines (two above and two below the model) could be used as an alternative to LV to assess wall interference. In addition, a one-step convergence procedure was demonstrated. By this procedure, intermediate iterations were simulated on the computer using influence coefficients to predict the effects of plenum pressure changes on velocities at the field and source levels. Only after this simulation predicted satisfactory convergence of measured and theoretical velocities at the field level were the cumulative plenum pressure changes actually applied and flow conditions re-measured.

A new test section was built that was very similar to that used in the first two-dimensional experiments except that the height was 11 rather than 13cm. The same 3 in. chord NACA 0012 airfoil was tested at Mach numbers up to 0.75 and angles of attack up to 4°. The tunnel blockage was 8.3% and the height-to-chord ratio was only 1.4. These tests showed that by using side-wall pressure measurements instead of LV to assess wall interference, testing time could be reduced by a factor of five. The one-step procedure further reduced test times by a factor of 2-3. Test times as short as eight minutes were achieved using the combination of sidewall pressure measurements and the one-step procedure. Five minutes of this was spent manually adjusting the plenum pressures.

6 EXPERIMENTS IN THE 2x2 ft. TWT

Shortly after completion of the first two-dimensional tests in the indraft wind tunnel, the decision was made to build a two-dimensional adaptive-wall test section for the Ames 2x2 ft. TWT. This test section was to be suitable for airfoil research. It would incorporate the key elements that had been demonstrated in the indraft wind tunnel: slotted top and bottom walls and solid, transparent sidewalls; upper and lower segmented plenums; LV for making the necessary flow measurements; and automatic control and on-line interference assessment by means of a dedicated minicomputer. Initially, empirical influence coefficients would be used to determine the proper plenum pressure changes; however, it was deemed likely that a better method would have to be developed. Improvements over the pilot test section were to include: more plenum compartments (32 per wall instead of 10); automatic rather than manual wall control; the capability to measure two components of velocity rather than just one; and outer-flow solutions that account for nonlinear effects. In addition, a new control system was required for existing pumps that would supply high- and low-pressure air to the test section.

The 2x2 ft. TWT is of the closed-return type. It is driven by four 1000hp electric motors and was designed to operate at Mach numbers up to 1.4 and at pressures between 0.1 and 3 atmospheres. For the adaptive-wall experiments, the wind tunnel was restricted to operations at one atmosphere because of cracks that were discovered in the pressure shell during installation of the test section.

Figure 14 illustrates the adaptive-wall installation. The test section was mounted inside a pressure vessel formed by circular bulkheads upstream and downstream and a cylindrical shell that sealed the space between the bulkheads. In the figure, the upstream bulkhead and the pressure shell are omitted so that the test section can be shown more clearly. The space between the bulkheads served as the low-pressure-air reservoir. It was pumped down to a pressure 1-2 psi below p_∞ by an auxiliary pump. A structural box-beam below the test section and a 4 in. -diameter pipe above the test section were pressurized by a second auxiliary pump and served as the high-pressure-air reservoirs.

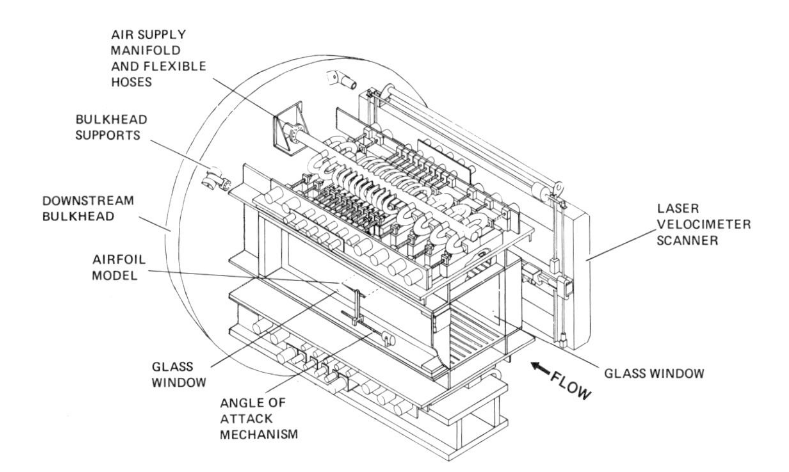

Figure 14. Adaptive-wall test section for the 2x2 ft. TWT.

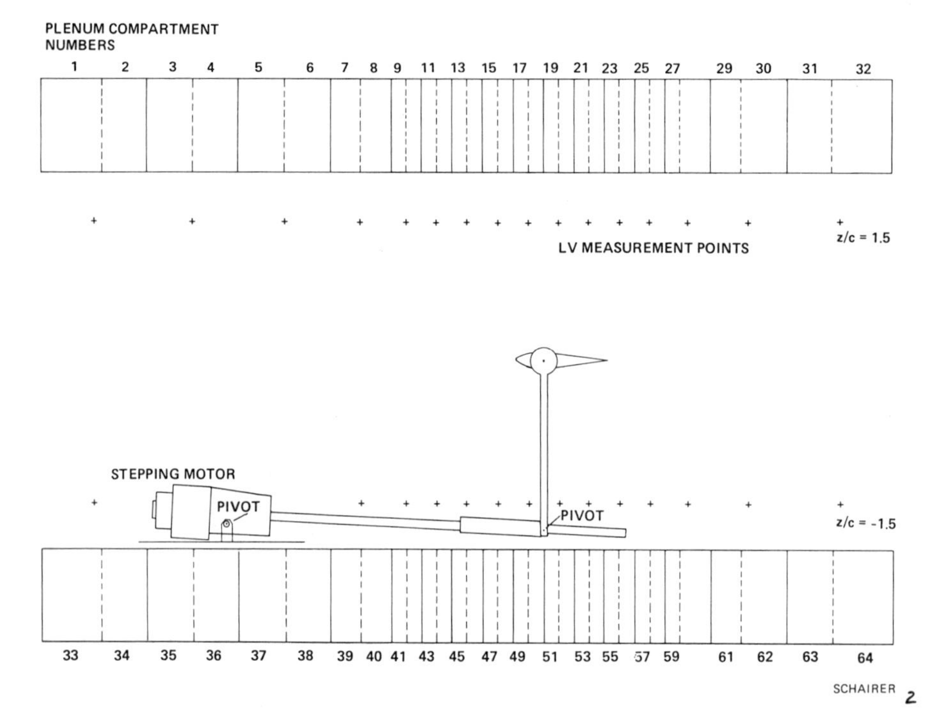

Figure 15. Side view of the 2x2 ft. TWT adaptive wall test section.

The test section was two feet high, two feet wide, and approximately five feet long. The top and bottom walls included nine equally spaced, open longitudinal slots (open-area ratio=0.14), and the sidewalls were two-inch thick Schlieren-quality glass windows. Separate plenums above and below the test section were each divided into 32 compartments by spanwise partitions, Figure 15. Pressures in the compartments were controlled by three-way, proportional slide valves that were connected to the high- and low-pressure reservoirs. Stepping motors drove the slide valves in the spanwise direction along a valve seat that formed the back of the plenum, Figure 16. Depending on the position of the valve, ports in the valve seat were either closed or open to one of the reservoirs, allowing air to be injected into or removed from the test section.

A fast-scanning, two-component LV was custom-built for the test section. It was designed to measure vertical and streamwise velocities at virtually any point in the plane midway between the test section side-walls. A unique, on-axis, forward-scatter optical design employed a retro-reflector behind the collecting lens to return scattered light to the transmitting lens much as if the light had been directly back-scattered, Figure 17. This design combined the much higher signal strength of a forward-scatter system and the superior tolerance to misalignment of a back-scatter system.

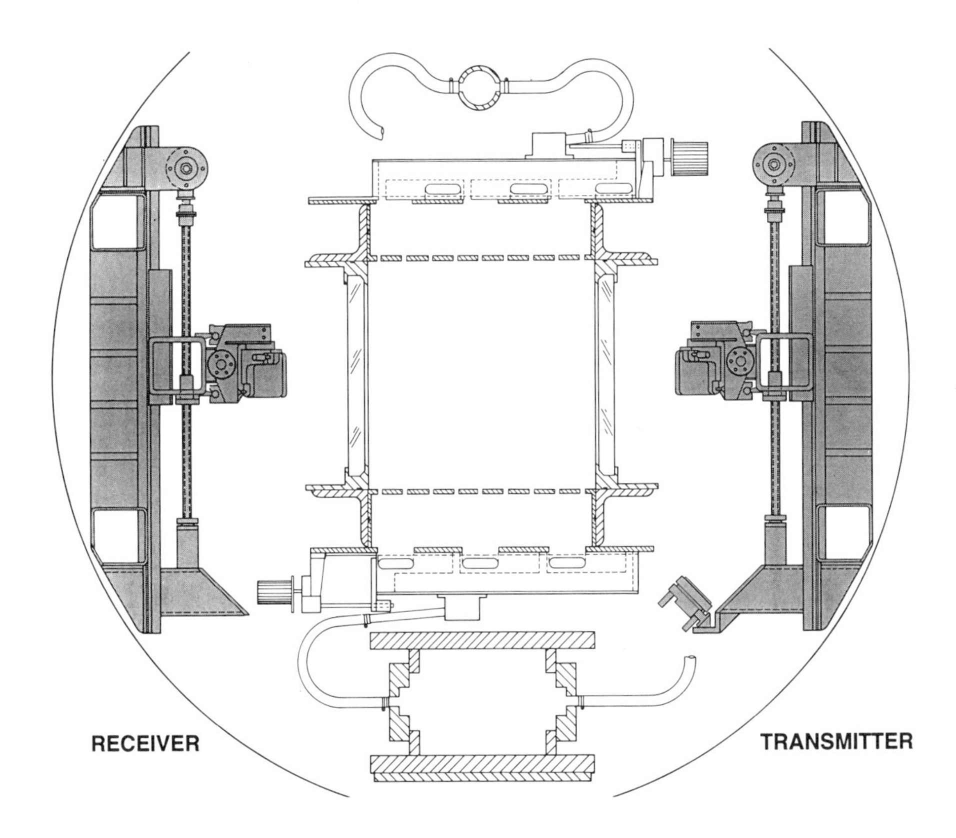

Figure 16. Lateral cross-section of the 2x2 ft. adaptive-wall test section.

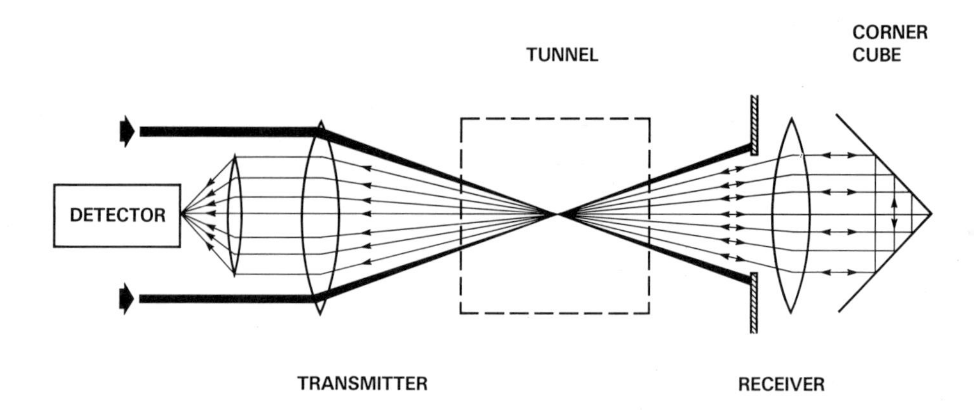

Figure 17. Schematic of LV optical design for the 2x2 ft. adaptive-wall test section.

The transmitting and collecting lenses were mounted on two-dimensional scanner platforms that were located on opposite sides of the test section and inside the pressure shell, Figure 16. As in the pilot experiments, the laser beams were directed to the transmitting lens by a series of turning mirrors mounted to the scanner, Figure 18. The laser beams were automatically aligned to the scanners using computer-controlled, motorized actuators to adjust the mirror angles and photodiodes behind the mirrors to measure the position of the laser beams. The moveable elements of the scanners could be retracted, Figure 19, permitting an unobstructed view of the model as would be required for other types of optical diagnostics, e.g., holography. The laser, beam-generating optics, receiving optics, and photodetectors were mounted on a stationary optics table in the wind-tunnel control room.

The model was a 6-in. chord NACA 0012 airfoil. It was supported by spherical bearings mounted in the glass windows. Angle of attack was set by identical stepping-motor-driven mechanisms on either side of the test section, Figure 15. The tunnel blockage was 3% (less than half of what it was in the indraft wind tunnel), and the height-to-chord ratio was 4.0.

In a departure from the method used in the pilot experiments, wall interference was assessed from measurements of two components of velocity. Two-component methods have several important advantages over the one-component method. Specifically, they require measurements at fewer locations, allowing the measurements to be made more quickly; they automatically define freestream Mach number and flow angle [30]; they are more sensitive to wall interference and yield more interference information (i.e., both components); and they allow assessing interference on a closed contour [31], thereby eliminating errors due to truncating infinite integrals.

Figure 18. Photograph showing how laser beams were directed into the 2x2 ft. adaptive-wall test section.

Several two-component methods were used to predict free-air velocities. The simplest was based on equations that were first published by Sears [5] and express the free-air relationship between axial and vertical velocity distributions along an infinite axial line:

$$u_f(x,z) = \frac{1}{\pi\beta} \int_{-\infty}^{\infty} \frac{w_f(\xi,z)}{x-\xi} d\xi \tag{2}$$

$$w_f(x,z) = \frac{-\beta}{\pi} \int_{-\infty}^{\infty} \frac{u_f(\xi,z)}{x-\xi} d\xi \tag{3}$$

Figure 19. Photograph showing LV scanners in retracted position.

These equations assume that the outer flow can be described by linear equations, but they make no assumptions about flow near the model. They are applied iteratively by substituting measured velocities for free-air velocities in the integrands. Solutions above and below the model are uncoupled. We assumed that velocity perturbations vanished beyond the ends of the test section.

In other cases, free-air velocities were computed using one-step/closed-contour equations [31,32]:

$$u_f(x,z) = u(x,z) + \frac{1}{2\beta\pi}\oint_c \frac{\beta^2(\eta - z)u(\xi,\eta) + (\xi - x)w(\xi,\eta)}{(\xi - x)^2 + \beta^2(\eta - z)^2} d\xi$$

$$- \frac{\beta}{2\pi}\oint_c \frac{(\xi - x)u(\xi,\eta) - (\eta - z)w(\xi,\eta)}{(\xi - x)^2 + \beta^2(\eta - z)^2} d\eta \tag{4}$$

$$w_f(x,z) = w(x,z) - \frac{\beta}{2\pi}\oint_c \frac{(\xi - x)u(\xi,\eta) - (\eta - z)w(\xi,\eta)}{(\xi - x)^2 + \beta^2(\eta - z)^2} d\xi$$

$$- \frac{\beta}{2\pi}\oint_c \frac{\beta^2(\eta - z)u(\xi,\eta) - (\xi - x)w(\xi,\eta)}{(\xi - x)^2 + \beta^2(\eta - z)^2} d\eta \tag{5}$$

In these equations, measurements are required along a closed rectangular contour surrounding the model. This eliminates the problem of extrapolating velocity distributions beyond the ends of the test section. The equations are more restrictive than the iterative equations because they are based on linear wall-interference theory, which assumes that both the inner and outer flows can be described by linear equations. Note that, unlike in Equations (2) and (3), velocities in the integrands of Equations (4) and (5) are measured, not free-air, and that solutions above and below the model are coupled.

For higher Mach number cases, free-air velocities were predicted using a zonal method that combined a finite-difference solution to the transonic small perturbation equation in a zone near the model and an analytic, linear solution at more distant points [25].

Free-air solutions were computed for both components of velocity. However, because of our previous experience and the intuitive relationship between plenum pressures and vertical velocities, wall adjustments were computed from differences between measured and theoretical upwash distributions. Figure 15 illustrates the points in the test section where axial and vertical velocities were measured.

Applying the method of influence coefficients to compute plenum pressure changes presented several difficulties. First, it was not possible to isolate the effect of pressure changes in a single plenum compartment because adjusting the slide valve to one compartment produced significant pressure changes in neighboring, passive, compartments. Therefore the influence coefficients were defined in terms of "self-induced" pressure changes (i.e., pressure changes in a compartment produced by adjusting that compartment's control valve), and the pressure changes computed from them were self-induced. In order to apply these changes, the slide valves had to be adjusted one at a time; otherwise, self-induced pressure changes would be confused with pressure changes induced by adjusting other control valves. Although the same constraint applied in the pilot experiments, it was less restrictive there because the control valves were adjusted by hand, usually by a single experimenter, and there were far fewer of them. In the 2x2 ft. TWT, however, where the means existed to automatically adjust all the control valves simultaneously, this constraint drastically increased the time required to adjust the walls. To help alleviate this problem, adjacent slide valves were paired and were simultaneously adjusted as if they were a single control, both when influence coefficients were measured and during the adaptive-wall procedure.

A second problem was that the influence matrix was ill conditioned and resulted in pressure-change solutions with very large elements of alternating sign. This problem was not encountered in the pilot experiments. It was resolved by computing approximate but more physically meaningful solutions using a least-squares method [33].

The model was tested at $\alpha = 0^\circ$ at Mach numbers up to 0.85 and at $\alpha = 2^\circ$ at Mach numbers up to 0.75. For all cases the model boundary layer was untripped.

Figure 20 illustrates how wall adjustments improved compatibility of flow measurements with free-air conditions for the highest-speed case that was run ($M = 0.85$, $\alpha = 0^\circ$). Since at zero incidence the flow was symmetrical about the tunnel centerline, flow measurements were only made above the model, and upper and lower plenums were adjusted symmetrically. Outer-flow solutions were computed using Equations (2) and (3). By the third cycle of wall adjustments, maximum suction had been applied to plenum compartments immediately above and below the model. The principal effect of the adjustments was to relieve blockage and thereby reduce peak axial velocities immediately above the model. This is reflected in the pressure distribution on the model, Figure 21, where the reduction in blockage decreased the effective Mach number at the

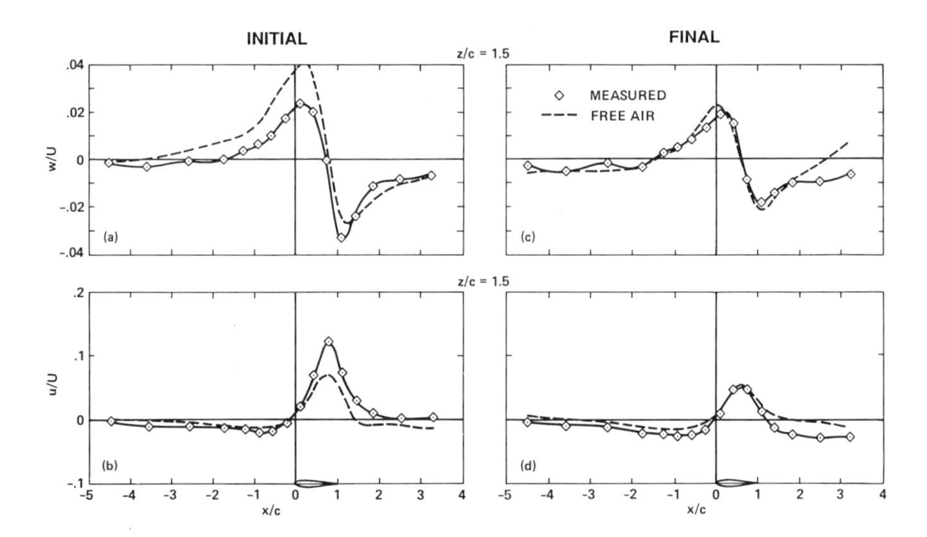

Figure 20. Effect of wall adjustments on axial and vertical velocities ($M = 0.85$, $\alpha = 0^{o}$).

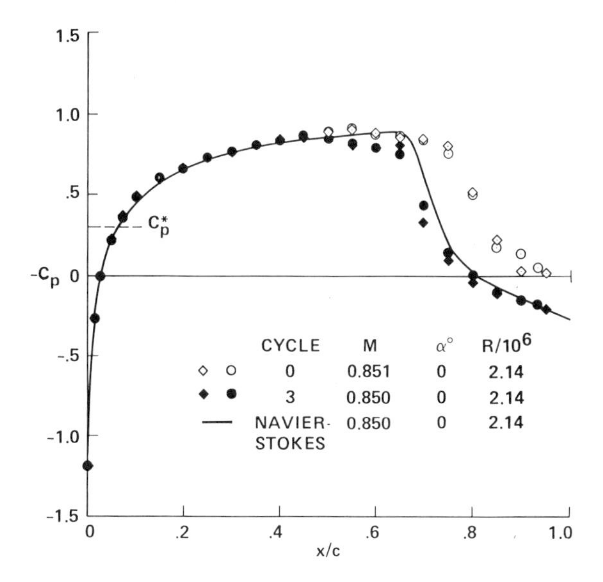

Figure 21. Effect of wall adjustments on model pressure distribution ($M = 0.85$, $\alpha = 0^{o}$).

model resulting in forward movement of the shock wave and elimination of flow separation near the trailing edge. Figure 21 compares the experimental pressure data with a numerical solution to the Navier-Stokes equations [34]. It suggests an over-correction for wall interference. This conclusion is supported by a linear WIAC (wall interference assessment and correction) analysis, Figure 22, by which data at the measurement level are used to compute wall-induced velocities along the tunnel centerline [32].

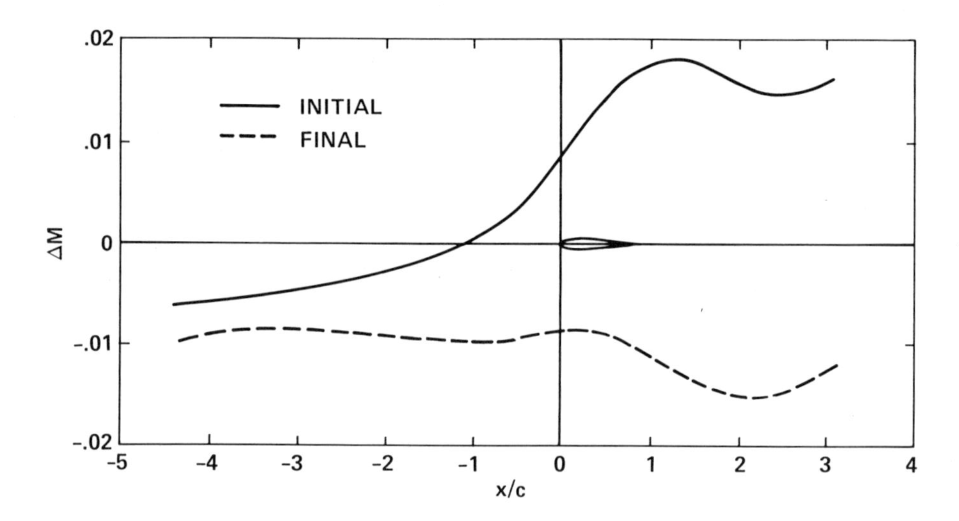

Figure 22. Effect of wall adjustments on wall-induced velocities along the tunnel centerline ($M = 0.85$, $\alpha = 0°$).

Data for a lifting case, $M = 0.70$ and $\alpha = 2°$, are presented in Figure 23. In this case free-air velocities were computed using the one-step/closed-contour formulae, Equations (4) and (5). The data show dramatic improvement in the free-air compatibility of the axial velocities above and below the model and smaller changes in vertical velocities. Again, the principal effect of the wall adjustments was to reduce blockage interference, thus decreasing axial velocities near the model. The shock wave on the model moved upstream, Figure 24, and wall-induced axial velocities along the tunnel centerline were virtually eliminated, Figure 25. Figure 25 shows, however, that there was significant residual wall-induced downwash.

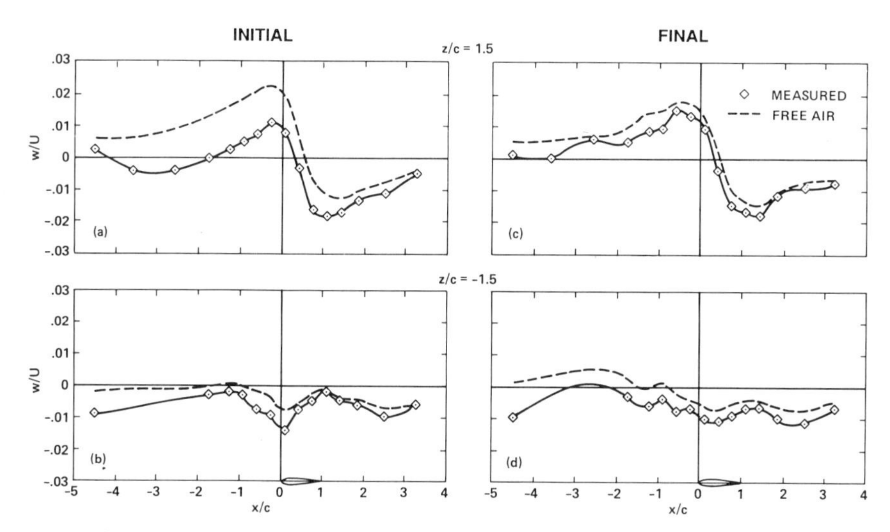

Figure 23a. Effect of wall adjustments on vertical velocities ($M = 0.70$, $\alpha = 2°$).

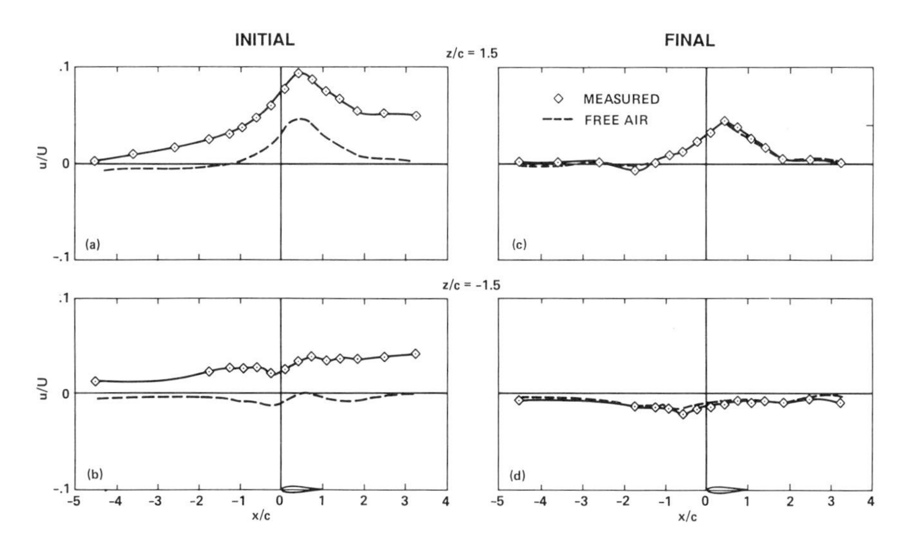

Figure 23b. Effect of wall adjustments on axial velocities ($M = 0.70$, $\alpha = 2°$).

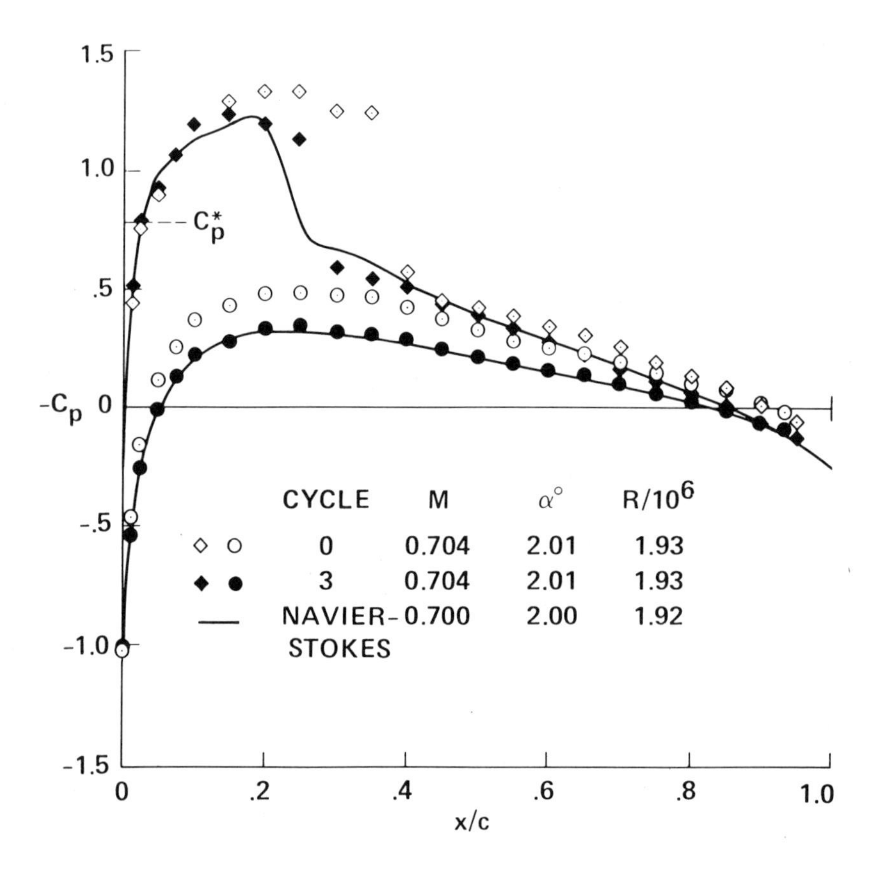

Figure 24. Effect of wall adjustments on model pressure distribution ($M = 0.70$, $\alpha = 2°$).

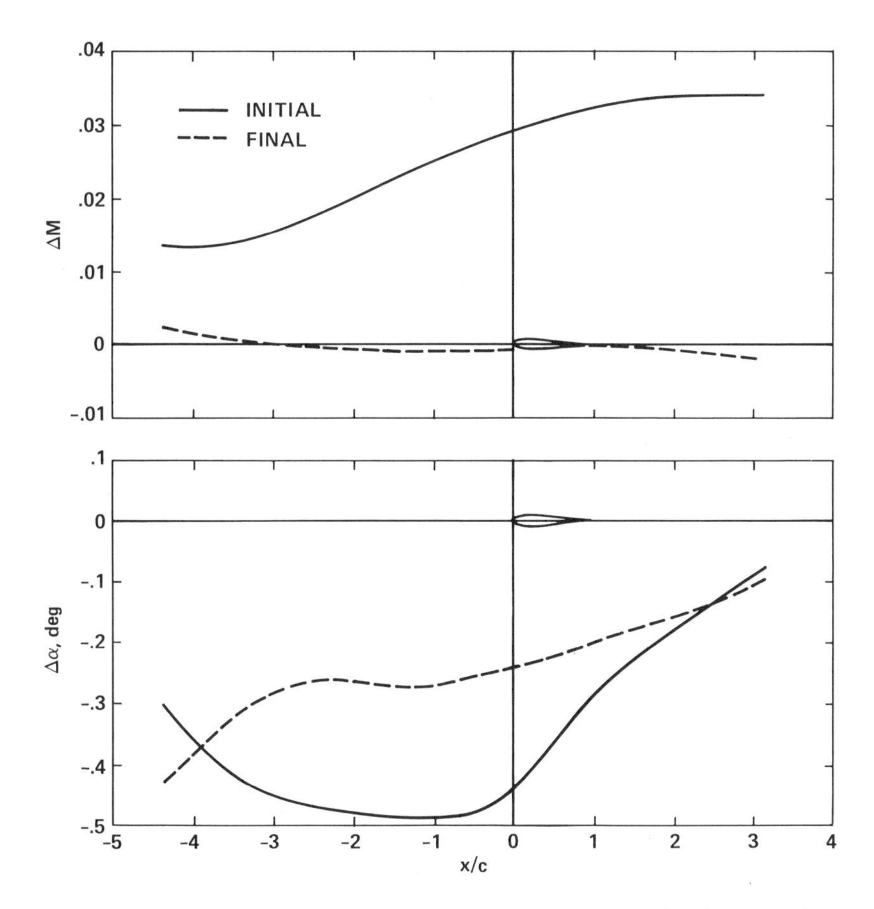

Figure 25. Effect of wall adjustments on wall-induced velocities along tunnel centerline ($M = 0.70$, $\alpha = 2^{\circ}$).

The cases presented here were typical of the limited success that was achieved with the 2x2-ft. test section. Wall interference was always significantly reduced but never to levels low enough to make corrections for residual interference unnecessary. As in the pilot experiments, the principal reason was our inability to accurately predict the effects of plenum pressure changes on the measured upwash distributions. In particular, the influence coefficients were not known with sufficient accuracy, and they were based on over-simplified assumptions. In addition, the quantity we chose to control, upwash, was far less sensitive to wall adjustments than was axial velocity. A better approach might have been to determine the plenum pressure changes in terms of axial velocity, at least for cases dominated by blockage interference.

The short time allocated for these experiments did not allow us to refine the wall-setting algorithm as we gained experience. Therefore, further development is clearly required before the test section would be useful for anything other than adaptive-wall research. In addition to the wall-control problems discussed above, the time required to adjust the slide valves (about 9 min/cycle) far exceeded what would be acceptable for production testing. The slide valves could be adjusted in a matter of seconds if an accurate method were developed that allowed them to be adjusted simultaneously. Laser velocimeter data-acquisition times also needed to be substantially reduced. Times approaching 1 min would be possible with improved seeding. Finally, the reliability and robustness of some of the wind tunnel systems would have to be significantly improved. In particular, the control system for the auxiliary pumps was inadequate, resulting in many unexpected pump shutdowns. In addition, the slide-valve controllers were very unreliable; never were all 64 operational at the same time.

7 HIGH REYNOLDS NUMBER CHANNEL 2

Two test sections with flexible top and bottom walls have been built for the Ames High Reynolds Number Channel 2 (HRC-2). These test sections were designed to allow direct comparison between experimental and numerical simulations of transonic flows. Therefore it was important that experimental wall boundary conditions be easily and accurately included in numerical simulations. This requirement called for impermeable, as opposed to ventilated walls, and flexible walls to allow testing at transonic speeds without choking. For two-dimensional tests, wall contours approximating free-air streamlines for a representative test condition have been determined a priori from numerical simulations of flow past the model. This wall shape was then used for experiments over a range of test conditions.

HRC-2 is a blow-down facility that is supplied with dry air from the Ames 3000 psi air system and exhausts either to the atmosphere or to a vacuum sphere. The test section is contained within a cylindrical test cabin that serves as the pressure vessel, Figure 26, and is vented to this cabin through porous panels in the sidewalls downstream of the model station. Mach number is set by adjusting the height of a sonic throat at the downstream end of the test section.

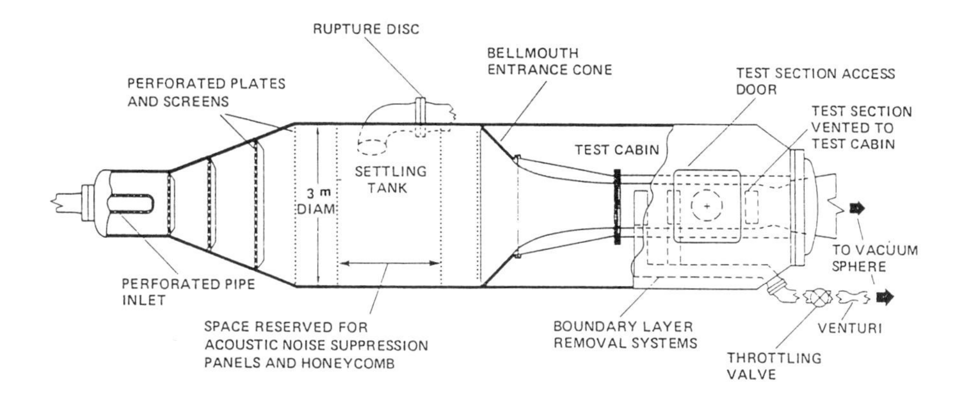

Figure 26. High Reynolds Number Channel 2.

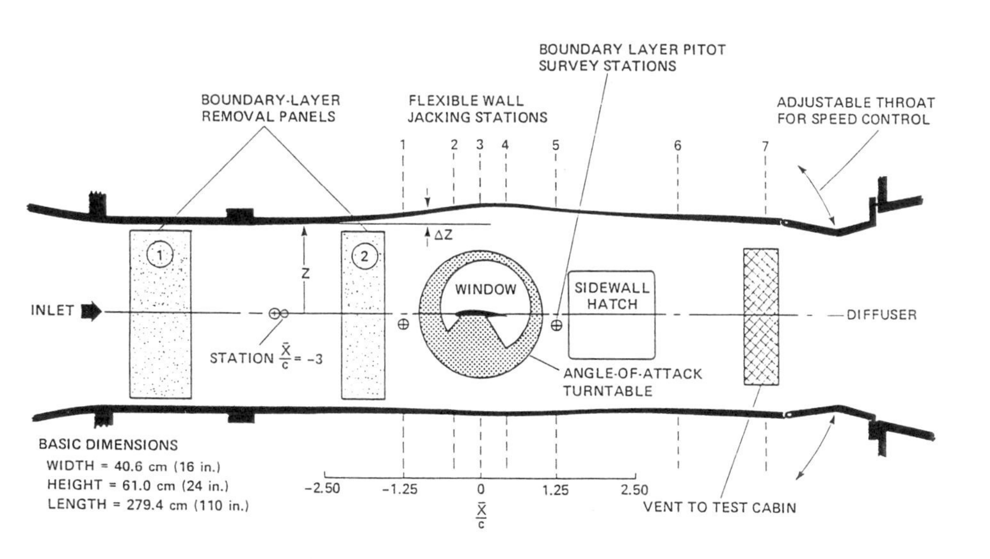

Figure 27. Test section of HRC-2.

Both flexible-wall test sections are 16 in wide, 24 in high, and about 8 ft. long. Both have porous panels in the sidewalls upstream of the model station through which sidewall boundary layers can be drawn off. The principal difference between them is that one has 7 manually actuated jacks per wall, Figure 27, whereas the other has 10 motorized jacks per wall. The test section with manual jacks has been used for two-dimensional airfoil research [35,36] and, more recently, for tests of a sidewall-mounted low-aspect-ratio wing [37]. For the three-dimensional test, the walls were set to diverge from the tunnel centerline at a constant angle of 0.24°. The test section with motorized jacks has never been used. In fact, its adjustable walls were recently disassembled.

8 DISCUSSION

It is unlikely that the adaptive-wall approach pursued at Ames will ever be applied in a production wind tunnel. With further development, better results than those presented here could undoubtedly be achieved in the 2x2 ft TWT. However, there is little hope that the time required to determine and apply the necessary wall adjustments would ever be short enough to be acceptable for production testing or testing with some other goal than eliminating wall interference. The Achilles heel of the slotted wall/segmented plenum approach is that, because of the complexity of flow at the walls, the interference assessment surface must lie between the model and the test section walls, and cannot coincide with the walls. This makes it very difficult to quickly and accurately measure flow conditions at the surface and to accurately predict the effect of wall adjustments on the flow there.

In contrast, researchers at other laboratories have demonstrated two-dimensional test sections with flexible impermeable walls in which the self-streamlining process occurs in less than one minute. Furthermore, these researchers have shown that two-dimensional free-air flows can be simulated at high transonic speeds even when the shock wave from the model extends to the tunnel walls [38]. Finally, the flexible-wall approach to the two-dimensional problem does not compromise optical access and may even improve flow quality and tunnel efficiency by eliminating ventilated walls – a major source of noise. Two-dimensional test sections with flexible walls have been installed in the Langley 0.3m cryogenic wind tunnel [39-40] and the ONERA CERT T-2 wind tunnel [41].

No satisfactory solution to the three-dimensional adaptive-wall problem has been demonstrated. The flexible-wall approach is not easy to apply in three dimensions because the walls must take on shapes with compound curvature. Attempts to solve this problem have included demonstration of a test section with rubber walls [42], one with eight flexible walls [43], and several investigations of how well three-dimensional interference can be eliminated in two-dimensional, flexible-wall test sections. The solutions that more completely eliminated wall interference (i.e., rubber walls and octagonal test section) result in extremely complex mechanical systems that compromise other aspects of the test section, for instance, optical access. Applying two-dimensional adaptive walls to three-dimensional flows involves considerable mechanical complexity without completely solving the wall-interference problem.

The problems that make adaptive walls difficult to apply in two-dimensional ventilated test sections are significantly harder to deal with in three dimensions. In particular, no fast and accurate methods have been demonstrated for measuring flow conditions on a surface removed from the tunnel walls. Laser velocimetry takes too long, even in two dimensions, and is much more difficult and expensive to apply in three dimensions. Pressure pipes, developed at AEDC, are intrusive, very sensitive to manufacturing defects, and pose significant mechanical problems.

Even if a viable technical solution to the three-dimensional adaptive-wall problem is developed, designers and users of transonic wind tunnels will be reluctant to abandon conventional ventilated test sections for adaptive walls. This was apparent at a meeting of wind tunnel users and adaptive-wall specialists at TU-Berlin in 1988 [44]. Most of the tunnel users at this meeting opined that wall interference in conventional three-dimensional test sections is a manageable problem for most test conditions. Wall interference in three-dimensional flows is generally less severe than in two dimensions and thus data can be more easily and often corrected for its effects. Therefore, the potential benefits of adaptive walls are less in three dimensions than in two, whereas the cost and complexity are much higher. Furthermore, many wind tunnel users are not comfortable with replacing ventilated walls and their forgiving interference properties with impermeable walls that, if improperly set, could produce large flow distortions.

An irony of adaptive-wall research has been that it has accelerated the development of competing technologies for coping with wall interference. In particular, wall interference assessment and correction (WIAC) methods have been developed that allow corrections for wall interference to be computed from flow measurements at or near the test section walls [32], i.e., from the same flow measurements that are required in adaptive-wall test sections. Therefore, to successfully implement WIAC methods it is necessary to overcome the adaptive-wall flow-measurement problem – something that is easy in solid-wall test sections, but much more difficult in ventilated test sections. A wind tunnel equipped to make such measurements but without adjustable walls may be an attractive compromise between data accuracy and test-section complexity.

After more than 20 years of adaptive-wall research the Sears concept of adaptive walls has yet to be applied in a major transonic wind tunnel. The closest thing to an industrial application is in the Russian wind tunnel T-128 where variable porosity wall panels can be adjusted to minimize wall interference. Wall settings in T-128, however, are determined by comparing numerical simulation of flow past the model with pressure measurements at the walls, and not by matching inner and outer flows. There are now discussions in NASA about the need for a major new transonic wind tunnel [45]. In light of the state-of-the-art of three-dimensional adaptive-wall technology and the decline in adaptive-wall research in the U.S., it seems unlikely that this wind tunnel, if built, will have adaptive walls.

9 CONCLUDING REMARKS

A decade of adaptive-wall wind-tunnel research at Ames Research Center has been summarized. This research included a series of adaptive-wall experiments in a transonic test sections with ventilated walls, beginning in a small, indraft wind tunnel and ending in the 2x2-ft. TWT. A concept for two-dimensional testing was shown to be technically viable; however, further development would be required before it could be applied in a production test section, and it is doubtful that it would ever be as efficient as the flexible-wall approach. Adaptive-wall research at Ames was discontinued before a satisfactory solution to the three-dimensional problem could be demonstrated. Making the necessary flow measurements quickly and accurately appears to be the biggest challenge. Since laser velocimetry is poorly suited to this task, some alternative method would have to be developed.

REFERENCES

1. Preston, J.H., and Sweeting, N.E.: The Experimental Determination of the Interference on a Large Chord Symmetrical Joukowski Aerofoil Spanning a Closed Tunnel, British Aeronautical Research Committee, R&M No. 1997, Dec. 1942.
2. Preston, J.H., Sweeting, N.E. and Cox, D.K.: The Experimental Determination of the Two-Dimensional Interference on a Large Chord Piercy 12/40 Aerofoil in a Closed Tunnel Fitted with a Flexible Roof and Floor, British Aeronautical Research Committee R&M No. 2007, Sept. 1944.

3 Lock, C.N.H. and Beavan, J.A.: Tunnel Interference at Compressible Speeds Using the Flexible Walls of the Rectangular High-Speed Tunnel, British Aeronautical Research Committee R&M No. 2005, Sept. 1944.

4 Becker, J.V.: The High-Speed Frontier, NASA SP-445, pp. 181, 1980.

5 Sears, W.R.: Self-Correcting Wind Tunnels, *Aeronautical J.,* **78**, pp. 80-89, 1974.

6 Goodyer, M.J.: The Self Streamlining Wind Tunnel, NASA TM-X-72699, Aug. 1975.

7 Goodyer, M.J.: A Low Speed Self Streamlining Wind Tunnel, AGARD-CP-174, Mar. 1976.

8 Chevallier, J.P.: Adaptive Wall Transonic Wind Tunnels, AGARD-CP-174, Mar. 1976.

9 Ganzer, U.: Wind Tunnels with Adapted Walls for Reducing Wall Interference, *Zeitschrift fur Flugwissenschaften and Weltraumforschung,* **3**, pp. 129-133, 1979.

10 Vidal, R.J., Erickson, J.C., Jr. and Catlin, P.A.: Experiments with a Self-Correcting Wind Tunnel, AGARD-CP-174, Mar. 1976.

11 Sears, W.R., Vidal, R.J., Erickson, J.C., Jr. and Ritter, A.: Interference-Free Wind-Tunnel Flows by Adaptive-Wall Technology, *J. of Aircraft,,* **14**, pp. 1042-1050, 1977.

12 Vidal, R.J., and Erickson, J.C. Jr.: Experiments on Supercritical Flows in a Self-Correcting Wind Tunnel, AIAA Paper 78-788 presented at the AIAA 10th Aerodynamic Testing Conference, San Diego, Cal., Apr. 1978.

13 Kraft, E.M., and Parker, R.L., Jr.: Experiments for the Reduction of Wind Tunnel Wall Interference by Adaptive-Wall Technology, AEDC-TR-79-51, Oct. 1979.

14 McDevitt, J.B., Polek, T.E. and Hand, L.A.: A New Facility and Technique for Two-Dimensional Aerodynamic Testing, *J. of Aircraft,* **20**, pp. 543-551, June 1983.

15 Wittliff, C.E., Nenni, J.P. and Erickson, J.C. Jr.: Recent Experience Using a Static Pipe to Measure Streamwise and Normal Velocity components, presented at the 55th Semiannual Meeting of the Supersonic Tunnel Association, Amsterdam, The Netherlands, Apr. 27-29, 1981.

16 Davis, S.S.: A Compatibility Assessment Method for Adaptive-Wall Wind Tunnels, *AIAA J.,* **19**, pp. 1169-1173, Sept. 1981.

17 Bodapati, S., Schairer, E.T. and Davis, S.S.: Adaptive-Wall Wind-Tunnel Development for Transonic Testing, *J. of Aircraft,,* **18**, pp. 273-279, Apr. 1981.

18 Schairer, E.T.: Experiments in a Three-Dimensional Adaptive-Wall Wind Tunnel, NASA TP-2210, Sept. 1983.

19 Mendoza, J.P.: A Numerical Simulation of Three-Dimensional Flow in an Adaptive-Wall Wind Tunnel, NASA TP-2351, Aug. 1984.

20 Morgan, D.G. and Lee, G.: Construction of a 2- by 2-Foot Transonic Adaptive-Wall Test Section at the NASA Ames Research Center, AIAA Paper 86-1089 presented at the AIAA 4th

Joint Fluid Mechanics, Plasma Dynamics, and Laser Conference, Atlanta, Georgia, May 1986.

21 Celik, Z. and Roberts, L.: An Experimental Study of an Adaptive-Wall Wind Tunnel, Joint Institute for Aeronautics and Acoustics TR-87, Aug. 1988.

22 Davis, S.S.: Applications of Adaptive-Wall Wind Tunnels, *J. of Aircraft,* **23**, pp. 158-160, Feb. 1986.

23 Schairer, E.T.: Two-Dimensional Wind-Tunnel Interference from Measurements on Two Contours, *J. of Aircraft,* **21**, pp. 414-419, June 1984.

24 Schairer, E.T.: Methods for Assessing Wall Interference in the 2- by 2-Foot Adaptive-Wall Wind Tunnel, NASA TM 88252, June 1986.

25 Schairer, E.T.: Nonlinear Effects in the Two-Dimensional Adaptive-Wall Outer-Flow Problem, *J. of Aircraft,* **27**, pp. 475-477, May 1990.

26 Schairer, E.T., Lee, G. and McDevitt, T.K.: A Two-Dimensional Adaptive-Wall Test Section with Ventilated Walls in the Ames 2- by 2-Foot Transonic Wind Tunnel, NASA TM 102207, Aug. 1989.

27 Schairer, E.T., Lee, G. and McDevitt, T.K.: Two-Dimensional Adaptive-Wall Tests in the NASA Ames Two- by Two-Foot Transonic Wind Tunnel, *J. of Aircraft,* **28**, Nov. 1991.

28 Vidal, R.K., Catlin, P.A. and Chudyk, D.W.: Two-Dimensional Subsonic Experiments with a NACA 0012 Airfoil, Calspan Report RK-5070-A-3, Dec. 1973.

29 Sleeman, W.C. Jr., Klevatt, P.L., and Linsley, E.L.: Comparison of Transonic Characteristics of Lifting Wings from Experiments in a Small Slotted Tunnel and the Langley High-Speed 7- by 10-Foot Wind Tunnel, NACA RM-L51F14, 1951.

30 Sears,W.R.: Adaptable Wind Tunnel for Testing V/STOL Configurations at High Lift, *J. of Aircraft,* **20**, pp. 968-974, Nov. 1983.

31 Everhart, J.L.: A Method for Modifying Two-Dimensional Adaptive Wind-Tunnel Walls Including Analytical and Experimental Verification, NASA TP-2081, Feb. 1983.

32 Mokry, M., Chan, Y.T. and Jones, D.J.: Two-Dimensional Wind Tunnel Wall Interference, AGARDograph No. 281, Nov. 1983.

33 Lawson, C.L. and Hanson, R.J.: *Solving Least Squares Problems,* Prentice Hall, 1974.

34 King, L.S.: A Comparison of Turbulence Closure Models for Transonic Flows about Airfoils, AIAA Paper 87-0418, Jan. 1987.

35 McDevitt, J.B. and Okuno, A.F.: Static and Dynamic Pressure Measurements on a NACA 0012 Airfoil in the Ames High Reynolds Number Facility, NASA TP-2485, June 1985.

36 Mateer, G.G., Seegmiller, H.L., Hand, L.A. and Szodruchm, J.: An Experimental Investigation of a Supercritical Airfoil at Transonic Speeds, NASA TM 103900, July 1992.

37 Olsen, M.R. and Seegmiller, H.L.: Low Aspect Ratio Wing Code Validation Experiment, AIAA Paper 92-0402 presented at the 30th Aerospace Sciences Meeting and Exhibit, Jan. 6-9, 1992, Reno, Nev.

38 Lewis, M.C.: Aerofoil Testing in a Self-Streamlining Flexible Walled Wind Tunnel, NASA CR-4128, May 1988.

39 Wolf, S.W.D. and Ray, E.J.: Highlights of Experience with a Flexible Walled Test Section in the NASA Langley 0.3-Meter Transonic Cryogenic Tunnel, NASA TM-101491, Sept. 1988.

40 Wolf, S.W.D.: Evaluation of a Flexible Wall Testing Technique to Minimize Wall Interferences in the NASA Langley 0.3-m Transonic Cryogenic Tunnel, AIAA Paper 88-0140 presented at the AIAA 26th Aerospace Sciences Meeting, Reno, Nev., Jan. 11-14, 1988.

41 Chevallier, J.P., Mignosi, A., Archambaud, J.P. and Seraudie, A.: T2 Wind Tunnel Adaptive Walls: Design, Construction, and Some Typical Results, *La Recherche Aerospatiale* (English Ed.) **4**, July/Aug. 1983.

42 Heddergott, A., Kuczka, D. and Wedemeyer, E.: The Adaptive Rubber Tube Test Section of the DFVLR Goettingen, Paper presented at the 11th International Congress on Instrumentation in Aerospace Simulation Facilities, Stanford, Cal., Aug. 26-28, 1985.

43 Ganzer, U., Igeta, Y. and Ziemann, J.: Design and Operation of TU-Berlin Wind Tunnel with Adaptable Walls, Presented at the 14th Congress of the International Council of the Aeronautical Sciences, Toulouse, France, Sept. 9-14, 1984.

44 Ganzer, U.: Adaptive Wind Tunnel Walls State of the Art 1988, Miniconference held at the Institut fur Luft- und Raumfahrt, TU Berlin, July 1988.

45 "Major Strides Needed in Subsonic Aircraft," *Aviation Week and Space Technology*, pp. 48-49, Jan. 1993.

Adaptive Wall Technology for Minimization of Wind Tunnel Boundary Interferences – Where Are We Now?

Stephen W. D. Wolf

ABSTRACT

The status of adaptive wall technology to improve wind tunnel simulations for 2- and 3-D testing is reviewed. This technology relies on the test section flow boundaries being adjustable, using a tunnel/computer system to control the boundary shapes without knowledge of the model under test. This paper briefly overviews the benefits and shortcomings of adaptive wall testing techniques. A historical perspective highlights the disjointed development of these testing techniques from 1938 to present. Currently operational transonic Adaptive Wall Test Sections (AWTSs) are detailed, showing a preference for the simplest AWTS design with two solid flexible walls. Research highlights show that quick wall adjustment procedures are available and AWTSs, with impervious or ventilated walls, can be used through the transonic range up to $M_\infty = 1.2$. The requirements for production testing in AWTSs are discussed, and conclusions drawn as to the current status of adaptive wall technology. In 2-D testing, adaptive wall technology is mature enough for general use, even in cryogenic wind tunnels. In 3-D testing, this technology has not been pursued aggressively, because of the inertia against change in testing techniques, and preconceptions about the difficulties of using AWTSs.

SYMBOLS

c	Chord
C_d	Drag coefficient
C_n	Normal force coefficient
C_p	Pressure coefficient
d	Body diameter
h	Test section height

M_∞	Free stream Mach number
P	Local static pressure
P_o	Stagnation pressure
U_∞	Free stream velocity
X	Streamwise location
Y	Local wall deflections from straight
α	Angle of attack
Δu	Induced streamwise velocity
Δw	Induced upwash velocity
ΔY	Local wall streamlining adjustment

1 INTRODUCTION

Progress in the science of aeronautics is measured by the improvements in the efficiency of flight vehicles. It is now generally accepted that these improvements are best achieved by using ground testing, CFD and flight testing technologies together. In effect, we must use a balanced combination of all available technologies to understand the remaining mysteries of aeronautics. In ground testing, this means we strive for better and better simulations of the "real" flow in our wind tunnel experiments. Consequently, wind tunnel improvements are, and will remain, the subject of considerable research effort in different parts of the world.

Ideally, for complete simulation of "real" flow conditions about a scale model within the confines of a wind tunnel, the values of test Reynolds number, Mach number, free stream turbulence level and the flow field shape must all be properly matched to full scale. Unfortunately, it is normal practice to test at the correct Mach number with the other three parameters seldom matched. Consequently, wind tunnel data still suffers from significant wall interference effects, particularly at transonic speeds. Traditionally, the wind tunnel community uses several well-known techniques to minimise wall interferences. Models are kept small compared with the test section size (sacrificing the test Reynolds number range). Ventilated test sections are used to relieve transonic blockage and prevent choking (but introduce other more complex boundary interferences). Post-test Wall Interference Assessment and Corrections (WIAC), of varying sophistication, are applied to the model data in an effort to remove wall interferences. Usually, all three techniques are used together in transonic testing. Alas, these techniques still fail to produce the high levels of data accuracy possible with modern testing techniques. In addition, these old techniques have led to expensive compromises in terms of test section size and drive power, which are not tolerable in today's economy.

A "modern" testing technique necessary to minimise wall interferences has existed, in a conceptual form, for about 55 years. This technique is a very intuitive solution to the problem, and involves minimising wall interferences at the very source of these disturbances. The technique adapts the test section boundaries to streamline shapes, so the test section walls become nearly invisible to the model under test. We know this concept as the **Principle of Wall Streamlining,** which was first used in 1938 as a means of relieving transonic blockage at NPL, England [1]. The concept effectively splits the real infinite flow field into two parts: the real flow field in the test section which contains all the viscous flow interactions with the model; and an imaginary flow field surrounding the test section and extending to infinity, as shown in Figure 1. The boundary between the two flow fields is a streamtube (ignoring the boundary layer growth on the test section walls).

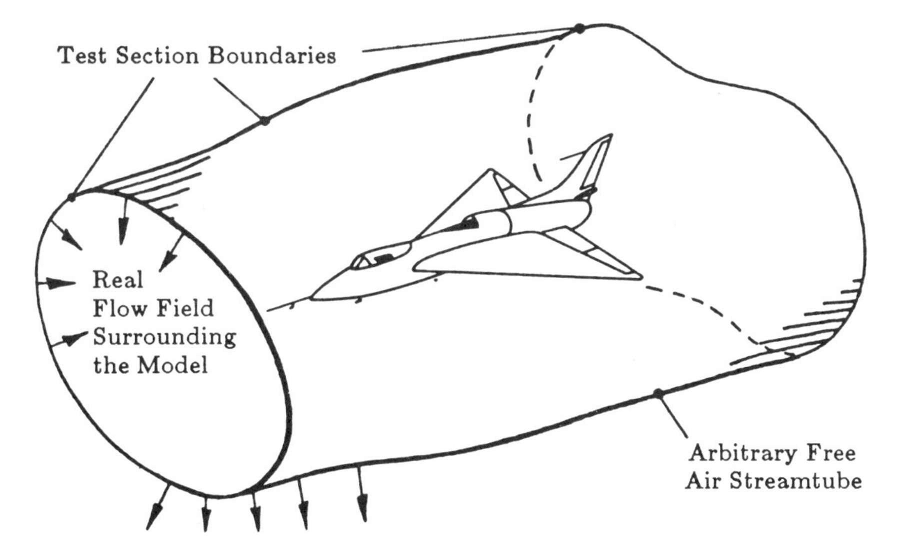

Figure 1. Principle of Wall Streamlining, general 3-D case.

The paper considers the benefits and shortcomings of adaptive wall testing techniques as a precursor to discussing the current status of the technology. A brief review of the development of adaptive walls shows the contribution of Professor Sears, whose work we are commemorating at this symposium.

Operational transonic AWTSs are detailed (which are currently used for both conventional and turbomachinery research) to demonstrate the current wave of enthusiasm. From these AWTSs, 2- and 3-D adaptive wall research is reviewed to illustrate the State of the Art. Finally, we consider the operational aspects of AWTSs, since the practicalities of adaptive walls play a critical factor in the more widespread use of this technology. In conclusion, an assessment of the accumulated adaptive wall experience is presented and possible directions for future developments are indicated.

2 ADAPTIVE WALL BENEFITS

Although the potential rewards for using adaptive wall testing techniques have been reported many times, a brief overview is appropriate. Adaptive walls offer several important advantages other than the major benefit of minimising wall interferences. With wall interferences minimised, we are free to increase the size of the model for a given test section. We can double the test Reynolds number and have a larger model to work with. Alternatively, we can shrink the test section and reduce the tunnel size and operating costs. Interestingly, the task of magnetically suspending models (to remove support interferences) becomes simpler in an AWTS because the supporting coils can be positioned closer to the model.

With solid adaptive walls (called flexible walls), the test section boundaries are smooth and impervious in contrast to the complex boundaries of a ventilated test section. This smoothness minimises disturbances to the tunnel free stream significantly improving flow quality. Of course, the considerable noise generated by flow through slots or perforations is eliminated. (A benefit that is gaining more and more significance in transition to turbulence experiments at transonic speeds.) Furthermore, this improved flow quality reduces the tunnel drive power required for a given test condition, with the model and test section size fixed. Flexible walls eliminate the plenum volume found in a conventional transonic wind tunnel, reducing settling times and minimising flow resonance, which is particularly important for blowdown tunnels.

With flexible walls, the boundary control is achieved by direct wall movement. This strong control of the test boundaries provides good data repeatability compared with the indirect and passive control of the boundaries in a conventional ventilated test section.

Adaptive walls can provide the aerodynamicist with real-time "corrected" data, even in the transonic regime, which presents another significant advantage to the wind tunnel user. Since the final results are known real-time, test programmes can be much more focused and test matrices can be minimised. Consequently, the use of adaptive walls can significantly reduce the number of data points and tunnel entries necessary to achieve test objectives.

It should be noted that the simulation of free-air conditions is one of six flow field simulations [2] that adaptive wall technology can provide. These simulations are: free air; cascade; open jet; closed tunnel; ground effect; and steady pitching. It is possible to use multiple simulations with the same model and AWTS. This feature of AWTSs can and has been shown to be a useful advantage for CFD validation work and tunnel versatility.

3 ADAPTIVE WALL SHORTCOMINGS

The simple, intuitive concept of adaptive walls is complicated by the need to continually adjust the test section boundaries for each test condition. This complication causes the AWTS hardware to be more complex than that of a conventional test section. However, this complexity is nothing new to a wind tunnel designer and should be considered as equivalent to the complexity of a flexible supersonic nozzle. As shown later, it would now seem that the AWTS complexity can be limited to only two adaptive walls, further reducing this shortcoming. In addition, it must be remembered that increasing the complexity of the test section hardware reduces the complexity of the correction codes. In effect the use of an AWTS reduces the computational overhead necessary to implement WIAC codes.

A sophisticated control system is required for an AWTS because wall streamlining is necessarily iterative (see Figure 2). The control system must be even more sophisticated if non-expert users are involved. However this shortcoming is not new and the AWTS control system is equivalent to a feedback control system for positioning other wind tunnel components like a sting support. Of course, sophisticated control systems are now very common with the advent of PC computers.

The number of data points per hour of tunnel run-time will be reduced with adaptive walls, because the wall adaptation will never be instantaneous. Computer control of the adaptation has greatly reduced this "wasted" tunnel time, but the associated time is still significant. However, the real-time AWTS data are the final data (as shown later) and therefore one must assess the importance of quality versus quantity.

Any AWTS must be designed with a good understanding of the expected testing requirements. The AWTS design can limit the test envelope as easily as other aspects of a wind tunnel system, such as drive power, diffuser design, etc. Fortunately, there are guidelines published to help design an AWTS for a given test envelope.

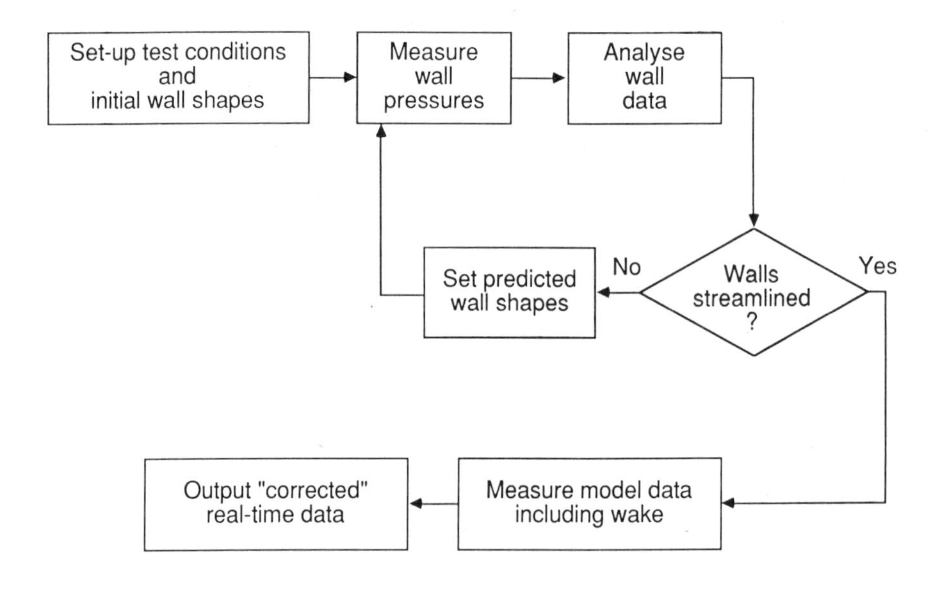

Figure 2. Wall streamlining procedure.

4 HISTORICAL OVERVIEW OF ADAPTIVE WALL RESEARCH

The adaptive wall testing techniques we know today are a rediscovery of the first solution to severe transonic wall interferences (i.e., choking). The National Physical Laboratory (NPL), UK, built the first adaptive wall test section in 1938, under the direction of Dr. H. J. Gough [1], about eight years before the first ventilated test section appeared. The pioneering research at NPL proved that streamlining the flexible walls of an AWTS was the first viable technique for achieving high speed (transonic) flows in a wind tunnel. In fact, NPL had transonic testing capability eight years before NACA. NPL researchers opted for minimum mechanical complexity in their AWTS and used only two flexible walls.

The absence of computers made wall streamlining a labour intensive process which was surprisingly fast, of the order 20 minutes. Sir G. I. Taylor developed the first wall adjustment procedure [3], which has since been validated in a modern AWTS [4]. NPL researchers went on to successfully use flexible walled AWTSs for general testing up until the early 1950s. They generated an extraordinary amount of 2- and 3-D transonic data [5] during this 14-year period, which is probably more than half of all the AWTS data produced to date. The NPL 20x8 in. (50.8x20.3cm) High Speed Tunnel even became the first adaptive wall wind tunnel to operate at low supersonic Mach numbers. Some early adaptive wall work was also carried out in Germany during the 1940s. Despite the building of the largest ever AWTS, with a 3m (118 in.) square cross-section, this effort came to nought.

The arrival of ventilated test sections at NACA Langley in 1946, provided a "simpler" approach to high speed testing. The adjustments to the test section boundaries are passive with ventilated walls, while the adaptive wall adjustments are active. The apparent simplicity of ventilated test sections led to the obsolescence of NPL's AWTSs and adaptive wall technology became forgotten in time.

After about twenty years (see Figure 3), interest in AWTSs was rekindled. In April 1971, an AGARD Fluid Dynamics Panel Specialist's meeting in Göttingen, Germany highlighted serious shortcomings in wind tunnel testing techniques at transonic speeds [6]. This serious concern led several researchers, in Europe and the USA, to independently rediscover the concept of adaptive wall testing techniques. One of the *fathers of Modern Adaptive Walls*, who presented and published his ideas in 1973, was of course Professor Sears [7]. The goal of these researchers was better free air simulations in transonic wind tunnels. The adaptive wall approach offered them an elegant way to simplify the wall interference problem. Adaptive wall adjustment procedures need only consider the flow at the test section boundaries (in the farfield), the complex flow field round the model need never be considered. Therefore, the adaptive wall concept allows us to simplify the "correction codes" at the expense of increasing the complexity of the test section hardware.

This renewed interest produced 5 adaptive wall research groups around the world. During the mid-1970s, these groups directed their initial research efforts towards low speed 2-D testing, because this provided a relatively quick way to re-invent the adaptive wall concept. Two methods of applying the adaptive wall principles were investigated. One was a modification of the conventional ventilated test section using variable porosity. The other was a complete re-invention of the NPL approach. In this early phase, notable work at Southampton University (England) demonstrated the AWTS versatility to create six flow field simulations and produced the first predictive wall adjustment procedure of Judd et al.[8]. In addition, the intuitive design principles for 2-D flexible walled AWTSs were quantified and optimized.

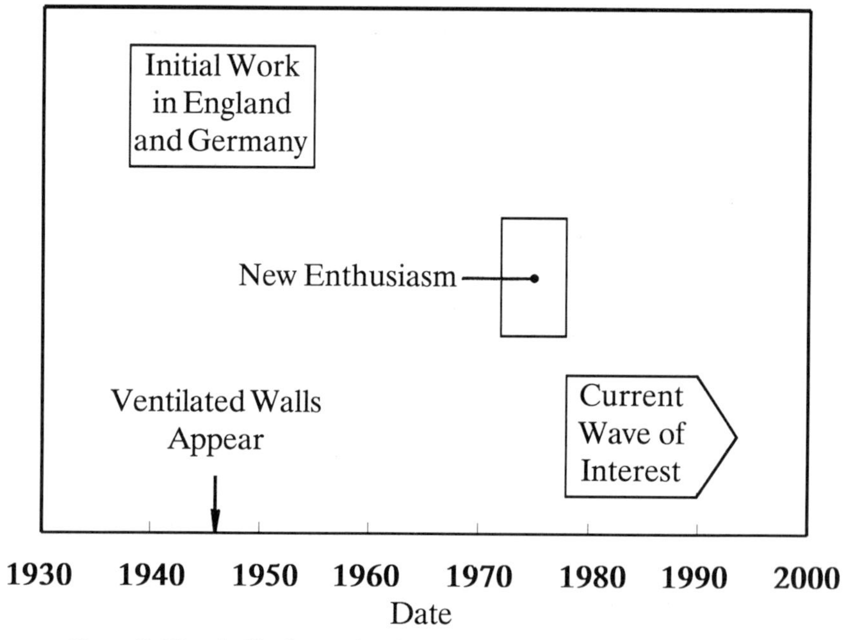

Figure 3. Historical background to the current status of adaptive wall technology.

In the mid to late 1970s, the successful low-speed research effort paved the way for transonic 2-D research, which introduced fast automatic wall adjustments (taking only seconds in some cases) with notable work at Southampton University, ONERA/CERT (France) and Technical University of Berlin (TU-Berlin, Germany). Rapid progress was made in the development of subsonic and transonic 2-D adaptive wall testing techniques, helped along by the growing availability of computers. During the 1980s, this progress led to successful 2-D tests at high lift and high blockage conditions, and the use of AWTSs in cryogenic wind tunnels (i.e., in the NASA Langley 0.3m Transonic Cryogenic Tunnel and the ONERA/CERT T2 tunnel). All these successes were achieved in flexible walled AWTSs. Parallel research using ventilated walled AWTSs (led by Professor Sears) was as vigorous at Calspan (Buffalo, NY) and AEDC, but alas this approach encountered some fundamental limitations. Boundary measurements proved difficult, and intrusive Calspan static pipes had to be used. Researchers also found it was impossible to achieve sufficient control of the ventilated adaptive walls, if relatively large model disturbances were present, and wall adjustment procedures were difficult to implement [9].

Initial research using AWTSs for 3-D testing began in the late 1970s, and concentrated on exotic and extremely complex AWTS designs like the rod-wall tunnel at AFFDL at Dayton, Ohio [10], the rubber-tube DAM AWTS at DFVLR Göttingen (Germany) [11] and the octagonal AWTS at TU-Berlin [12]. These designs proved that 3-D AWTSs using intuitive wall adjustments could minimise boundary interferences, but confirmed that these complex AWTS designs are impractical for general use. Fortunately, the research with the rod-wall tunnel did show that 2-D adaptive walls could be successfully used in 3-D testing. This important finding resulted in numerous 3-D subsonic and low transonic tests in the 1980s. Research at the Von Karman Institute (Belgium) and DFVLR Göttingen produced the first predictive wall adjustment procedure for 2-D AWTSs used in 3-D testing. This procedure was developed by Wedemeyer and Lamarche [13]. Meanwhile, studies of the residual interferences at Southampton University established the usefulness of 2-D AWTSs in 3-D testing up to $M_\infty = 0.97$. Other notable work at TU-Berlin, DFVLR Göttingen and Northwestern Polytechnical University (NPU - China) [14] involved preliminary 3-D tests at $M_\infty = 1.2$, in which oblique shock wave reflections were successfully attenuated by local flexible wall bending. Again, all these advances were achieved in flexible walled AWTSs. Parallel research with ventilated walled AWTSs for 3-D tests occurred at NASA Ames and AEDC in the 1980s. This work was eventually abandoned with the failure to develop fast wall adjustment procedures, and to achieve adequate boundary control when significant model disturbances are present. However, since 1986, researchers at TsAGI (Russia) have been performing production testing in the large, 2.75m (9 ft.) square, ventilated walled AWTS of their transonic T-128 tunnel. This automated AWTS has a simplified wall adjustment procedure that relies extensively on the use of WIAC codes to remove wall interferences from the data (see sub-section 5.12) [15].

Unfortunately, adaptive wall research has been significantly slowed in recent years by low priority funding and the demise of active research groups in the United States and Germany. The development of 3-D transonic adaptive wall testing techniques in the 1990s currently rests with 4 organizations: DLR Göttingen, ONERA, NPU, and Southampton University.

Research has continued into advanced 2-D testing techniques with the goal of extending the useful speed range up to low supersonic Mach numbers and improving cascade simulations. Initial supersonic 2-D testing at NPL was followed some 40 years later by preliminary research at M_∞ = 1.2 performed at ONERA during the mid-1980's. More recently there have been significant strides at Southampton University [16], again at M_∞ = 1.2. Cascade simulations, pioneered at Southampton University in 1974 using a single blade at low speeds, are currently only performed in Genoa University (Italy) at transonic speeds.

We find that AWTSs are now available for commercial use at NASA Langley (2-D only), ONERA/CERT (2- and 3-D), and TsAGI (3-D only). There are plans to build new transonic cascade AWTSs at Genoa University and DLR Göttingen, and new transonic flexible walled AWTSs in China (for the CARDC 0.6m high speed tunnel) and Germany (for the DLR TWG 1m transonic wind tunnel and the DLR transonic cryogenic Ludwieg tube). Furthermore, the French cryogenic T'3 transonic tunnel, with a flexible walled AWTS, is now being recommissioned at the Von Karman Institute in Belgium.

5 TRANSONIC AWTSs CURRENTLY OPERATIONAL

In this section features of the currently operational AWTSs are summarized. The facilities are discussed in alphabetical order by organization.

5.1 Aerodynamic Institute, RWTH Aachen, Germany

The test section of the Transonic- and Supersonic Tunnel (TST) at RWTH Aachen was equipped with flexible walls in 1985/6. The AWTS is 40cm (15.75 in.) square and 1.414m (4.64ft) long. The top and bottom walls are flexible and mounted between two parallel sidewalls. The flexible walls are made from 1.3mm (0.051 in.) thick spring steel. Each wall is supported by 24 motorized jacks (see Figure 4).

The TST is an intermittent tunnel capable of operation at M_∞ between 0.2 and 4, with run times between 3 to 10 seconds. The AWTS has only been used for 2-D testing up to about M_∞ = 0.8 [17]. Usually 3 or 4 tunnel runs are required for each data point at transonic Mach numbers. Boundary measurements are static pressures measured along the flexible walls. Wall adaptation calculations and automatic wall adjustments are made between tunnel runs.

Empty test section calibrations reveal Mach number discrepancies less than 2%, where the model is usually mounted, at M_∞ = 0.82. Lower Mach numbers produce lower discrepancies.

Mach number is controlled, up to low transonic Mach numbers, by a downstream sonic throat. The average accuracy of the wall contours, measured by potentiometers at each wall jack, is ±0.1mm (±0.004 in.).

Figure 4. Transonic and Supersonic Tunnel (TST), RWTH Aachen, Germany.

5.2 DLR - Institute of Experimental Fluid Mechanics, Göttingen, Germany

During 1987-8, researchers at DLR modified the 2-D supersonic nozzle of the DLR High Speed Wind Tunnel (HKG) into an AWTS. The top and bottom nozzle walls are made of highly flexible 4mm (0.157 in.) thick steel plates. The shape of each wall is set by 17 pairs of equally spaced hydraulic jacks (see Figure 5).

The AWTS consists of an initial contraction followed by a 2.2m (7.22 ft.) straight section. This straight portion, in which the model is mounted, is nominally 0.67m (2.2 ft.) high and 0.725m (2.38 ft.) wide. Each wall of the test section is equipped with 3 rows of pressure taps for boundary measurements. The wall adjustment procedure of Wedemeyer/Lamarche [13] is used to minimise interferences along the tunnel centerline.

This AWTS is used for evaluation of 2-D wall adaptation in 3-D testing. Researchers have tested sting mounted 3-D models, both lifting and non-lifting, up to about M_{∞}=0.8 [18].

Figure 5. DLR high speed wind tunnel (HKG) AWTS/supersonic nozzle, Göttingen, Germany.

5.3 Genoa University, Italy

The Department of Energy Engineering at Genoa University operates two adaptive wall cascade tunnels. These tunnels are the only current examples of the use of AWTSs in non free-air flow simulations. Both tunnels have a cross-section of 0.2m (7.87 in.) high and 5cm (1.97 in.) wide. One is the Low Deflection Blade Cascade Tunnel (LDBCT), which became operational in 1982. The other is the High Deflection Blade Cascade Tunnel (HDBCT) which became operational in about 1985.

The LDBCT can test up to 12 blades, at M_∞ = up to 2.0, with flow deflections up to about 35°. The AWTS has two flexible walls and two solid transparent sidewalls. The flexible walls are 1.58m (5.18 ft.) long and each is shaped by 36 manual jacks (see Figure 6). Wall streamlining is performed upstream and downstream of the cascade.

Figure 6. The Low Deflection Blade Cascade Tunnel (LDBCT) at Genoa University, Italy.

The HDBCT has a similar configuration except the AWTS is 1.6m (5.25 ft.) long and wall adaptation is performed only downstream of the cascade. The top flexible wall is supported by 13 manual jacks and the bottom flexible wall by 26 manual jacks (See Figure 7). The AWTS can accommodate flow deflections up to 140°. Up to 13 blades can be fitted in the cascade, with test M_∞ up to 1.18 reported [19].

Both AWTS need only approximate wall adaptation procedures due to the large number of blades used in the cascade. The smooth flexible walls have provided remarkably good flow quality for cascade research. The LDBCT is also used for probe calibration [19].

Figure 7. The High Deflection Blade Cascade Tunnel (HDBCT) at Genoa Univ., Italy.

5.4 NASA-Ames Research Center, California

The Thermo-Physics Facilities Branch at NASA Ames has an AWTS for use in their intermittent High Reynolds Number Channel-2 (HRC-2) facility. The AWTS was constructed in 1981. The AWTS is fitted with two flexible walls and two parallel solid sidewalls. The AWTS has a rectangular cross-section which is 0.61m (24 in.) high and 0.41m (16 in.) wide. The AWTS is 2.79m (9.15 ft.) long. The Ames AWTS has seven manually adjusted jacks supporting each flexible wall (see Figure 8). An automated version of the AWs was built and then dismantled without ever being used.

The flexible walls are made of 17-4 PH stainless steel plates and are 2.53 m (8.32 feet) long. The flexible walls are 15.9mm (0.625 in.) at the ends tapering to 3.17mm (0.125 in.) in the middle for increased flexibility. The downstream ends of the flexible walls each house a pivot joint which attaches to a variable sonic throat for Mach number control. Sidewall Boundary Layer Control (BLC) is available by installing porous plates in the sidewall, upstream of the model location. Mach number variations along the test section, due to BLC suction, were removed by suitable wall adaptation based on simple influence coefficients [20].

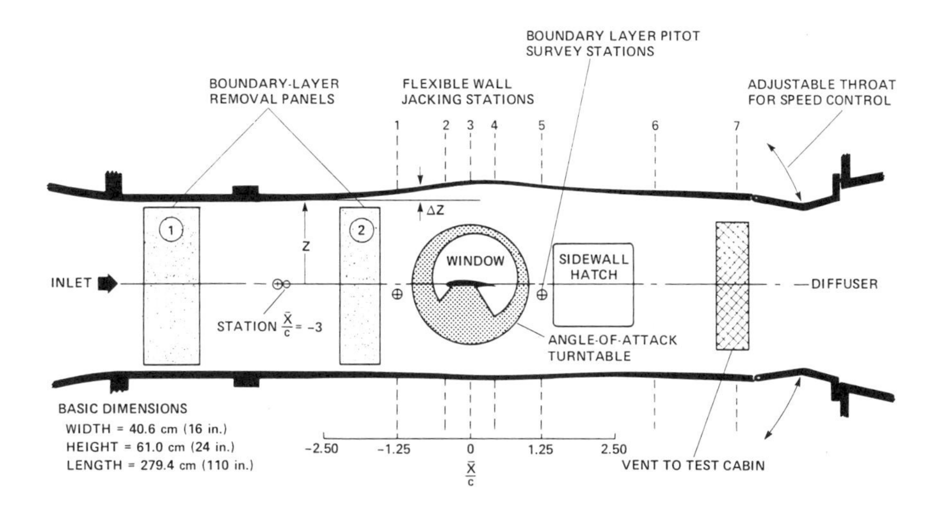

Figure 8. NASA Ames High Reynolds Channel–2 (HRC-2) AWTS.

More recently, the Ames AWTS has been used for 2-D and 3-D CFD code validation. No wall adjustment procedure is used. The flexible walls are simply set to predetermined shapes depending on the investigation underway, and move apart to prevent choking. Studies of LDA wake measurements behind 2-D airfoils have also been carried out. CFD validation 3-D tests with a sidewall-mounted half-model have been performed by Olsen with fixed diverging walls [21].

5.5 NASA Langley Research Center, Virginia

The NASA Langley 0.3m Transonic Cryogenic Tunnel (TCT) was fitted with an AWTS during 1985. The AWTS has two flexible walls mounted between two solid parallel sidewalls. The flexible walls are made of 304 stainless steel, 3.17mm (0.125 in.) thick at the ends, and thinning down to 1.57mm (0.062 in.) thick in the middle for increased flexibility.

The cross-section of the AWTS is 0.33m (13 in.) square and the AWTS is 1.417m (55.8 in.) long. The flexible wall shapes are controlled by 18 motorized jacks per wall. The downstream ends of the flexible walls are attached, by sliding joints, to a 2-D variable diffuser (formed by flexible wall extensions) between the AWTS and the rigid tunnel circuit. The shape of the variable diffuser is controlled by 6 motorized jacks. The wall jacks are designed with stepper motors whose power is insufficient to cause permanent damage to the flexible walls.

The AWTS functions over the complete operating envelope of the continuous running TCT [22]. The test gas is nitrogen. The AWTS has been operated continuously over an 8 hour work shift at temperatures below 120 K. In addition, the AWTS is contained in a pressure vessel for operation up to stagnation pressures of 90 psia (6 bars). The jack motors and position sensors are located outside the pressure shell in a near-ambient environment (see Figure 9). Sidewall boundary layer control is available by fitting porous plates in the sidewalls, upstream of the model location. Boundary layer suction has been successfully used in 2-D testing with normal wall adaptation. 2-D wake measurements are taken using a traversing pitot/static probe mounted in one of three positions downstream of the airfoil.

The wall adjustment procedure of Judd et al. [23] is used for 2-D testing. The 2-D test envelope includes normal force coefficients up to 1.54 (and through stall) and M_∞ up to 0.82, with a model blockage of 12%. The use of a large model combined with cryogenic test conditions has allowed testing at a record breaking Reynolds number of 72.4×10^6. Boundary measurements are static pressures measured along the centerline of the flexible walls at the jack locations. Wall streamlining takes on average less than 2 minutes and is paced by slow wall movements. A generalized and documented non-expert system [24] is used for AWTS operation within known 2-D test envelopes. Up to 50 data points (each with wall streamlining) have been taken during a 6 hour period.

Researchers have carried out tests at M_∞ up to 1.3, using sidewall mounted 3-D wings. For 3-D testing at Mach numbers below 0.8, the wall adjustment procedure of Rebstock [25] has been used to minimise wall interferences along a pre-set streamwise target line anywhere in the test section. Boundary measurements are static pressures from 3 rows of pressure taps on each flexible wall and a row of taps on the centerline of the sidewall, opposite the model. Downstream flexible wall curvature is automatically minimised by rotation of the tunnel centerline. For low supersonic tests, the adapted wall shapes are based on wave theory and form a 2-D supersonic nozzle ahead of the model.

The flexible walls are set to a nominal accuracy of ± 0.127mm (± 0.005 in.). No aerodynamic effect of AWTS shrinkage, due to cryogenic operation, has been reported. M_∞ is controlled by a feedback control system (based on a PC computer) to better than 0.002 during wall streamlining.

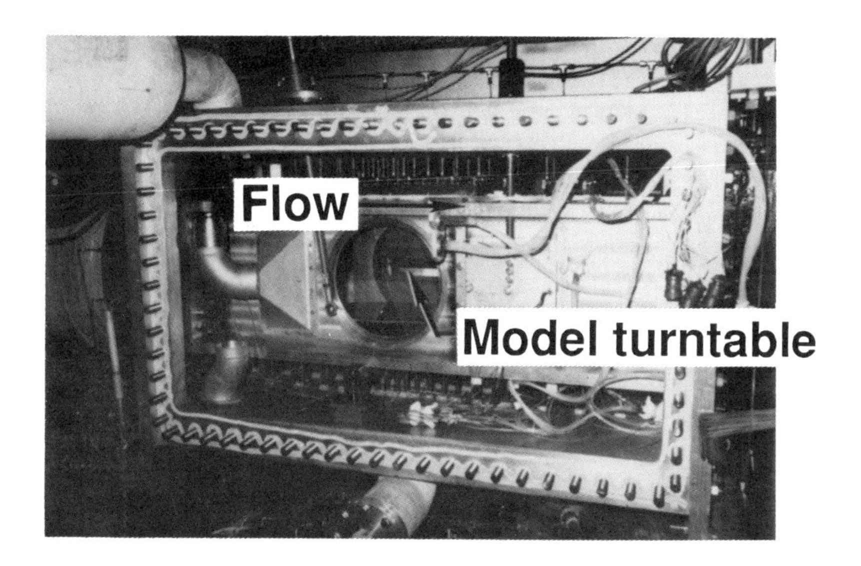

AWTS with Pressure Shell Sidewall Removed

Ice Buildup on AWTS Pressure Shell after a Long Cryogenic Run

Figure 9. NASA Langley 0.3m Transonic Cryogenic Tunnel.

5.6 Northwestern Polytechnical University (NPU), Xian, China

Researchers fitted an AWTS to the NPU WT52 supersonic wind tunnel in May/June 1990. The test section was originally 30cm (11.8 in.) square and is now 21.4 cm (8.42 in.) high, 30cm (11.8 in.) wide and 1.08m (42.52 in.) long. The floor and ceiling are impervious flexible walls, each positioned by 16 manually operated jacks (see Figure 10). Three rows of static pressure taps are provided along each flexible wall. The tunnel can operate at M_∞ up to 1.3 using suitable nozzle blocks.

Figure 10. Northwestern Polytechnical University high speed WT52 AWTS, Xian, China.

The flexible walls are made of 1mm (0.039 in.) thick bronze alloy. In the model region, the wall jack spacing is 6cm (2.36 in.). Each flexible wall is connected to the diffuser by a porous plate to balance the pressure across the flexible walls.

Wall adaptation around a 2-D airfoil was successfully achieved in WT52 at $M_\infty = 0.8$, using a correction method based on Transonic Small Perturbation (TSP) theory. The model blockage was 8% and the test section height 30cm (11.8 in.). In 3-D testing, much work has been done to verify the use of two flexible walls in low supersonic testing. Cone-cylinder models have been tested with blockage of 1% and 2%. Successful engineering and analytical adjustments were made to the flexible walls to alleviate bow shock reflections [14]. Further 3-D testing with a wing-body model has been completed up to $M_\infty = 0.87$ at zero angle of attack, with the test section height reduced to increase the blockage to 5.2%. Preliminary results show a significant reduction of wall interference with wall streamlining. This work continues as part of a joint research programme with DLR Göttingen.

5.7 ONERA/CERT, Toulouse, France

The AWTS fitted in the intermittent ONERA/CERT T2 Transonic Cryogenic Tunnel has been operational since 1978. This AWTS became the first cryogenic AWTS in 1981, when the T2 tunnel was modified to operate cryogenically for 1 to 2 minutes at a time. This French AWTS is 0.37m (14.57 in.) high, 0.39m (15.35 in.) wide and 1.32m (51.97 in.) long. The AWTS has two flexible walls and two solid parallel sidewalls (see Figure 11). The flexible walls are made of 1.5mm (0.059 in.) thick Invar steel plates. The shape of each flexible wall is controlled by 16 hydraulic jacks attached to wall ribs. These ribs are electron beam welded to the outside of the flexible walls. The hydraulic jacks move the flexible walls very rapidly at about 6mm (0.24 in.) per second. The wall jacks have enough power to damage the flexible walls in bending. During a cryogenic run, the flexible walls rapidly reach the low test temperatures, while the jack mechanisms remain at near ambient temperatures. Sidewall BLC is available for 2-D testing by placing porous plates around the airfoil/sidewall junctions. BLC suction is routinely used with wall adaptation. In 2-D tests, a pitot/static rake, mounted on a sting support downstream of the wing, is used for wake measurements.

A wall adjustment procedure developed by Chevallier et al. is used for 2-D testing. This procedure is tunnel dependent and has no documented test envelope for non-expert use. Computer controlled wall streamlining in about 10 seconds is possible. However, two short tunnel runs are normally required per data point for 2-D tests at about $M_\infty=0.8$. Boundary measurements are static pressures measured equidistant along the centerline of the flexible walls.

For 3-D testing, the Wedemeyer/Lamarche wall adjustment procedure is used. Both lifting and non-lifting models have been tested up to $M_\infty=0.97$ [26]. Researchers have carried out several production-type studies for Airbus Industries, including riblet studies on 3-D models (see Figure 11). Boundary measurements are static pressures measured along 3 rows on each flexible wall and a single row on one sidewall.

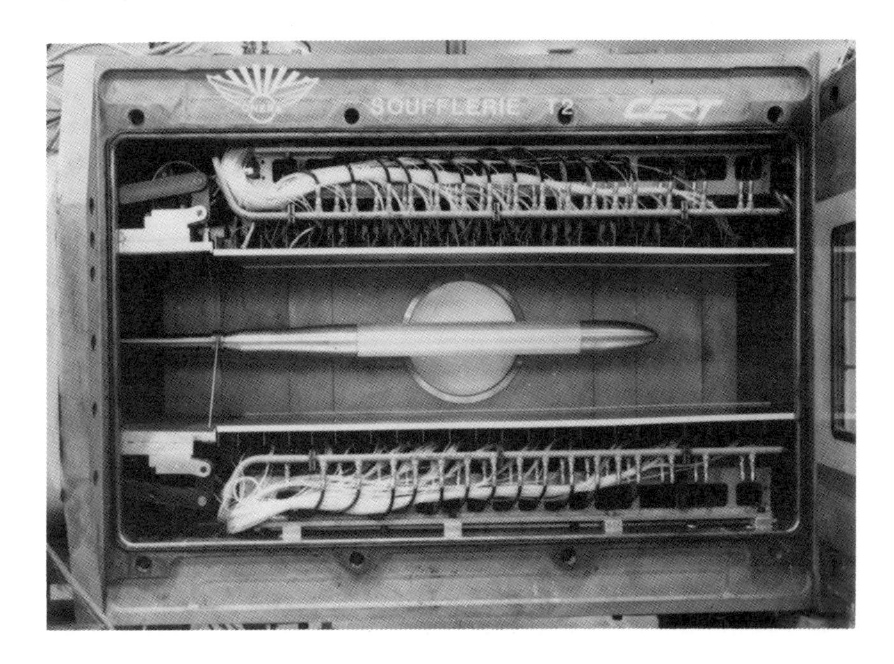

Figure 11. ONERA/CERT T2 transonic cryogenic AWST, Toulouse, France.

The shape of the flexible walls can be measured to 0.05mm (0.002 in.). The wall curvature is checked before any wall movement is initiated. Mach number is control by a downstream sonic throat, which acts as a fairing between the AWTS and the fixed diffuser. In general, the Mach number is not held constant during each wall adaptation process. The pressure fluctuations measured in the test section are a low 0.18% [27].

5.8 ONERA, Chalais-Meudon, France

The ONERA S3Ch transonic wind tunnel was fitted with an AWTS in 1992 and 3-D testing is about to commence. The tunnel can operate over a Mach number range from 0.3 to 1.3. The AWTS is 80cm (31.49 in.) square and 2.2m (86.61 in.) long. The floor and ceiling of the test section are impervious flexible walls mounted between solid parallel sidewalls. The adaptation of the floor and ceiling is controlled by 15 equally spaced wall jacks, 12.8cm (5 in.) apart (see Figure 12). Each wall jack is driven by a computer controlled stepper motor. The flexible walls are made from duralumin alloy and are 3mm (0.12 in.) thick.

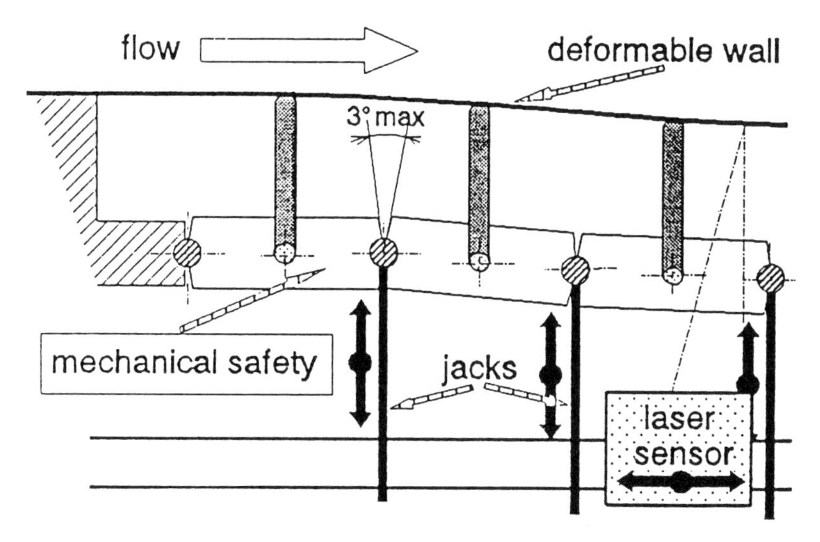

Figure 12. ONERA S3Ch transonic wind tunnel AWTS, Chalais-Meudon, France.

The unique design features of this flexible walled AWTS are as follows: the wall jacks have mechanical stops to prevent damage to the flexible walls; the wall jacks are equally spaced; and only one movable no-contact laser position sensor is used to measure the shape of each flexible wall. The wall can be moved ±50mm (±1.97 in.) from straight, with a position accuracy of ±0.1mm (±0.004 in.).

After considerable effort, it was decided that the boundary measurements for 3-D testing will consist of 4 streamwise rows of 50 static pressures along each flexible wall [28]. The wall adjustment procedure has been developed by Le Sant with a variable target line for test versatility. The AWTS boundary measurements will allow wall interferences to be evaluated everywhere in the test section after wall streamlining. With full computer control of the AWTS, wall streamlining is expected in about 1 minute per test condition.

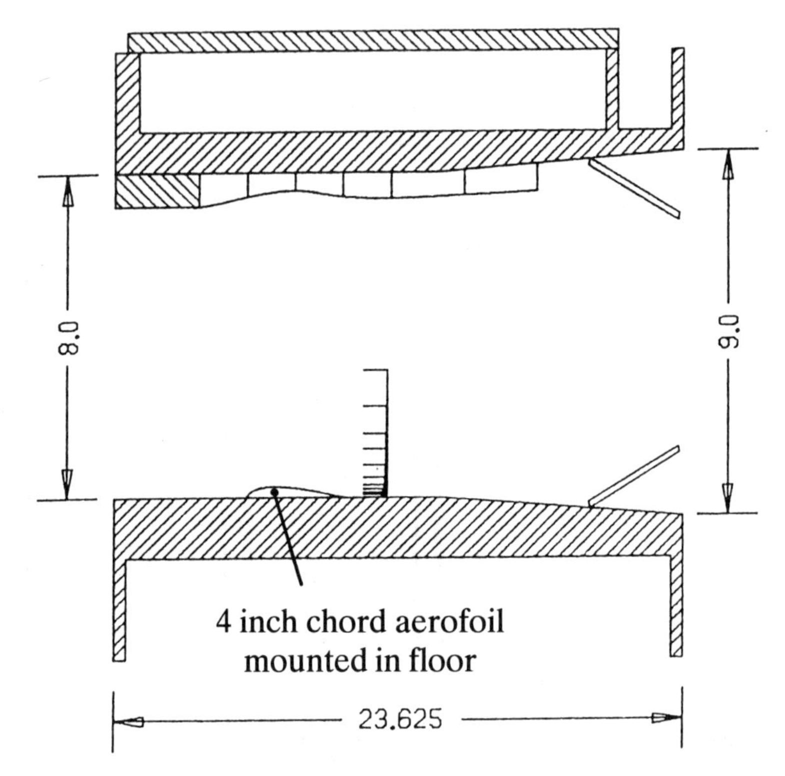

Figure 13. Rensselaer Polytechnical Institute 3 x 8 transonic wind tunnel, Troy, New York.

5.9 Rensselaer Polytechnic Institute, New York

Since the mid-1980s, the Rensselaer Polytechnic Institute has operated two AWTSs for rotorcraft research, in particular the study of 2-D airfoils with passive boundary layer control. The RPI 3 x 8 transonic wind tunnel is fitted with a rectangular AWTS, 20.3cm (8 in.) high, 7.6cm (3 in.) wide, and 0.6m (23.62 in.) long. The top wall is flexible and supported by 6 jacks. The other three walls are solid. The 2-D airfoil is mounted in the bottom wall with a boundary layer removal slot ahead of the leading edge. A relatively large airfoil with a 10.16 cm (4 inch) chord has been tested in this AWTS at M_∞ up to 0.86 (see Figure 13) [29]. Testing of such large models would be impossible at these Mach numbers without an AWTS. Interestingly, oscillatory flow field simulations have also been carried out in the RPI 3 x 8 AWTS with the top wall shape fixed [30].

The RPI 3 x 15 transonic tunnel has a similar AWTS arrangement except the test section height is increased to 38 cm (15 inches). Also the top wall is not flexible and different wall shapes are set in the AWTS by using interchangeable wooden wall inserts. Tests of 14% thick airfoils at M_∞ up to 0.9 are reported [29].

Researchers use a simple wall adjustment procedure in these AWTSs. One-dimensional wall influence coefficients are used to remove the blockage effects associated with testing a large airfoil in these small test sections. Boundary measurements are static pressures measured along the test section walls.

5.10 Southampton University, Hampshire, England

The Transonic Self-Streamlining Tunnel (TSWT) at Southampton University is one of the first fully automated AWTSs. Built in 1976/7, TSWT has a 15cm (6 in.) square test section which is 1.12m (3.67 ft.) long. The floor and ceiling are flexible and made from woven man-made fibre (Terylene). The flexible walls are 5mm (0.2 in.) thick at the ends tapering to 2.5mm (0.1 in.) thick in the middle for increased flexibility. Each flexible wall is supported by 19 motorized jacks (see Figure 14). A sliding joint attaches the downstream ends of the flexible walls to a 2-D variable diffuser (which is two plates, each controlled by a single motorized jack). The wall jacks are designed with stepper motors whose power is not sufficient to damage the flexible walls. The two sidewalls are solid and parallel.

The wall adjustment procedure of Judd et al [8] for 2-D testing was developed in TSWT, and is used routinely for all 2-D tests, up to speeds where the flow at the flexible walls is just sonic. Wall streamlining is generally achieved in less than 2 minutes. If the walls become sonic, a Transonic Small Perturbation (TSP) code is included in the Judd procedure and 2-D testing has been successfully carried out up to $M_\infty = 0.96$ [31]. For low supersonic 2-D testing at up to $M_\infty=1.2$, a wall adjustment procedure based on wave theory is used to generate a simple 2-D supersonic nozzle in the AWTS, upstream of the model [16]. Since 1978, researchers have used TSWT to build up a substantial database on 2-D testing in an AWTS, with blockage ratios up to 12% and test section height to model chord ratios down to unity [2].

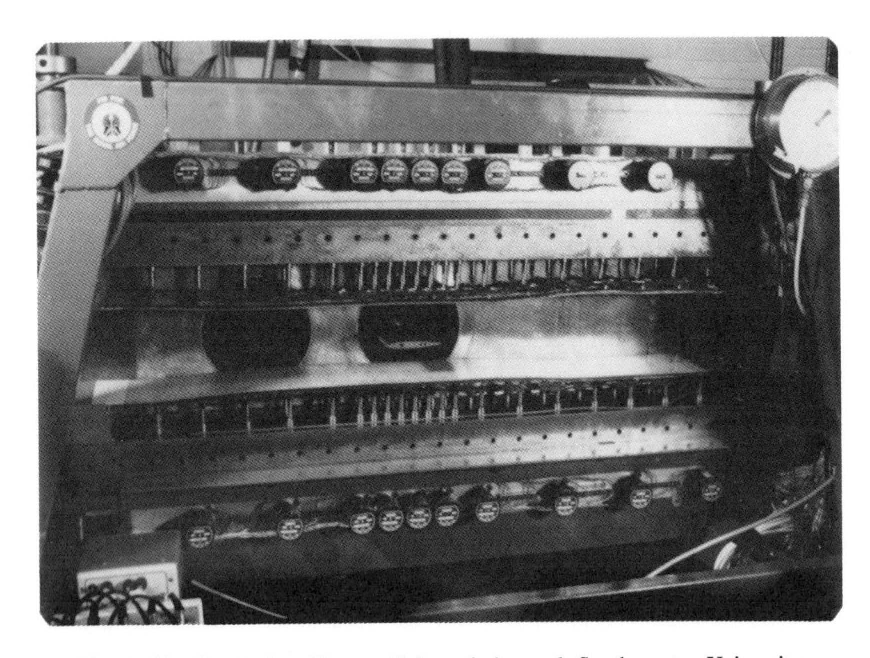

Figure 14. Transonic self-streamlining wind tunnel, Southampton University, England.

In addition, TSWT has been used for 3-D tests with sidewall and sting mounted models with blockage ratios up to 4%. A wall adjustment procedure developed by Goodyer et al is used for 3-D test up to about $M_\infty = 0.9$ [16,31]. Boundary measurements are static pressures measured along five rows on each flexible wall and a single row on one sidewall.

The wall shapes are measured by potentiometers at each wall jack to an accuracy of ±0.127mm (±0.005 in.). M_∞ is controlled by automatic throttling of the inducing air pressure. M_∞ variation up to .002 is typical during a test at M_∞=0.8. Calibration of TSWT with an empty test section reveals a standard deviation in M_∞ variation of about 0.003 at M_∞=0.8.

5.11 Technical University of Berlin, Germany

During 1980, an octagonal AWTS was built at the Technical University of Berlin (TU-Berlin) to study the use of adaptive walls in 3-D testing. This AWTS is currently available for undergraduate teaching. This unusual test section is 15cm (5.9 in.) high, 18cm (7.09 in.) wide and 83cm (32.68 in.) long [12]. The test section is formed by eight flexible walls supported by a total of 78 jacks powered by individual DC motors (see Figure 15). The flexible walls are made of thin steel plates. The corners are sealed by spring steel lamellas so the test section boundary is impermeable and continuous.

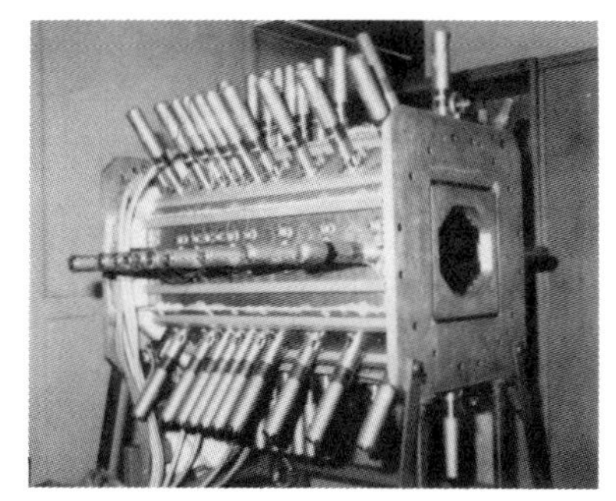

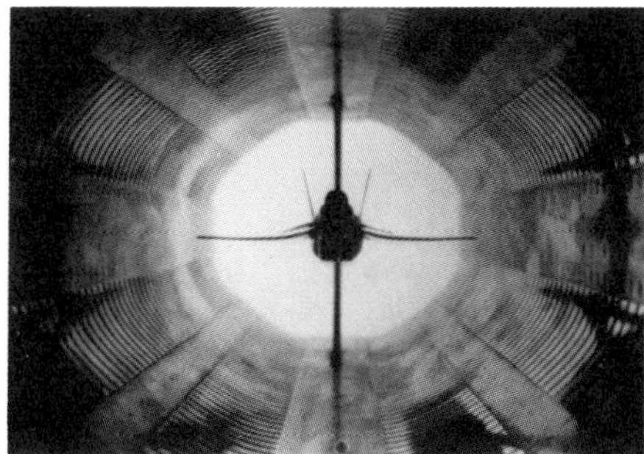

Figure 15. Octagonal flexible walled AWTS III, Technical University of Berlin.

This 3-D AWTS uses a wall adjustment procedure developed by Rebstock et al. Wall adaptation is possible in two iterations at M_∞ up to 0.95. Model blockage ratios up to 1.3% have been successfully tested, both with lifting and non-lifting sting mounted models. Boundary measurements are static pressures measured along the centerline of each flexible wall. Low supersonic tests of non-lifting bodies indicate that bow shock reflections from the flexible walls can be deflected away from the model [32].

5.12 TsAGI (Central Aero-Hydrodynamic Institute), Zhukovsky, Russia

Since 1986, TsAGI has been operating the largest AWTS anywhere for industrial testing. The massive Russian T-128 tunnel has a test section size of 2.75m (9 ft.) square and 8m (26.25 ft.) long, and is enclosed within a building (see Figure 16a). The T-128 has five interchangeable test sections and can operate at M_∞ up to 1.7, over a dynamic pressure range up to 10.15 psi (0.71 bar). Test section No. 1 is an AWTS with all four walls perforated (see Figure 16b). The turbulence level in the test section is quoted as 0.5%. Each wall of the AWTS is made up of 32 segments. The porosity of each of the 128 segments can be varied between 0 and 18%. Each segment is made up of two porous plates (one on top of the other). These two plates are moved relative to one another by computer control, to achieve a desired porosity over the segment (see Figure 17).

Figure 16a. TsAGI T-128 transonic wind tunnel, Zhukovsky, Russia.

A simplified wall adjustment procedure has been developed by Neyland for 3-D testing at transonic speeds, which requires knowledge and representation of the model under test [15]. Moreover, wall streamlining is only used for the few test conditions where WIAC is deemed inaccurate, mostly around $M_\infty=1$. Boundary measurements are static pressures measured along rows on the centerline of each wall. These pressures are assumed accurate for low wall porosities necessary for testing near sonic speeds. These simple boundary measurements made T-128 the first major transonic tunnel to be fully instrumented for WIAC use. The maximum reported blockage ratio for 3-D testing is about 3%, as shown on Figure 16b. This high blockage ratio is beyond the capabilities of other reported variable porosity AWTSs. The T-128 tunnel is a world-class industrial wind tunnel, which has received recent notoriety because the Boeing Company is one of the major users.

Figure 16b. TsAGI T-128 variable porosity AWTS with a 3% Buron model.

Figure 17. TsAGI T-128 AWTS variable porosity segments showing sliding plate assembly, Zhukovsky, Russia.

6 AN OVERVIEW OF AWTS DESIGNS

In 2-D testing, only two walls need to be adaptable and a simple AWTS is sufficient. Meanwhile in 3-D testing, the logical desire to control the AWTS boundaries in 3-D has led to a variety of AWTS designs. Moreover, some approximation in the shape of the test section boundaries is inevitable. The magnitude of this approximation has been the subject of much research. The most favoured number of adaptive walls for a 3-D AWTS is now two, which provides an acceptable design compromise. From practical considerations, this compromise is between size/correctability of residual wall interferences (after wall streamlining), hardware complexity, model accessibility, and the existence of rapid wall adjustment procedures.

There are strong theoretical [13] and experimental [31] indications that the simpler the AWTS design the better the testing technique (see sub-section 7.2). A simple design reduces both the complexity of calculating the residual wall interferences and the complexity of the tunnel hardware, and gives better model access as a bonus. A major factor in the design of new AWTSs will undoubtedly be the trade-off between the complexity of the boundary adjustments and the quality/cost of the residual wall interference corrections. The Russian T-128 tunnel represents the case where WIAC is used instead of adaptive walls for all but a limited number of test conditions.

Published data clearly shows that flexible walled AWTSs provide testing capabilities superior to that of similar sized variable porosity AWTS designs. We can summarize the effectiveness of flexible walls thus:

a) Flexible walls can be rapidly streamlined.
b) Flexible walls provide more powerful and direct adaptation control of the test section boundaries, necessary for large models and high lift conditions.
c) Flexible walls provide simple test section boundaries for adaptation measurements, residual wall interference assessments and setting of test conditions.
d) Flexible walls improve flow quality providing reduced tunnel interferences and reduced tunnel disturbances which lower operating costs.
e) No plenum is required around the test section reducing tunnel volume.

Interestingly, of the 16 transonic AWTSs now operational worldwide, only two AWTSs do not have flexible walls (see Table 1 at the end of this chapter). In the low speed regime, there are seven AWTSs currently operational which are all fitted with flexible walls. So the current total of AWTSs is 23, of which, 21 have flexible walls and 16 have only two flexible walls.

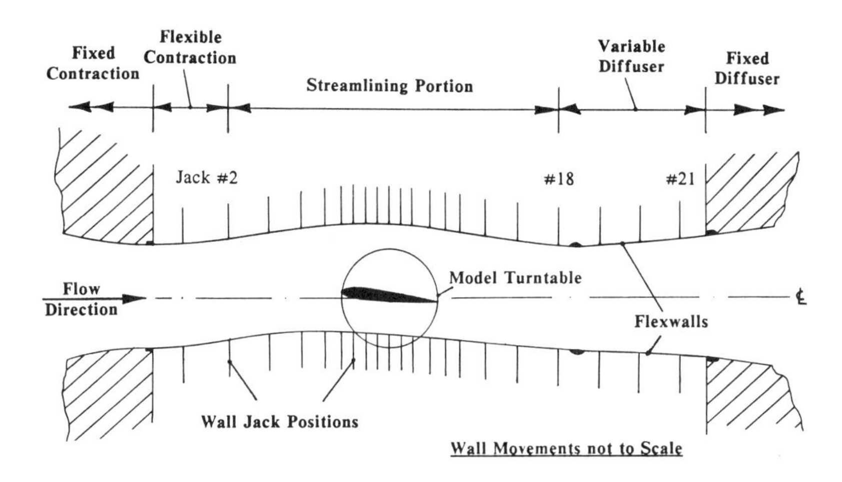

Figure 18. Design layout of the NASA-Langley 0.3m TCT adaptive wall test section.

The optimum 2-D AWTS has two flexible walls supported by jacks closely grouped in the vicinity of the model. A good example of this optimum design is the AWTS fitted in the NASA Langley 0.3m TCT, shown on Figure 18. The flexible walls (made of thin metal) are anchored at the upstream ends, and the downstream ends are attached by a sliding joint to a variable 2-D diffuser (refer to sub-section 5.5). The AWTS requires a square cross-section for optimum 2-D testing (i.e. maximizing Reynolds number capability). For 3-D testing, a rectangular cross-section, which is wider than it is tall, seems better for minimising wall induced gradients with 2-D wall adaptation [13]. However, we find that only 3 of the AWTSs used for 3-D testing have this desirable cross-section. To offset this situation, researchers have found that swept target lines for zero interference can compensate for less than optimum AWTS cross-sections [16].

7 REVIEW OF AWTS RESEARCH

Research into adaptive wall testing techniques, with both variable porosity and flexible wall AWTS designs, has concentrated on the following goals:

1) To define fast adaptive wall testing techniques for different test regimes.

2) To identify acceptable measurement tolerances.
3) To find the optimum AWTS design for different applications.
4) To find if any fundamental limitations to the adaptive wall concept exist.

However, we now know that a variable porosity AWTS is much less effective than a flexible walled AWTS. Therefore, only flexible wall research will be considered in this review.

Since 1938, researchers have made significant reductions to the time associated with wall streamlining. A major factor in this progress has been the development of rapid wall adjustment procedures for flexible walled AWTSs. (The term *rapid* refers to minimisation of the number of iterations necessary in any wall adaptation procedure.) Early empirical type methods (requiring eight iterations) have given way to analytical methods (requiring one or two iterations) as computer support has improved. These analytical methods now use both linear and nonlinear theory, and require no prior knowledge of the model. Nevertheless, simple empirical methods are still appropriate where the use of large models (relative to the test section size) is not important, as found in some of the AWTSs (particularly the cascade AWTSs) described in Section 5.

For 2-D free air simulations, the linear wall adjustment procedure of Judd et al. [23] (Southampton University) is now well established for reasons of speed, accuracy, simplicity (non-experts can easily use the method on any PC computer), and adaptability for general use with any flexible walled AWTS. A non-linear version is also available for use in 2-D testing where the flow at the walls is sonic [31]. For free air simulations in 3-D testing, researchers use the linear methods of Wedemeyer/Lamarche [13] (DLR), Rebstock [25] (NASA), Goodyer et al. [6] (Southampton), and Le Sant [28] (ONERA). However, all these 3-D methods are still under development. Supersonic 2- and 3-D testing is possible using the method of characteristics (wave theory) to predict the wall shapes necessary to generate supersonic flow [14,16,32].

Another time-saving feature of modern AWTSs is automatic wall streamlining. Researchers have shown that computer controlled movement of the adaptive walls and automatic acquisition of wall data dramatically reduce the time attributed to wall streamlining, from weeks to seconds! In addition, researchers have found that fast wall streamlining requires a good practical definition of when the walls are streamlined. We call this definition the *streamlining criterion* (the point at which we stop wall adaptation). The criterion is directly related to the accuracy of the tunnel/wall measurements (discussed later). For 2-D free air simulations, the best approach appears to be a quantitative approach which is to set, as the streamlining criterion, an acceptable maxima for the residual wall interferences [2]. This approach is used at the Southampton University, NASA

Langley and the University of Naples (Italy). At present there are only qualitative streamlining criteria in 3-D testing, whereby the walls are streamlined when the model data is unaffected by subsequent iterations of the wall adjustment procedure. On-line residual wall interference codes are available but require development for 3-D testing techniques in AWTSs [31,33,34].

Researchers have probed the limits of 2-D adaptive wall testing techniques. These limits are related to the aerodynamic theory and to mechanical hardware used in the wind tunnel tests. The use of sidewall BLC is only a factor in altering the wall curvature requirements. In 2-D testing, the operating envelope of an AWTS is bounded by the limitations of the test section geometry, the wall adjustment procedure and the instrumentation. These are the same factors that need to be defined in the design phase of a new AWTS. Researchers have provided many design guidelines to eliminate wall hardware problems, so far encountered, from future AWTS designs. With good design, only theoretical assumptions should restrict the operating envelope for 2-D testing.

In 3-D testing, the situation is far less clear, as no AWTS operating envelopes are well defined. Research has been spread thinly over many AWTS designs and numerous model configurations. The result is that the favoured AWTS design for 3-D testing has become the simplest design (as described earlier), because hardware complexity does not produce significant improvements.

Researchers have examined the effects of measurement accuracy on AWTS operation, particularly for flexible wall designs [2]. With flexible walls, the position of each wall can be measured only at a finite number of points. The measurement accuracy at each of these points is of the order ±0.127mm (±0.005 in.) in current AWTSs. The relative position of these measurement points, along each wall, can be optimized for 2-D flexible walled AWTS designs (as shown on Figure 18). Operationally, flexible walled AWTSs have proved tolerant to wall jacks being disconnected due to hardware failures [22]. Interestingly, because the wall position accuracy requirements are proportional to (l/h), the measurement accuracy requirements reduce significantly for a large AWTS. This situation is already proven in large supersonic nozzle systems operational today, and should encourage use of large AWTSs amongst potential operators.

We have found the flexible wall adaptation procedures to be tolerant to uncertainties in the wall pressures. This important feature is due to the smearing effect of the wall boundary layers. However, at high Reynolds numbers (when the wall boundary layers are thin) or with near sonic flow at the adaptive walls, this tolerance to measurement uncertainties reduces. The uncertainties in the wall pressures can be caused by wall imperfections or fluctuations in the tunnel test conditions. Again, large AWTSs should provide more tolerance to these uncertainties. However, we do know that if the model perturbations at the adaptive walls are small (as found in 3-D testing), the accuracy of the wall pressures needs to be better than when the model perturbations are large (as found in 2-D testing).

Furthermore, the allowance necessary for the boundary layer growth on the test section flexible walls is dependent on the accuracy of the wall pressures. In theory, each test condition should require a different boundary layer allowance (i.e., a change in test section cross-sectional area). In practice, researchers have shown that a series of say four *Aerodynamically Straight* wall contours are sufficient to provide uniform Mach number distributions, through an empty AWTS, for M_∞ up to 0.9 [2,22]. In addition, we do not need to make an allowance for the wall boundary layer thinning due to the presence of the model itself, until the flow on the flexible walls is sonic. Most AWTS operators monitor this boundary layer thinning real-time. Researchers have demonstrated that the adaptive wall testing techniques are tolerant to simple boundary layer allowances. In the NASA-Langley 0.3m TCT, for example, approximate *Aerodynamically Straight* contours are used which have simply linear divergence. This situation is a result of unacceptable wall waviness in the experimentally determined wall contours, with an empty test section due to wall imperfections. The quality of TCT data is unaffected by the use of these approximate wall boundary layer allowances.

The application of adaptive walls to different testing regimes is ongoing. Some interesting examples are: high lift 3-D testing of V/STOL aircraft at low speeds in an AWTS at the University of Arizona [35], under the supervision of Professor Sears; swept wing studies and minimum test section height studies with height to chord ratio of less than 0.5:1 in a low speed 2-D AWTSs at the Southampton University (See Figure 19); laminar flow studies using very large chord models (order 1m or 39.37 in.) in low-speed flexible walled AWTSs at NPU, China and FFA, Sweden; automotive testing at Sverdrup, Tennessee, and Southampton University using large blockage models (order 10%) [36]. Of the six AWTS flow field simulations (investigated by Goodyer [37]), adaptive wall research has tended to concentrate on free-air and cascade simulations, because these are of most interest. Nevertheless, we now find closed tunnel simulations are proving to be very useful for CFD code validation [21].

Figure 19. Minimum test section height studies in the low speed Self-Streamlining Wind Tunnel (SSWT), Southampton University, England.

7.1 2-D Testing Experience in AWTSs

Validation data shows that real-time 2-D data from AWTSs is essentially free of top and bottom wall interferences [38,39]. We have found no problems with testing an airfoil through stall (no wall shape induced model hysteresis present). Routine testing is possible up to drag rise Mach numbers ($M_\infty = 0.8\text{-}0.85$). Data repeatability from day to day is excellent (order 0.001 in C_n and 0.0005 in C_d) but, in common with all wind tunnel measurements, calibration procedures affect long term repeatability.

We have observed that the wake of a 2-D model in an AWTS shows minimal spanwise variation. We can speculate that the use of large models (with blockage up to 12%) intrinsically minimises secondary flows at the airfoil-sidewall junction, particularly when sidewall boundary layers are thin in high Reynolds number testing. This observation may explain why sidewall BLC does not significant effect wing performance in a relatively small AWTS. There are strong indications that the flow in an AWTS can be an excellent simulation of a 2-D flow field. If we ever need to use sidewall BLC in an AWTS, then researchers have found that no special testing procedures are necessary. The wall adjustment procedure simply senses the boundary layer mass removal as a model change.

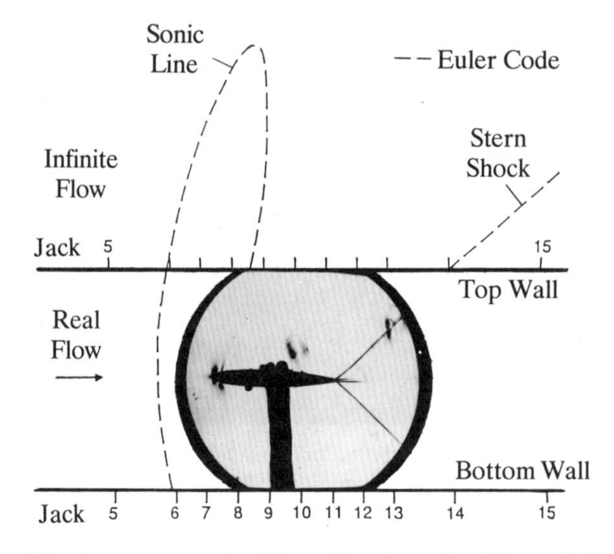

Figure 20. Low supersonic 2-D testing in a flexible walled AWTS, Southampton University TSWT data, NACA 0012-64 Airfoil (M_∞=1.2, $\alpha \approx 1.0°$).

Researchers have found many limitations to the various 2-D adaptive wall testing techniques, none of which are fundamental. These limitations are associated with wall movement (hardware), model size (theoretical assumptions) and M_∞ (theory sophistication). Researchers have made 2-D tests close to $M_\infty = 1.0$, and up to $M_\infty = 1.2$, as shown in Figure 20 [16]. In supersonic tests, researchers used local wall curvature (aided by wall boundary layer smearing) to remove oblique shock reflections on to the model. This work disproves long standing preconceptions that flexible walled AWTSs cannot be used through the speed of sound. However, the usefulness of 2-D testing in the supersonic regime is probably only academic, providing experience leading to production-type supersonic 3-D testing.

7.2 3-D Testing Experience in AWTSs

Limited 3-D validation tests support the claim that wall interferences are minimised in AWTSs [38,39]. However, the wall interferences present before any wall streamlining tend to be already small. Consequently, the effectiveness of AWTSs to minimise severe wall interferences in 3-D testing has not been studied.

This situation is due to the low blockage of the 3-D models so far tested in AWTSs. We can increase the model disturbances in the test section by using larger models or testing only at high speeds. Unfortunately, the roughly square cross-section of current AWTSs restricts the size of non-axisymmetric lifting models. Researchers have found that they must use low aspect ratio models to increase the model blockage above the normally accepted value of 0.5 percent. (This is because the model span must be limited to about 70 percent of the test section width to minimise interactions between the tip vortices and the tunnel sidewalls.) Consequently, there is a need for a new generation of 3-D AWTSs with rectangular cross-sections, which are wider than they are tall. These new AWTSs will help find the maximum 3-D model blockage which can be successfully tested in an AWTS (5.2% is the maximum reported).

Numerous 3-D AWTS designs have been studied (as discussed earlier). In fact, researchers have spent considerable time and effort to develop a wide range of complex 3-D AWTS designs, when it now appears the simpler 2-D design is adequate. (In hindsight, this effort appears unnecessary but the contribution to overall knowledge is nevertheless important.) An example of the promise of simple AWTSs in 3-D testing is shown on Figure 21. Data from residual interference codes (based on the work of Ashill and Weeks) are presented as contour plots of blockage and upwash wall interferences on a simple cropped delta wing, mounted on a sidewall of the 2-D AWTS in the Southampton University TSWT. Notice how the blockage interference patterns, with straight walls, are normal to the flow and 2-D in nature. We can see 2-D wall streamlining, with a straight target line, eliminates the blockage interference. Similarly, the upwash interference pattern with the walls straight exhibits some two-dimensionality, and again 2-D wall streamlining significantly reduces the upwash. If necessary, these remaining gradients could be further reduced by use of swept target lines in the wall adjustment procedure.

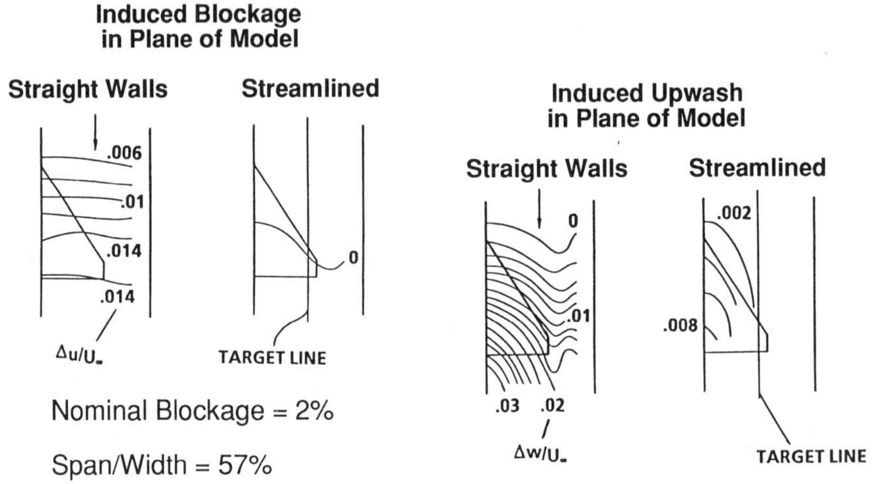

Induced Blockage in Plane of Model

Straight Walls — Streamlined

.006 .01 .014 .014 — $\Delta u/U_\infty$ — 0 — TARGET LINE

Nominal Blockage = 2%

Span/Width = 57%

Induced Upwash in Plane of Model

Straight Walls — Streamlined

0 .01 .03 .02 — $\Delta w/U_\infty$ — .002 .008 — TARGET LINE

Figure 21. Boundary interference minimization with 2-D wall adaptation in 3-D tests, Southampton University TSWT data, ($M_\infty = 0.7$, $\alpha = 8^o$).

No fundamental limits to Mach number have been found when using flexible walled AWTSs in 3-D testing. Preliminary tests at $M_\infty = 1.2$ show that bending AWTSs' flexible walls in 2-D or 3-D can eliminate oblique shock reflections on to the model (see Figure 22). The smearing of the shock/boundary-layer interaction does much to ease the curvature requirements on the flexible walls, as found in 2-D testing. The example shown in Figure 22 has been repeated in two other AWTSs. Nevertheless, in supersonic testing, the quality of the model data after wall streamlining needs to be better defined and robust wall adjustment procedures are required.

The wall adjustment procedures for 3-D testing have taken advantage of the fast and large capacity mini-computers available for real-time 3-D flow computations. Fast wall adjustment procedures are available up to about $M_\infty = 0.97$. Adaptive wall and WIAC research continues to identify when wall streamlining can be stopped and corrections applied with confidence. The various wall adjustment procedures for 3-D testing attempt to minimise wall interferences along pre-determined target lines (as discussed earlier). For example, the Rebstock method [25] minimises interferences along a pre-set streamwise target line anywhere in the test section. The Goodyer method has shown that swept target lines are very effective at reducing wall induced gradients [16] in square AWTSs. In addition, the Rebstock and Goodyer methods minimise wall curvature by introducing a uniform angle of attack error throughout the test section. Currently, we do not know where best to place the target line to eliminate the wall induced gradients for different model configurations. We also do not know where the concept of a uniform angle of attack error will break down.

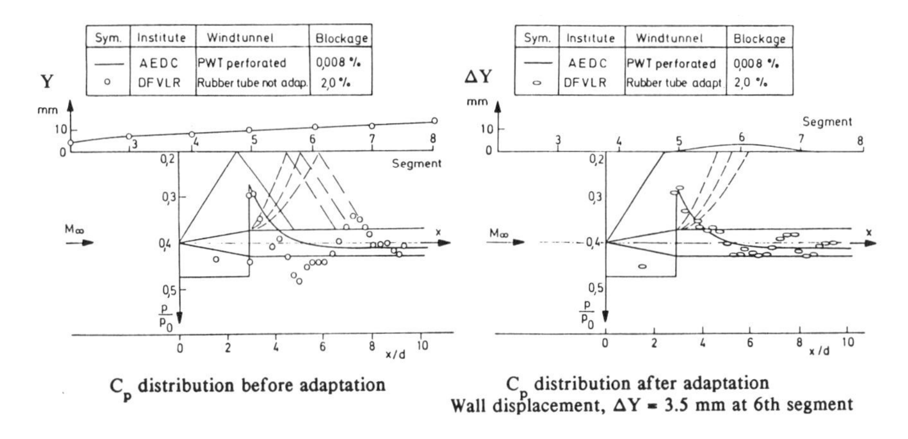

Figure 22. Supersonic tests in the DFVLR (now DLR) rubber tube AWTS, $M_\infty = 1.2$.

The type of wall pressure measurements necessary to adequately assess the residual wall interferences is also an unknown. The exploitation of real-time residual interference assessment and WIAC codes is now critical to progress in 3-D adaptive wall testing techniques. This has come about because we now realize that 3-D wall interferences cannot be eliminated with even the most complex AWTS system [28,33].

Hardware limitations currently restrict AWTS test envelopes (in particular model lift) for reported 3-D tests in small AWTSs. These hardware limitations have arisen because the AWTS design criteria was inadequate, or the AWTSs were originally designed for only 2-D testing. Unfortunately, these limitations have hampered 3-D adaptive wall research. It would appear that this situation is caused by low priority funding.

Despite these problems, some routine AWTS use in 3-D testing has been demonstrated in the ONERA T-2 and TsAGI T-128. This situation indicates that even first-generation adaptive wall testing techniques can be used for production 3-D testing if the demand is present.

8 PRODUCTION REQUIREMENTS

The production requirements for an adaptive wall testing technique are the same as for any modern testing technique. Firstly, the technique must be easy to use. Consequently, we need to make the complexities of the AWTS invisible to the tunnel operators (similar to operating large flexible supersonic nozzles). Secondly, the technique must not require excessive tunnel time for wall streamlining. So we require the AWTS wall movements to be quick. Thirdly, the technique must have a known test envelope for successful use. Therefore, we must ensure the testing technique is well researched, so that we know the limitations and restrictions to be avoided during normal operations. Fourthly, the technique must be financially and politically acceptable.

How can the adaptive wall testing technique meet the production requirements listed above? First, lets consider the complexity of an adaptive wall testing technique. We must design the associated test section hardware so the wall shapes can be continually changed. We also need an interaction between the AWTS and a computer system to set the wall to streamline shapes. If we make the AWTS of simple design then access to the model is unaffected. Furthermore, if we make the wall adjustments automatic via a user-friendly computer system, the operator need only issue *Go/Stop* commands (see Figure 23). Consequently, the complexity of the testing technique is invisible to the operator. The tunnel operator's contact with the AWTS becomes similar to setting angle of attack or changing tunnel test conditions.

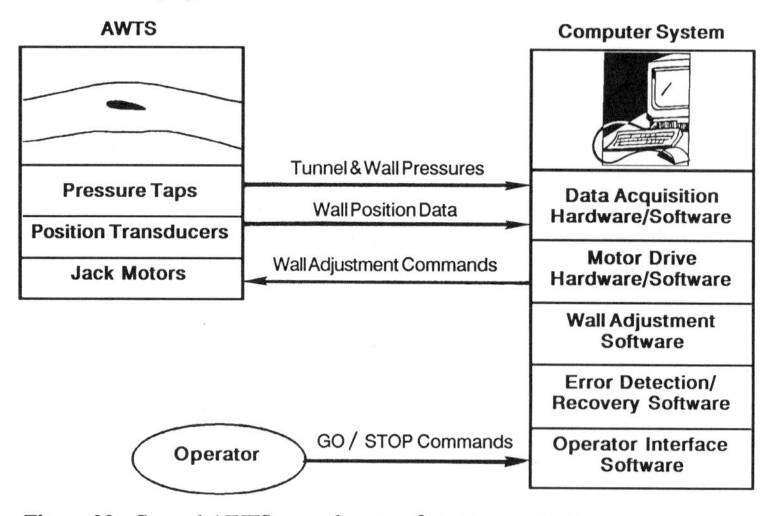

Figure 23. General AWTS control system for non-expert use.

Second, lets consider the time factor. Adjusting the walls of the test section takes time. How much time depends on the AWTS hardware (jack type) and the wall adjustment procedure. We can design the wall jacks to be very responsive. The wall adjustment procedure can find the streamline shapes in one or two iterations. The result is that wall streamlining can be quick. Researchers at ONERA have already demonstrated wall streamlining in 10 seconds for 2-D testing. Computer advances will make this possible for 3-D testing in the future. Another time factor is the elimination of post-test corrections and lengthy test programmes, because real-time AWTS data are the final data. We show the importance of this fact on Figure 24. In this example, we compare real-time transonic 2-D lift data from a deep slotted walled test section with equivalent real-time data from a shallow flexible walled AWTS, at the same test conditions with transition fixed. The differences in maximum C_n and the lift-curve slope are alarming. With the final data known real-time, AWTS operators can and should save overall tunnel run time.

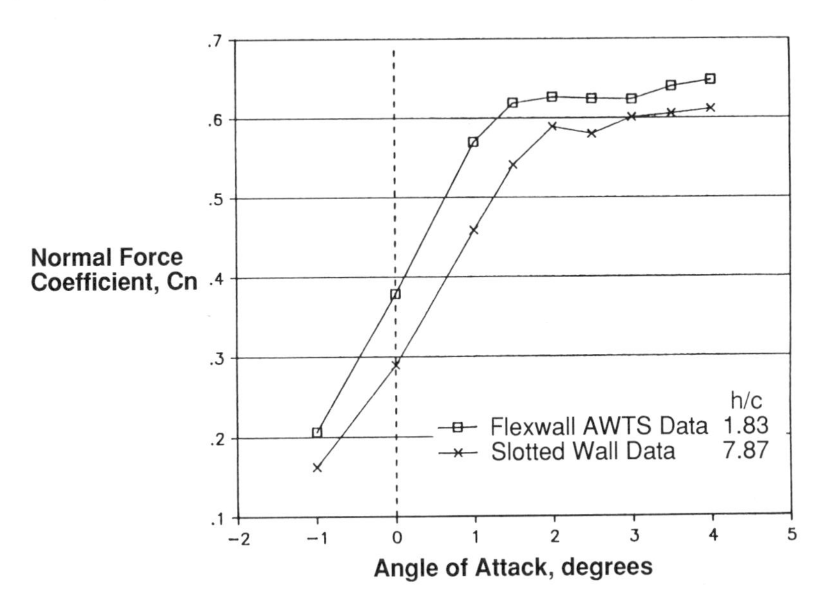

Figure 24. The importance of corrected real-time 2-D data. NASA Langley 0.3m TCT airfoil data, $M_\infty = 0.765$, transition fixed.

Third, lets consider the test envelopes for AWTSs. Researchers have defined the test envelope for various 2-D adaptive wall testing techniques (described earlier). So we can direct non-expert users away from these known limitations. Alas, in 3-D testing, we are still learning what the limitations are, but the experience base is growing.

Fourth, let us consider the cost and political factors. The simplest AWTS design can be incorporated in existing wind tunnels by the replacement of only two walls. Also, the plenum, which surrounds ventilated transonic test sections, can provide adequate volume, within the pressure vessel, for the jack mechanisms. These factors will reduce the overall hardware costs. Furthermore, the advent of inexpensive, but powerful, PC computer systems means that an AWTS control system can cost considerably less now. In addition, an AWTS control system can be integrated with other tunnel systems, which do not need to operate at the same time as wall streamlining, such as the sting positioning system. Other favourable cost factors are the reduction of tunnel operating costs possible by using a smaller AWTS (as much as 75% smaller than the original), and by removing test section ventilation through use of flexible walls.

Third, lets consider the test envelopes for AWTSs. Researchers have defined the test envelope for various 2-D adaptive wall testing techniques (described earlier). So we can direct non-expert users away from these known limitations. Alas, in 3-D testing, we are still learning what the limitations are, but the experience base is growing.

Fourth, let us consider the cost and political factors. The simplest AWTS design can be incorporated in existing wind tunnels by the replacement of only two walls. Also, the plenum, which surrounds ventilated transonic test sections, can provide adequate volume, within the pressure vessel, for the jack mechanisms. These factors will reduce the overall hardware costs. Furthermore, the advent of inexpensive, but powerful, PC computer systems means that an AWTS control system can cost considerably less now. In addition, an AWTS control system can be integrated with other tunnel systems, which do not need to operate at the same time as wall streamlining, such as the sting positioning system. Other favourable cost factors are the reduction of tunnel operating costs possible by using a smaller AWTS (as much as 75% smaller than the original), and by removing test section ventilation through use of flexible walls.

Politically, minimisation of risk is of major concern when considering the use of a new testing technique. It is always easier to copy what has gone before, but if we do not take risks, progress invariably suffers. Surprisingly, even providing simple boundary measurements for WIAC, with little or no risk involved, has met with resistance from tunnel managers. Naturally, adaptive walls introduce a level of risk which is all too often assessed on out-dated misconceptions, such as:

1) Adaptive walls are too complex for industrial use.
2) Adaptive walls are only effective in small scale facilities.
3) Model flow field calculations are required.
4) A rubber tube with a large number of jacks is required for 3-D testing.
5) Sonic flow at the AWTS flexible walls is a fundamental limit to test M_∞.

In reality, if the adaptive walls remain structurally sound, only the quality of the real-time data is put at risk by using adaptive walls. The tunnel test envelope can only be restricted by mechanical problems. Of course, there is the natural tendency to ignore improvement unless there is a significant reason for making that improvement. Clearly, the problem of wall interferences is perceived by many as more of a nuisance than a serious problem that needs to be overcome. This is certainly the current situation in the United States.

Interestingly of the three wind tunnels with AWTSs which come closest to being production-type tunnels, non-experts can only use one. The Langley 0.3m TCT has the only *User Friendly* AWTS control system that allows non-expert 2-D testing within defined test envelopes. However, the TsAGI T-128 is currently the only major transonic wind tunnel where effort is being made to improve experimental testing techniques, and non-expert use is probably close at hand.

The extensive Langley experience with adaptive walls has identified several special requirements for production use of AWTS. An accurate technique for setting the flexible walls to a straight datum is required. The wall position transducers should have high priority calibration status on a par with pressure instrumentation. The stability of the test Mach number should be checked during wall pressure scans. Finally, robust control software, which does not perpetuate data errors from test to test, should be used. These requirements may seem trivial but are crucial and often overlooked.

9 THE FUTURE OF AWTSs?

The vast and successful 2-D AWTS testing experience will continue to be a catalyst for the development of 3-D adaptive wall testing techniques. There is now a need to build more AWTSs with two flexible walls specially for 3-D testing, to probe design principles and testing technique limitations. Currently, six research groups around the world are actively pursuing the development of 3-D adaptive wall testing techniques. This effort needs to focus on current research problems in 3-D testing (such as low supersonic testing) to bridge the gap between academic and industrial interests. Furthermore, this research needs to emphasize the importance of adaptive walls in transition research and CFD code validation, to gain popular support for the many advantages of AWTSs. We must consolidate limited resources on developing flexible wall testing techniques and WIAC, which together offer the best chance for success.

The current status of adaptive wall technology is ongoing and positive. There are four new AWTSs being designed and built at this time. If production testing is the ultimate goal, then we have finished developing 2-D adaptive wall testing techniques for free-air simulations [39]. The last gathering of adaptive wall researchers at ICAW '91 in June 1991 set a precedent for discussing adaptive wall and WIAC interests together [40]. The comments from this meeting repeated those from previous such meetings, notably:

a) Flexible walls are usable over the transonic range up to $M_\infty = 1.3$.
b) AWTSs can significantly reduce wall induced gradients in 3-D testing.
c) The choice of where to place the target line for interference elimination in 3-D testing is critical to success.
d) WIAC is effective at benign test conditions where CFD is almost as effective as the wind tunnel.

An International Working Group was established in June 1991 to promote progress in WIAC and adaptive wall testing techniques, utilizing the readership of the *Adaptive Wall Newsletter.* The group will help coordinate research effort on mutual problems of wall interferences. Currently, the most pressing concern amongst group members is how best to tackle low supersonic 3-D testing. In this era of shrinking research budgets, this group will serve progress very well. The next group meeting is expected in December 1993, when plans will be laid for a programme to validate 3-D testing techniques.

Much has been achieved since the start of the modern era of adaptive walls in 1971. However, we find that AWTSs are incorporated in only one of the major transonic wind tunnels built worldwide in the last 20 years. However, several of these new wind tunnels do include provisions in the test section design for the eventual use of adaptive wall technology, during the life of the wind tunnel. (The European Transonic Wind Tunnel is a prime example of this practice because of a 1988 management disagreement as to which AWTS design to use.) Some projects, like the refurbishment of the NASA 12-foot wind tunnel, are managed in such a manner as to preclude any inclusion of modern testing techniques regardless of the benefits. The inclusion of WIAC and/or adaptive walls in new wind tunnel projects requires an advocate. To quote Orville

Wright, *"I had thought that truth must eventually prevail, but I have found silent truth cannot withstand error aided by continual propaganda."* Without advocates, preconceptions, based on limited practical experience or an over-exposure to the use of ventilated test sections, will continue to delay the proper utilization of adaptive wall technology.

We can summarize the current status as the development of a "new" technology to a point where this technology could be made very useful to the aerodynamicist (both theoretician and experimentalist) given the right priority. I am certain that if adaptive wall research had been given the same priority as the development of complex transonic "correction codes", we would have production 3-D adaptive wall testing techniques available for general use right now.

Lack of progress and loss of aerospace prowess in the United States may cause more risk-taking in ground testing in the future. The recent statement by the NASA Administrator, Daniel Goldin, that NASA aeronautical facilities should be improved and *"... we've got to be prepared to spend what it takes"* is good news indeed [41]. NASA aeronautical research is still highly respected around the world and leads by example. During the last five years we have seen a major scaling down of NASA's research into experimental testing techniques, which has sent a message to the world that our wind tunnels are just fine as they are. This new NASA initiative, caused by US aerospace companies seeking out better ground testing facilities outside the United States, could have a profound influence on wind tunnel design worldwide, by encouraging more risk-taking and progress. I certainly hope so.

Now that the expectations of CFD have become more realistic, the relationship between wind tunnel and computer has become much more mature and stronger. The adaptive wall concept embodies a near-perfect combination (marriage even) of experimental and theoretical aerodynamics (wind tunnel and computer) to improve our understanding of aerodynamics in the future. Our wind tunnels are not perfect and adaptive wall technology is available to help us aspire to higher levels of data accuracy [42].

10 CONCLUSIONS

1. Adaptive wall testing techniques, particularly those which utilize flexible walls, offer major advantages over conventional techniques in transonic testing.
2. We can significantly improve transonic data quality by using adaptive wall technology available for use now.

3. Computer advances have removed any impractical aspects of adaptive wall technology.
4. Non-expert use of AWTSs for routine 2-D testing has been demonstrated, even in cryogenic wind tunnels.
5. We can now design new AWTSs with no hardware restrictions to the 2-D and 3-D operating envelopes.
6. In 2-D testing, adaptive wall testing techniques are well proven and are already in use for production-type transonic testing in cryogenic wind tunnels.
7. Adaptive wall technology can significant reduce wall-induced gradients in 3-D testing with just two adaptive walls.
8. The next major step for adaptive wall technology is the development of testing techniques for general 3-D transonic testing, particularly at low supersonic speeds.
9. For general acceptance of adaptive wall testing techniques in 3-D testing, advocates are required to overcome the inertia against change in testing techniques, and preconceptions about the practicalities of AWTSs.
10. The current status of adaptive wall technology is ongoing and positive.

REFERENCES

1 Bailey, A. and Wood, S.A.: Further Development of a High-Speed Wind Tunnel of Rectangular Cross-Section. British ARC R&M 1853, September 1938.

2 Wolf, S.W.D.: *The Design and Operational Development of Self-Streamlining Two-Dimensional Flexible Walled Test Sections*, Ph.D. Thesis, Southampton University, also as NASA CR-172328, March 1984, N84-22534.

3 Hilton, W.F.: *High-Speed Aerodynamics*. Longmans, Green and Co., pp. 425-429, 1951.

4 Lewis, M.C.: An Evaluation in a Modern Wind Tunnel of the Transonic Adaptive Wall Adjustment Strategy Developed by NPL in the 1940's, NASA CR-181623, February 1988.

5 Holder, D.W.: The High-Speed Laboratory of the Aerodynamics Division, N.P.L., Parts I, II and III. British ARC R&M 2560, December 1946.

6 Facilities and Techniques for Aerodynamic Testing at Transonic Speeds and High Reynolds Number. AGARD CP-83, April 1971.

7 Sears, W.R.: Self Correcting Wind Tunnels. 16th Lanchester Memorial Lecture, May 1973. In: *Aeronautical J.*, **78**, pp. 80-89, Feb/Mar 1974, A74-27592.

8 Judd, M., Goodyer, M.J. and Wolf, S.W.D.: Application of the Computer for On-site Definition and Control of Wind Tunnel Shape for Minimum Boundary Interference. Presented

at AGARD Specialist Meeting on "Numerical Methods and Wind Tunnel Testing," June 1976, AGARD CP-210, October 1976, Paper 6, N77-11975.

9 Sears, W.R., Vidal, R.J., Erickson, J.C. and Ritter, A.: Interference-Free Wind Tunnel Flows by Adaptive-Wall Technology. *J. of Aircraft,* **14**, pp 1042-1050, November 1977.

10 Harney, D.J.: Three-Dimensional Testing in a Flexible-Wall Wind Tunnel, AIAA Paper 84-0623 (Technical Papers A84-24203, pp 276-283), AIAA 13th Aerodynamic Testing Conference, March 1984.

11 Heddergott, A., Kuczka, D. and Wedemeyer, E.: The Adaptive Rubber Tube Test Section of the DFVLR Göttingen, ICIASF 1985 Record, IEEE 85CH2210-3, A86-38244, pp. 154-156, 11th International Congress on Instrumentation in Aerospace Simulation Facilities, August 1985.

12 Ganzer, U., Igeta, Y. and Ziemann, J.: Design and Operation of TU-Berlin Wind Tunnel With Adaptable Walls. ICAS Paper 84-2.1.1, Proceedings of the 14th Congress of the International Council of the Aeronautical Sciences, Vol. 1, A84-44926, pp. 52-65, September 1984.

13 Wedemeyer, E. and Lamarche, L.: The Use of 2-D Adaptive Wall Test Sections for 3-D Flows, AIAA Paper 88-2041, AIAA 15th Aerodynamic Testing Conference, A88-37943, May 1988.

14 He, J.J., Zuo, P.C., Li, H.X. and Xu, M.: The Research of Reducing 3-D Low Supersonic Shock Wave Reflection in a 2-D Transonic Flexible Walls Adaptive Wind Tunnel, AIAA Paper 92-3924, 17th AIAA Aerospace Ground Testing Conference, A92-56755, July 1992.

15 Neyland, V.M.: Adaptive Wall Wind Tunnels with Adjustable Permeability Experience of Exploitation and Possibilities, A91-52779, Paper A3, Proceedings of International Conference on Adaptive Wall Wind Tunnel Research and Wall Interference Correction, Xian, China, June 1991.

16 Lewis, M.C.; Taylor, N.J.; and Goodyer, M.J.: Adaptive Wall Technology for Three-Dimensional Models at High Subsonic Speeds and Airfoil Testing Through the Speed of Sound, pp. 42.1-42.12, Proceedings of the European Forum on Wind Tunnels and Wind Tunnel Testing Techniques, Southampton, England, September 1992.

17 Romberg, H.-J.: Two-Dimensional Wall Adaptation in the Transonic Wind Tunnel of AIA, *Experiments in Fluids*, **9**, No. 3, pp. 177-180, May 1990, A90-38497.

18 Holst, H.; and Raman, K.S.: 2-D Adaptation for 3-D Testing, DFVLR-IB 29112-88 A 03, June 1988.

19 Pittaluga, F.; and Benvenuto, G.: A Variable Geometry Transonic Facility for Aerodynamic Probe Calibration, pp. 1-1/23, Proceedings of 8th Symposium on Measuring Techniques for Transonic and Supersonic Flow in Cascades and Turbomachines, Genoa, Italy, October 1985.

20 McDevitt, J.B. and Okuno, A.F.: Static and Dynamic Pressure Measurements on a NACA 0012 Airfoil in the Ames High Reynolds Number Facility, NASA TP-2485, January 1985, N85-27823.

21 Olsen, M.E. and Seegmiller, H.L.: Low Aspect Ratio Wing Code Validation Experiment. AIAA Paper 92-0402, 30th AIAA Aerospace Sciences Meeting, January 1992.

22 Wolf, S.W.D.: Application of a Flexible Walled Testing Technique Section to the NASA Langley 0.3-m Transonic Cryogenic Tunnel, ICAS Paper 88-3.8.2, vol. 2, pp. 1181-1191, Proceedings of the 16th Congress of the International Council of the Aeronautical Sciences, August-September 1988, A89-13620.

23 Wolf, S.W.D. and Goodyer, M.J.: Predictive Wall Adjustment Strategy for Two-Dimensional Flexible Walled Adaptive Wind Tunnel, A Detailed Description of the First One-Step Method, NASA CR-181635, January 1988, N88-19409.

24 Wolf, S.W.D.: Wall Adjustment Strategy Software for Use With the NASA Langley 0.3-Meter Transonic Cryogenic Tunnel Adaptive Wall Test Section, NASA CR-181694, August 1988, N89-13400.

25 Rebstock, R. and Lee, E.E.: Capabilities of Wind Tunnels with Two Adaptive Walls to Minimize Boundary Interference in 3-D Model Testing, Proceedings of the NASA Langley Transonic Symposium, NASA CP-3020, Vol. 1, Part 2, pp 891-910, April 1988, N89-20942.

26 Archambaud, J.-P. and Mignosi, A.: Two-Dimensional and Three-Dimensional Adaptation at the T2 Transonic Wind Tunnel of ONERA/CERT, AIAA Paper 88-2038, AIAA 15th Aerodynamic Testing Conference, May 1988, A88-37940.

27 Archambaud, J.-P., Michonneau, J.F. and Prudhomme, S.: Use of Flexible Walls to Minimise Interferences at T2 Tunnel, Paper 15, European Forum on Wind Tunnels and Wind Tunnel Testing Techniques, Southampton, England, September 1992.

28 Le Sant, Y. and Bouvier, F.: A New Adaptive Test Section at ONERA Chalais-Meudon, pp. 41.1-41.14, Proceedings of the European Forum on Wind Tunnels and Wind Tunnel Testing Techniques, Southampton, England, September 1992.

29 Nagamatsu, H.T. and Trilling, T.W.: Passive Transonic Drag Reduction of Supercritical and Helicopter Rotor Airfoils, A88-51785, Proceedings of the 2nd AHS International Conference on Rotorcraft Basic Research, February 1988.

30 Mitty, T.J., Nagamatsu, H.T and Nyberg, G.A.: Oscillatory Flow Field Simulation in Blow-Down Wind Tunnel and the Passive Shock Wave/Boundary Layer Control Concept, AIAA Paper 89-0214, AIAA 27th Aerospace Sciences Meeting, January 1989.

31 Lewis, M.C., : *Airfoil Testing in a Self-Streamlining Flexible Walled Wind Tunnel,* Ph. D. Thesis, Southampton University, also as NASA CR-4128, May 1988, N88-22865.

32 Rill, S.L. and Ganzer, U.: Adaptation of Flexible Wind Tunnel Walls for Supersonic Flows, AIAA Paper 88-2039, AIAA 15th Aerodynamic Testing Conference, May 1988, A88-37941.
33 Mokry, M.: Residual Interference and Wind Tunnel Wall Adaption, pp. 175-193, Proceedings of the NASA Langley CAST-10-2/DOA 2 Airfoil Studies Workshop Results, NASA CP-3052, September 1988, N90-17655.
34 Holst, H.: Determination of 3-D Wall Interference in a 2-D Adaptive Test Section Using Measured Wall Pressures, Proceedings of the 13th International Congress on Instrumentation in Aerospace Simulation Facilities, September 1989.
35 Sears, W.R.: Adaptable Wind Tunnel for Testing V/STOL Configurations at High Lift. *J. of Aircraft*, **20**, pp. 968-974, November 1983.
36 Whitfield, J.D., Jacobs, J.L., Dietz, W.E. and Pate, S.R.: Demonstration of the Adaptive Wall Concept Applied to an Automotive Wind Tunnel, AIAA Paper 82-0584, AIAA 12th Aerodynamic Testing Conference, March 1982.
37 Goodyer, M.J.: The Self Streamlining Wind Tunnel, NASA TM-X-72699, August 1975, N75-28080.
38 Tuttle, M.H.; and Mineck, R.E.: Adaptive Wall Wind Tunnels - A Selected, Annotated Bibliography, NASA TM-87639, August 1986, N86-29871.
39 Hornung, H.G. (ed.): *Adaptive Wind Tunnel Walls: Technology and Applications*, AGARD AR-269, April 1990.
40 He, J.J. (ed.): ICAW papers A91-52777 to A91-52805, Proceedings of the International Conference on Adaptive Wall Wind Tunnel Research and Wall Interference Correction, Xian, China, June 1991, A91-52776.
41 Goldin Defines Policy to Reinvigorate Aeronautics Research, Infrastructure, *Aviation Week and Space Technology,* December 14/21, 1992.
42 Steinle, F. and Stanewsky, E.: Wind Tunnel Flow Quality and Data Accuracy Requirements, AGARD AR-184, 1982.

Table 1 Currently Used Adaptive Wall Test Sections

Organization	Tunnel	X-Section (h x w) m	Length. m	Approx. Max. Mach No.	Approx. Max. R_c (millions)	Walls	Adaptation Control	Remarks
Aachen Aero. Institute[2]	TST	0.4 Square	1.414	4.0	2.8	2 Flexible 2 Solid	24 Jacks/Wall	Issue 10
DLR[3]	DAM	0.8 Circular	2.40	1.2	...	Rubber Tube	64 Jacks Total	Issue 5
DLR[3]	HKG	0.67 x 0.725 Rectangular	4.0	>1.2	...	2 Flexible 2 Solid	17 Jacks/Wall	Issues 7, 14
Genova University[2]	High Defl. Cascade	0.2 x 0.05 Rectangular	1.58	2.0	1	2 Flexible 2 Solid	36 Jacks/Wall	Issue 7
Genova University[2]	High Defl. Cascade	0.2 x 0.05 Rectangular	1.6	>1.18	1	2 Flexible 2 Solid	13 Jacks-Celing 26 Jacks-Floor	Issue 7
NASA Ames[2+3]	HRC-2 AWTS1	0.61 x 0.41 Rectangular	2.79	>0.8	30	2 Flexible 2 Solid	7 Jacks/Wall	
NASA Langley[2+3]	0.3-m TCT	0.33 Square	1.417	>1.3	120	2 Flexible 2 Solid	18 Jacks/Wall	Issues 1-5, 7, 8, 13
N P Univ.[2+3] Xian, China	WT52	0.21 x 0.3 Rectangular	1.08	1.2	...	2 Flexible 2 Solid	16 Jacks/Wall	Issue14
N P Univ.[2+3] Xian, China	Low Speed	0.256 x 0.238 Rectangular	1.3	0.12	0.50	2 Flexible 2 Solid	19 Jacks/Wall	Issues 2, 5 9, 14, 15
N P Univ.[2] Xian, China	LTWT	1.0 x 0.4 Rectangular	6	0.23	4	2 Flexible 2 Solid	15+ Jacks/Wall	
ONERA/CERT[2+3]	T2	0.37 x 0.39 Rectangular	1.32	>1.0	30	2 Flexible 2 Solid	16 Jacks/Wall	Issue 2
ONERA[3]	S3Ch	0.8 Square	2.2	1.3	...	2 Flexible 2 Solid	15 Jacks/Wall	
RPI[2] Troy, NY	3 x 8	0.20 x 0.07 Rectangular	0.6	0.86	...	1 Flexible 3 Solid	6 Jacks	
RPI[2] Troy, NY	3 x 15	0.39 x 0.07 Rectangular	0.6	0.8	...	4 Solid	Multiple Top Wall Inserts	
Southampton University[2+3]	SSWT	0.152 x 0.305 Rectangular	0.914	0.1	0.38	2 Flexible 2 Solid	17 Jacks/ Wall	Variable T.S. Height
Southampton University[3]	AWT	0.305 Square	0.914	0.1	...	3 Flexible 1 Solid	? Jacks/Wall	
Southampton University[2+3]	TSWT	0.15 Square	1.12	>1.0	2.5	2 Flexible 2 Solid	19 Jacks/Wall	Issue 1
Sverdrup Technology[3]	AWAT	0.305 x 0.61 Rectangular	2.438	0.2	...	3 Multi-Flexible Slats 1 Solid	102 Jacks-Ceiling 15 Jacks/Sidewall	Issue 4
Tech. University Berlin[3]	III	0.15 x 0.18 Octangonal	0.83	>1.0	...	8 Flexible	78 Jacks Total	Issue 6
TsAGI[3], U.S.S.R.	T-128	2.75 Square	8.0	1.7	9	4 Porous	32 Control Panels per Wall	Issues 11, 13
Umberto Nobile[2]	FWWT	0.2 Square	1.0	0.6	3.5	2 Flexible 2 Solid	18 Jacks/Wall	
VKI[2], Belgium	T'3	0.10 x 0.12 Rectangular	?	0.2	...	2 Flexible 2 Solid	? Jacks/Wall	
Univ. Waterloo[2], Canada	UWFWWT	0.91 x 0.61 Rectangular	6.55	0.1	...	2 Flexible 2 Solid	48 Jacks/Wall	

[2] - 2D Testing Capability, [3] - 3D Testing Capability, [2+3] - 2D and 3D Testing Capability   February 1993 SWDW

Note: The Remarks column refers to issues of the ***Adaptive Wall Newsletter*** which contain related articles.

List of Symposium Registrants

Mr. Ashkenas, Irv
600 Harbor St.
STI Hawthorne CA
Unit 2, Marina del Rey CA 90291

Professor Atassi, Hafiz
Center for Applied Math.
U of Notre Dame
Notre Dame, IN 46556

Prof. Balsa, Tom
Aero. & Mech. Eng. Dept.
U of Arizona
Tucson, AZ 85721

Mr. Bates, Ron
Aero. & Mech. Eng. Dept.
U of Arizona
Tucson, AZ 85721

Mr. Bean, George
P.O. Box 3707
Boeing
Seattle, WA 98124

Dr. Chan, Cholik
Aero. & Mech. Eng. Dept.
U of Arizona
Tucson, AZ 85721

Mr. Chavez, F.
Mech. & Aero. Eng.
ASU
Tempe, AZ 85287

Prof. Chen, Tony
Aero. & Mech. Eng. Dept.
U of Arizona
Tucson, AZ 85721

Prof. Cheng, H. K.
2335 Westridge Rd.
USC
Los Angeles, CA 90049

Dr. Chow, Kwok
Math. Dept.
U of Arizona
Tucson, AZ 85721

Mr. Colvin, Jim
Aero. & Mech. Eng. Dept.
U of Arizona
Tucson, AZ 85721

Mr. Crockett, Richard
Aero. & Mech. Eng. Dept.
U of Arizona
Tucson, AZ 85721

Prof. Crow, Steve
Aero. & Mech. Eng. Dept.
U of Arizona
Tucson, AZ 85721

Dr. Davis, Sandy
MS 260-1
NASA-Ames
Moffett Field, CA 94035

Mr. De-Silva, Sirilath
Aero. & Mech. Eng. Dept.
U of Arizona
Tucson, AZ 85721

Mr Diamond, Jeff
Aero. & Mech. Eng. Dept.
U of Arizona
Tucson, AZ 85721

Dr. Dulikravich, George
233 Hammond Bld.
Penn State U
State College, PA 16802

Dr. Farassat, Feri
Acoustics Division
NASA-Langley
Langley, VA 23681

Prof. Fasel, Hermann
Aero. & Mech. Eng. Dept.
U of Arizona
Tucson, AZ 85721

Professor Ffowcs Williams, Sohn
Dept. Engineering.
Cambridge Universtiy
England, CB2 1PZ

Dr. Flax, Al
2101 Constitution Ave N. W.
NAE/Wash. DC
Washington, DC 20418

Dr. Fung, K.-Y.
Aero. & Mech. Eng. Dept.
U of Arizona
Tucson, AZ 85721

Dr. Geffen, Nima
School of Math.Sciences
U of Tel-Aviv
Tel-Aviv, Israel

Dr. Goldstein, Marv
MS 3-17
NASA-Lewis
Cleveland, OH 44135

Mr. Grabowsky, Wally
26812 Venado Dr.
Aerospace Corp
Mission Viejo, CA 92675

Dr. Hartunian, Dick
2714 Graysby Ave.
Aerospace Corp
San Pedro, CA 90732

Mr. Heine, Christoph
Aero. & Mech. Eng. Dept.
U of Arizona
Tucson, AZ 85721

Prof. Ho, Chih-Ming
MANE
UCLA
Los Angeles, CA 90024-1597

Mr. Ingmire, Gordon
Aero. & Mech. Eng. Dept.
U of Arizona
Tucson, AZ 85721

Dr. Jacobs, Jeff
Aero. & Mech. Eng. Dept.
U of Arizona
Tucson, AZ 85721

Mr James, Richard
Aero. & Mech. Eng. Dept.
U of Arizona
Tucson, AZ 85721

Mr. Jones, Mike
Aero. & Mech. Eng. Dept.
U of Arizona
Tucson, AZ 85721

Dr. Kerschen, Ed
Aero. & Mech. Eng. Dept.
U of Arizona
Tucson, AZ 85721

Dr. Korkegi, Bob
4418 Springdale St. N.W.
NRC/Wash. DC
Washington, D.C. 20016

Professor Kroo, Ilan
Dept. of Aero. & Astro.
Stanford
Stanford, CA 94305

Dr. Liepmann, Hans
4036 Robin Hill Road
CALTECH
Pasadena, CA 91011

Prof. Lighthill, Sir James
Gower Street WCI
U College London
London E6BT, England

Prof. Liu, Danny
Mech. & Aero. Dept.
ASU
Tempe, AZ 85287

Prof. Liu, Joe
Division of Engineering
Brown University
Providence, RI 02912

Prof. Lukasiewicz, Luke
46 Whippoorwill Dr.
Carleton U
Ottawa Ont. K1J 7H9, Canada

Mr. Man, Raymond
Aero. & Mech. Eng. Dept.
U of Arizona
Tucson, AZ 85721

Prof. Marble, Frank
1691 San Pasqual St.
CALTECH
Pasadena, CA 91106

Mr. Meitz, Hubert
Aero. & Mech. Eng. Dept.
U of Arizona
Tucson, AZ 85721

Dr. Messiter, Art
Aerospace Eng.
U of Michigan
Ann Arbor, MI 48109

Mr. Miller, Scott
Aero. & Mech. Eng. Dept.
U of Arizona
Tucson, AZ 85721

Mr. Mitchell, Dominique
Aero. & Mech. Eng. Dept.
U of Arizona
Tucson, AZ 85721

Dr. Mori, Yasuo
Setagaya-ku
Tokyo Inst Tech
Tokyo 157, Japan

Dr. Nagamatsu, Henry
1046 Cornelius Ave.
R.P.I.
Schenectady, NY 12309

Professor Neyland, Vera
140160 Zhukovsky
TSAGI Moscow
Moscow Region, RUSSIA C.I.S.

Dr. Ordway, Donald
P.O.Box 416
Sage Action
Ithaca, NY 14851

Mr. Posnansky, Hernan
2165 Quadelupe Suite B228
ZONA TECH
Mesa, AZ 85283

Mr. Reba, Ramons
Aero. & Mech. Eng. Dept.
U of Arizona
Tucson, AZ 85721

Dr. Rosenzweig, Marty
27 Portugese Bend Rd.
Rolling Hills CA
Rolling Hills, CA 90274

Dr. Rott, Nicholas
1685 Bryant St.
Stanford
Palo Alto, CA 94301

Dr. Rubbert, Paul
P.O. Box 3707
Boeing
Seattle, WA 98124

Prof. Rubin, Stanley
10695 Deershadow Lane
U of Cincinnati
Montgomery, OH 45242

Dr. Schaff, Al
8143 Billowvista Dr.
AMETEK Elsegundo CA
Playa del Rey, CA 90293

Mr. Schairer, Ed
MS 229-1
NASA-Ames
Moffett Field, CA 94035

Dr. Schairer, George
4242 Hunts Point Road
Boeing
Bellevue, WA 98004

Mr. Schuster, Bill
Aero. & Mech. Eng. Dept.
U of Arizona
Tucson, AZ 85721

Prof. Scott, Larry
Aero. & Mech. Eng. Dept.
U of Arizona
Tucson, AZ 85721

Dr. Seebass, Dick
1477 Patton Dr.
U of Colorado
Boulder, CO 80303

Prof. Shevell, Dick
151 Stockbridge Ave
Stanford
Atherton, CA 94027

Dr. Smerdon, Ernie
College of Engineering
U of Arizona
Tucson, AZ 85721

Mr. Sung, In-Kyung
Aero. & Mech. Eng. Dept.
U of Arizona
Tucson, AZ 85721

Mr. Tourbier, Dietmar
Aero. & Mech. Eng. Dept.
U of Arizona
Tucson, AZ 85721

Dr. Treanor, Chuck
140 Segsbury Dr.
CALSPAN
Williamsville, NY 14221-3425

Dr. Triffet, Terry
4717 E. Ft. Lowell Rd.
U of Arizona/SERC
Tucson, AZ 85712

Mr. Tsu, Jack
1316 6th St.
Northrop
Manhattan Beach, CA 90266

Professor van Dommelen, Leo
College of Engineering
Florida State U
Tallahassee, FL 32309

Mr Vionnet, Carlos
Aero. & Mech. Eng. Dept.
U of Arizona
Tucson, AZ 85721

Dr. Waddell, Jack
P.O. Box 185
Boeing
Silesia, MT 59080

Dr. Wolf, Steve
MS 260-1
NASA-Ames
Moffett Field, CA 94035

Prof. Wygnanski, Wygi
Aero. & Mech. Eng. Dept.
U of Arizona
Tucson, AZ 85721

Mr. Wernz, Stefan
Aero. & Mech. Eng. Dept.
U of Arizona
Tucson, AZ 85721

Mr. Wygle, Brian
Aero. & Mech. Eng. Dept.
U of Arizona
Tucson, AZ 85721

Mr. Yang, Yongsheng
Aero. & Mech. Eng. Dept.
U of Arizona
Tucson, AZ 85721

Mr. Yao, Z. X.
Mech. & Aero. Eng.
ASU
Tempe, AZ 85287

Mr. Yi, Jianwen
Aero. & Mech. Eng. Dept.
U of Arizona
Tucson, AZ 85721

Ms Zhang, Hu
Aero. & Mech. Eng. Dept.
U of Arizona
Tucson, AZ 85721

Dr. Zhou, Mingde
Aero. & Mech. Eng. Dept.
U of Arizona
Tucson, AZ 85721

Symposium Participants at Lunch Break, Hotel Park Tucson, Tucson, Arizona